Practical Density Measurement and Hydrometry

Series in Measurement and Technology

Series Editor: **M Afsar**, Tufts University, USA

The Series in Measurement Science and Technology includes books that emphasize the fundamental importance of measurement in all branches of science and engineering. The series contains texts on all aspects of the theory and practice of measurements, including related instruments, techniques, hardware, software, data acquisition and analysis systems. The books are aimed at graduate students and researchers in physics, engineering and the applied sciences.

Other books in the series

Evaluating the Measurement Uncertainty:
Fundamentals and Practical Guidance
I Lira

Uncertainty, Calibration and Probability:
The Statistics of Scientific and Industrial Measurement
C F Dietrich

SQUIDs, the Josephson Effects and Superconducting Electronics
J C Gallop

Physical Acoustics and Metrology of Fluids
J P M Trusler

Series in Measurement Science and Technology

Practical Density Measurement and Hydrometry

S V Gupta

UNIDO Consultant in Metrology
Former Scientist, National Physical Laboratory,
New Delhi

Institute of Physics Publishing
Bristol and Philadelphia

British Library Cataloguing-in-Publication Data

A catalogue record for this book is available from the British Library.

ISBN 0 7503 0847 8

Library of Congress Cataloging-in-Publication Data are available

Series Editor: **M Afsar**, Tufts University, USA

Commissioning Editor: John Navas
Production Editor: Simon Laurenson
Production Control: Sarah Plenty
Cover Design: Victoria Le Billon
Marketing: Nicola Newey and Verity Cooke

Published by Institute of Physics Publishing, wholly owned by The Institute of Physics, London

Institute of Physics Publishing, Dirac House, Temple Back, Bristol BS1 6BE, UK

US Office: Institute of Physics Publishing, The Public Ledger Building, Suite 1035, 150 South Independence Mall West, Philadelphia, PA 19106, USA

Typeset in LaTeX 2_ε by Text 2 Text, Torquay, Devon
Printed in the UK by Antony Rowe Ltd, Chippenham, Wiltshire

Dedicated to my wife, Mrs Prem Gupta, and my children

Contents

Foreword

Dr S V Gupta has been scientist-in-charge of mass, volume, density and viscosity measurements at the National Physical Laboratory, New Delhi for four decades starting from the mid-fifties. With such a long and continuous association with precision measurements of density and other parameters, Dr Gupta is in a position to write authoritatively on Practical Density Measurements. As far as I know there does not exist any other book which deals with measurement of density in such great detail. The book would therefore be of great use to metrologists, students, teachers, scientists and technologists working in industry. With the introduction of ISO 9000 quality standards, industry will find this book very practical and beneficial.

The unique feature of this book is that, in one place, one finds all the information that is needed for accurate measurement of density of solids and liquids. It tries to bring out assiduously the basic physical principles involved in the measurements as well as the techniques of the measurement, along with the details of the measuring equipment/apparatus.

The book is a result of dedicated work and effort by Dr Gupta. This volume will go a long way in demystifying the measurement of density of solids and liquids.

S K Joshi
Former Director-General
Council of Scientific and Industrial Research
New Delhi

While metrology is an essential activity for every country, books catering to learners at undergraduate and graduate levels and at the same time to professional measurement scientists are rare. There is increasing demand in industries, including those in developing countries like India, for specialized knowledge for quality standards. The introduction of the ISO 9000 quality standard has made this requirement more urgent and more demanding.

This book, written by an expert with long experience in weights and measures in several institutions in India, will fill a longstanding gap in one major area: practical density measurements and hydrometry. There is currently no written material in a comprehensive and consolidated form on this topic, reflecting improvements in methodology and technology.

The author starts with the principles of physics, and the technology, involved to measure accurately the density of solids and liquids to required precision. It has a detailed discussion on national and international density standards, on different methods used to measure density, comparison of these methods, and on thermal dilation of solids. For industrial use, several areas have been specially highlighted: correction tables for calibrating volumetric glassware and capacity measures at different temperatures and for different materials; determination of density of ceramics and glasslike materials; methods for density determination for insulating materials, refractory materials, leather, PVC etc; calibration of hydrometers.

Dr Gupta is uniquely qualified in the preparation of such a volume. For nearly four decades he was associated with the areas of mass, volume, density and viscosity measurements at the National Physical Laboratory, New Delhi—the custodian of primary standards in India. During this period his focus was on the introduction of increasing sophistication in measurements and of new techniques. As Director of weights and measures, Government of India, during the late seventies and as an expert advisor in Cyprus, Syria, Kuwait, Oman and Vietnam he gathered experience in the use and application of measurement standards.

As a former Director of the National Physical Laboratory and a former Director-General of the Council of Scientific and Industrial Research, I am glad that such a book is now being published. I have always felt that the science of standards must extend beyond the boundaries of a standard laboratory, and enter into the curricula and research efforts of academic institutions and industries with in-house measurement facilities. A wide base of expertise is a prerequisite for rapid growth in measurement standards. This book will be an important asset to such groups in India and elsewhere.

A P Mitra, FRS
Former Director-General
Council of Scientific and Industrial Research
New Delhi
and
Former Director
National Physical Laboratory
New Delhi

Preface

Raw from university, when I joined the National Physical Laboratory in Delhi and was placed in the Division of Weights and Measures, I had no iota of an idea about measurement standards. At school I had been taught something very elementary about weighing when studying the principle of levers and that a two-pan balance is a lever of the first kind. From the very first day it was clear to me that this meagre knowledge was not going to serve the purpose for which I was employed so I started to search for literature regarding weighing, balances and hydrometers. I was given some useful verbal guidance as to the practice followed at the National Physical Laboratory, Teddington, UK. However, no explanations were offered, so I remained starved of the reasoning behind a particular method. Then I came across the *The Dictionary of Applied Physics* Volume IV, edited by Glazebrook. From this I gained a good idea of hydrometry. The basics of balances I learnt from NPL's *Notes on Applied Sciences*, published by HMSO. Similarly information about volumetric measurement was dug out from some British Standards. All this was in the late 1950s. Since then no proper book dealing with the subject of mass, volume or density measurement has been published. The density of water as measured by Chappuis at BIPM and Thiesen *et al* at PTR, Germany was taken to be final. The controversy of isotopic composition and dissolution of air arose a little later but again no compilation of the relevant literature was ever carried out. This is one reason I undertook the task of collecting the necessary information.

Measurement science with respect to mass, volume and density is not specifically taught at any level of education—perhaps because of the lack of appropriate books. Therefore, while compiling the information, I have taken special care to write a text that can be taught at the undergraduate level. Here I would specifically like to mention chapter 6 'Special methods for density measurements'. Creating a density or temperature gradient in a long column of liquid, which hitherto was supposed to be a source of error, has been properly utilized, to compare the densities of two objects with very similar values. Variations in the densities of liquids due to changes in temperature or pressure have similarly been used to compare the density of objects. Chapter 2 'Special interferometers', i.e. those used to find the dimensions of solid objects and also the coefficients of linear expansion, will also provide suitable material for teaching at undergraduate level.

The principle of measurements remains the same—it is technology which has played a vital role in improving measurement repeatability. The establishment of solid-based density standards and their inter-comparison among various national laboratories has considerably improved reproducibility in density measurements (chapter 1).

Identifying and characterizing the properties of water and arriving at a definition for Standard Mean Ocean Water has given a definite meaning to the density of water. Recent determinations of water density, discussed in chapters 3 and 4 not only laid to rest the controversy about the maximum density of water and the temperature at which it becomes a maximum but have also highlighted the importance attached to and the amount of effort put into this project. My effort to arrive at a harmonized relation connecting water density with its temperature will also help the scientific community. The water density table in the range 5–41 °C in steps of 0.1 °C in terms of ITS 90 will equally benefit users.

The density of mercury and its updated status with a new density table giving a higher decimal place will benefit metrologists who use mercury as the standard for generating pressure and as a liquid of known density for volumetric measurement.

The use of ISO 9000 has boosted the need to calibrate measuring equipment so many laboratories for calibrating hydrometers are being set up. All these laboratories will study chapter 7 (which deals with hydrometers) to their advantage. Furthermore, the manufacturers of hydrometers will gain insight into the subject. Hydrometers have different names in the petroleum, sugar and dairy industries, so the methods for calibrating density, Brix and milk hydrometers will be of interest to these industries. Similarly all breweries will be equally interested in the calibration of alcoholometers, especially against density hydrometers.

The compilation of methods for determining the density of materials used in industry—chapters 8 and 9—will help the user industry and popularize the methods suggested by ASTM.

It is my pleasant duty to thank quite a number of people, who have guided and helped me in completing this book. Professor A R Verma, former Director of the National Physical Laboratory, New Delhi, Dr R S Davis of the International Bureau of Weights and Measures (BIPM), Paris and Dr John Navas of the Institute of Physics Publishing UK, who have all constantly guided and helped me in this endeavour, deserve my profound thanks. Dr V N Bindal, Dr Ashok Kumar, Mrs Reeta Gupta, Tripurari Lal and his colleagues at the National Physical Laboratory, New Delhi, are thanked for their continuing support. The work carried out at the National Physical Laboratory, New Delhi, mentioned in this book was teamwork, so every colleague deserves my appreciation and thanks. My thanks are also due to Dr J R Anand who has conceived and drawn quite a number of the illustrations. Thanks also to Mr Manmohan Gupta and Mr Ansul Agrawal for attending to my day-to-day needs in computer software, and to John Navas, Sarah Plenty and Simon Laurenson at Institute of Physics Publishing.

Chapter 1

Solid-based density standards

Symbols

ω	Solid angle
Σ	Summation
π	Ratio of the length of the circumference to the diameter
λ	Wavelength of the radiation used
ω_{ij}	Solid angle subtended by an elementary area formed at the centre b bounded by
d_0	Diameter of sphere at $0\,°C$
d_t	Diameter of sphere at $t\,°C$
$g\,cm^{-3}$, $g\,ml^{-1}$, $kg\,m^{-3}$	A unit of density
m^3, cm^3	A unit of volume
ppm	Parts per million = one part in 10^6
V_t	Volume at $t\,°C$
S_{ij}	Difference between the jth diameter of the ith meridian and its equatorial diameter d_{i1}.

1.1 Introduction

Density measurements are required in practically all areas of measurement science. The most sophisticated measurements are required for the density of silicon crystals and hence Avogadro's constant; whereas only a mediocre accuracy is required for checking the net contents of commercial items like ice cream and edible oils. The requirement of accuracy can vary from better than 1 ppm to only 1%. In industry, density measurements are required in many areas; for example to make a correct alloy or mixtures with desired properties, density determination is vital. In industry, many other properties of a material are, in fact, inferred from its density. Measurement of the density of liquids has become all the more important because water and mercury are used as reference liquids. Precise knowledge of the density of water is not only vital for the calibration of volumetric glassware and capacity measures but is also equally important in oceanography. Mercury is used as a confining liquid in investigations of pressure–volume and pressure–temperature relations of other fluids. The accuracy of a primary standard barometer depends upon the accuracy with which the density of mercury, with which it is filled, is known. So the density of mercury, its measurement technique and its values at various temperatures on the latest 1990 international temperature scale are of great interest to the scientific community engaged in pressure and force measurements.

1.2   Density

Only the densities of solids and liquids are discussed in this book.   The measurement of the density of porous solid materials, powders etc will be taken up in chapter 9. The density of liquids used in industry will be discussed in chapter 8. However, the density of gases and their measurement techniques are altogether different, so these will not be discussed in this monograph.

1.2.1   Definition and units

Density is defined as the mass per unit volume of a substance at a specified temperature and pressure. Although the pressure on the density of a solid has practically no effect, the density of a liquid is affected to a limited extent.

The unit of density in the International System of units is $kg\,m^{-3}$ but its sub-multiples, $g\,cm^{-3}$ and $g\,ml^{-1}$, are also commonly used, ml and cm^3 being taken as synonymous. As the density of every material depends upon its temperature, the temperature should qualify the density value. The density of a liquid is affected by pressure to a certain extent and that of solids to a still lesser degree, so the pressure should be mentioned in all high-precision measurements of density.

1.2.2   Water as the primary standard of density

All density measurements would essentially require the measurement of two parameters, namely mass and volume. The mass of a body can be adequately found with the help of a balance by the usual established procedures. The volume may be measured by dimensional measurement, which is more problematic. Therefore, this method is employed only in special circumstances: either when the accuracy requirement is very high, say for the purpose of maintaining a solid-based density standard, or where accuracy requirements are small. However, the hydrostatic weighing method is most often used to measure the volume of a solid body. The volume so measured will be with reference to the density of water or that of the liquid used in the hydrostatic weighing. Similarly capacity measures and other volumetric glassware should be calibrated by taking water as the density standard. Hence water acts as a primary standard of density more often than not.

This establishes the need and importance of knowing the density of water and improving the accuracy in its measurement.

1.2.3 Advantages and disadvantages of water as a primary standard of density

Water has the following advantages when used as a density standard.

(1) It is generally easily available.
(2) It is one of the least expensive materials.
(3) It is easy to handle.
(4) It is non-corrosive so it may be handled by an inexperienced worker.
(5) It remains liquid in most commonly used temperature ranges.
(6) It is a self-cleaning liquid and workers do not need to clean their hands etc.
(7) It has no smell and is a friendly liquid.
(8) It is a non-toxic substance.
(9) Its density is known to a fairly high degree of accuracy for any routine work requiring accuracy of one part in one hundred thousand.
(10) It is non-inflammable.

However, to be used as a primary standard of density, water has the following disadvantages:

(1) The density of water becomes uncertain within a few parts per million because of the unknown concentration of dissolved gases, normally unknown isotopic compositions and contamination. For these reasons, the density of water may become uncertain even within a few parts per hundred thousand.
(2) High surface tension causes serious problems in hydrostatic weighing. The surface tension of water can vary due to surface contamination at the air–water interface, which increases the uncertainty.

1.2.4 Primary standard of density

Solid bodies of known mass and geometric shape, with volumes determined in terms of the primary standard of length, are replacing water as the primary standard of density. In this procedure, the volume of a solid object of known geometrical shape is determined by linear measurements using interferometric techniques. This volume, if necessary, is then transferred to other solid bodies of more stable materials like silicon or zerodur by hydrostatic weighing. The use of fluorocarbon as the hydrostatic liquid improves the accuracy of transfer as it has

(1) low surface tension,
(2) high density, 1.6 to 1.9 times that of water, and
(3) low viscosity.

These bodies with known mass and volume are then used as standards for determining the density of solids and liquids by the hydrostatic weighing method.

1.3   Solid-based density standards

The concept of solid-based density standards probably started from the then National Bureau of Standards in the USA, now known as the National Institute of Standards and Technology (NIST), USA.

1.3.1   Advantages of using solid-based density standards

There are three distinct advantages to choosing a solid-based density standard in preference to water:

- Its volume and mass are directly traceable to the base units of length and mass, i.e. the metre and kilogram.
- It is possible to compare the densities of two solids to a precision better than what is achievable by comparing the volumes of solids in water.
- The density of an unknown liquid may be determined much more easily with the help of a solid standard of known density than by comparing the density of liquid with that of water.

1.3.2   Choice of form of solid-based density standards

To measure the volume of a solid from dimensional measurements, it is necessary to have an artefact of known geometry. The cube and sphere are two simple geometrical shapes the volumes of which can be expressed in terms of their respective linear dimensions.

From a maintenance point of view, the sphere is preferred over the cube— the sharp edges of a cube are not only difficult to make but are also much more difficult to keep intact.

Dimensional measurement appears to be simpler for a cube as, in this case, any interferometer which is good for calibrating slip gauges may be used. However, the uncertainty in volume may be larger due to the presence of imperfect or unnoticed broken edges which, in the case of a cube, number 12.

So most of the national laboratories maintain solid-based density artefacts in the form of spheres. The NIST initiated the work in this direction in the early 1970s.

1.3.3   Desirable qualities of materials used for solid-based density standards

The most important properties of a material to be used for a solid-based density standard are:

- its material properties should not change with time;
- it should be homogeneous;
- it should be sufficiently conducting both electrically and thermally;
- it should form a sufficiently hard surface, which is inert to atmospheric gases;

- it should have a low coefficient of expansion;
- it should be inert to water and most other liquids;
- it should not be too expensive; and
- it should be such that it can be easily machined, ground and polished so that bodies of the desired form and surface are worked out.

1.3.4 Materials for solid-based density standards

The ultra-low expansion (ULE) glass used by the Commonwealth Scientific and Industrial Research Organization (CSIRO), Australia and other laboratories does not meet the requirements of electrical and thermal conductivity. Zerodur, a ceramic material, meets most of the requirements except that it is suspected to shrink with time. However, silicon meets most of the requirements:

- It has a high temporal stability.
- It is a very homogeneous material, so minor chipping or abrasion in use causes an equal proportional loss of mass and volume, so that the density remains constant.
- It has sufficient electrical conductivity so electrostatic charges do not pose a problem while weighing. ULE glass and quartz suffer from this drawback.
- It has a highly stable surface, as a silicon oxide film is formed within a minute or so. The change thereafter occurs at a very slow rate, which causes a change in density by no more than 0.1 ppm per year. The density of silicon oxide is nearly equal to that of the parent material. However, its thickness can now be calculated and its effect can be estimated.
- Its coefficient of linear expansion is low—only $2.5 \times 10^{-6}\,°C^{-1}$.
- It is easily machined, ground and polished.
- It is easily available commercially and not too expensive.

1.4 Artefacts of solid-based density standards in various laboratories

1.4.1 National Institute of Standards and Technology, USA

NIST, formerly known as the National Bureau of Standards, introduced solid-based density standards. Bowman *et al* [1] have given a detailed account of the work concerning these density standards which was achieved in two steps.

Step I. They acquired a set of four steel ball bearings, with diameters in different planes differing by not more than 1 ppm, from a local firm. Dr J B Saunders, from NBS itself, designed and had fabricated an interferometer [2] especially for the purpose of measuring the diameters of each sphere. He also estimated the variation in the diameter, which was found to be less than 63 nm. Johnson [3] showed that the true volumes of such spheres differ insignificantly from the volumes calculated on the basis of the average diameter. Therefore they

Table 1.1. Diameter, SD and volume of steel spheres at NIST.

Sphere	Diameter (μm)	SD of mean (μm)	Volume at 20 °C (cm^3)
A	63 500.1116	0.0070	134.067 062
B	63 500.0841	0.0040	134.066 888
C	63 499.9851	0.0060	134.066 261
D	63 500.1882	0.0041	134.067 547
E	63 500.1064	0.0054	134.067 028
F	63 500.0964	0.0064	134.066 966

calculated the volume of each sphere from knowledge of the average diameter. In fact four spheres, designated as A, B, C and D, were used. Later on it was felt that the measurements would be better designed with six spheres, so instead of taking two more fresh spheres, the observations were repeated for spheres A and B and these were termed spheres E and F. The mean diameters and their standard uncertainty together with volumes of spheres at 20 °C are given in table 1.1.

Step II. The volume of the spheres was transferred to silicon crystals using the hydrostatic method. An improved procedure for high-precision density determination by hydrostatic weighing developed by the group was used [4]. Four crystals, designated X2, X3, X4 and X5, were fabricated. Silicon crystals were cut from a 6.5 cm diameter bole in slices about 2.5 cm thick. Each crystal was ground to about 205 g and finally trimmed to about 200 g by etching with hydrofluoric acid (HF). The etching exposed the original crystal lattice surface and removed all surface damage due to grinding and cutting.

Determination of the volume ratios of four objects [5]—ratio of volumes

If W_K, W_L are the apparent masses of objects K, L which have true masses M_K, M_L and volumes V_K, V_L and ρ is the density of the hydrostatic liquid, then

$$W_K = M_K - \rho V_K \tag{1.1}$$
$$W_L = M_L - \rho V_L \tag{1.2}$$

giving

$$V_K / V_L = (M_K - W_K)/(M_L - W_L). \tag{1.3}$$

For simplicity, let us denote the ratio of the volumes of the two bodies by their respective designation, i.e. V_K/V_L simply as K/L.

Let the four loads under consideration be designated K, L, M and N. To eliminate any linear drift, the hydrostatic weighing of four bodies is carried out in the following order: K–L–K–M–K–N–K–L–M–L–N–L–M–N–M. Furthermore, an equal interval of time must elapse between each weighing. This means that to obtain the experimental values of the volume ratios of the four bodies in all possible combinations, 15 hydrostatic weighings should be carried out. Instead of finding the differences of apparent weights of four objects, six volume ratios of four objects taken in all possible combinations of two at a time are determined. This comparison format is called the 4–1 series. The six ratios are K/L, K/M, K/N, L/M, L/N and M/N.

K–L–K will give the volume ratio of K/L and similarly M–N–M will give volume ratio of M/N and so on. Furthermore it should be noted that $(K/L)(L/M)(M/K)$ would be exactly equal to unity if the experiment were to be performed without any error. These observed ratios might, therefore, be adjusted by the least-squares method so that these products have unit value. The formula for the least-squares estimate for K/L is of the form:

$$(K/L) = [\{(K/L)\}^2\{(K/M)(M/L)\}\{(K/N)(N/L)\}]^{1/4}. \qquad (1.4)$$

The case of four crystals and two spheres at a time

In this case let M and N represent two pairs of crystals. Two pairs of crystals out of four crystals can be taken in three ways. If the ratio of the volumes of the four crystals is also determined, then this 4–1 series will comprise four sets of operations. For two given spheres A and B, the four crystals, designated as X2, X3, X4 and X5, can be grouped in the following four ways.

Set no	K	L	M	N
I	A	B	X2 + X3	X4 + X5
II	A	B	X2 + X4	X3 + X5
III	A	B	X2 + X5	X3 + X4
IV	X2	X3	X4	X5

Here each set will give six volume ratios. Hence there will be 24 observation equations.

It may be noted that in each set $(K/L)(L/M)(M/K)$, $(K/L)(L/N)(N/K)$, $(K/M)(M/N)(N/K)$ and $(L/M)(M/N)(N/L)$ would be exactly equal to unity if the experiment were to be performed without any error. This way each set will give a least-squares estimate of six ratios; adding a constraint equation that the sum of the volumes of the spheres A and B is given, we obtain 25 equations. Solving these 25 equations by the least-squares method for six variables gives the volumes of the two spheres A and B and four crystals X2, X3, X4 and X5. It should be noted that the least-squares method has been used twice for one experimental datum, which may be questionable.

Two spheres out of six spheres can be taken in three ways. Similarly, the determination of the volumes of four crystals and two spheres is carried out by taking spheres C and D as standards in the second group while spheres E and F are taken as standards for the third group of observations. So there would be three groups each with 25 equations. Checks on the volumes of the spheres

Table 1.2. Mass, volume, density and SD of silicon discs at NIST.

Crystal designation	Mass (g)	Volume (cm^3)	Density (g cm^{-3})	SD of mean (g cm^{-3})
X2	200.420 673	86.049 788	2.329 1245	0.000 0007
X3	199.763 695	85.767 851	2.329 1209	0.000 0006
X4	200.010 787	85.874 033	2.329 1184	0.000 0004
X5	199.932 651	85.840 491	2.329 1182	0.000 0003

calculated by the least-squares method through hydrostatic weighing and the volumes of spheres determined by measuring their respective diameters would give the overall accuracy of the whole experiment. The actual difference between the volumes of spheres calculated from the diameter and obtained from the hydrostatic weighing varied from 0.13 to 0.37 ppm only.

This method was used by NIST, USA. The mass, volume and density of crystals with their standard deviation from the mean are given in table 1.2.

Sample holder

A special sample holder is shown in figure 1.1. The sample holder consists of a column C and a disc. The disc has four holes in a circle and at right angles to each other. On each hole, four objects can be placed concentrically. The pan P is suspended with a very fine wire W from the pan of the balance. The column C with the help of a guiding bush G can be moved in exact steps of 90° in either direction and also vertical directions. The system moves in such a way that any particular object can be loaded on the pan of the balance. For loading a particular object, it is moved in such a way that the object is just above the immersed balance pan. Then the holder moves vertically downward and the object comes onto the balance pan. To unload the object, it is moved vertically upward taking the object with it thus removing the object from the pan. The sample holder can also be connected to a computer and can thus be controlled through it, so that each of the four objects can be chosen in all possible combinations. Normally two standards S and two objects O are taken whose volumes may be compared in all possible ways. A similar arrangement has been used in the balance HK1000 supplied to NPL, New Delhi.

1.4.2   National Measurement Laboratory (CSIRO) Australia

The National Measurement Laboratory of CSIRO, Sydney, Australia [6] developed a sphere of ultra-low expansion (ULE) glass. This is made from two hemispherical shells joined together with some glue. The scattering at the glued portion reduces the reflection from inside the sphere. The great circle where the

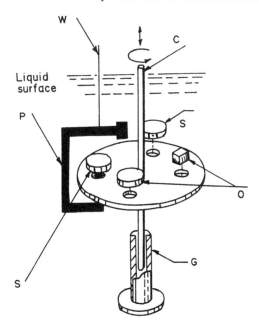

Figure 1.1. Sample holder.

two hemispheres join is called the equator. A scale has been marked along the equator, which helps to position the sphere precisely at the required positions. The diameters of the equator are directly measured with the interferometer at five points with an angular distance of 36°. In addition the relative values of the diameters have been measured with a Talyrond roundness machine at 18 points along each of the five meridians. The points are equally spaced with an angular separation of 10°. All diameters along a meridian are measured relative to its respective equatorial diameter. The first Talyrond measurement on each meridian is at the equatorial point. In this way all 648 Talyrond measurements can be carried out.

Measurements are also carried out at different temperatures, so that the diameter and volume of the sphere can be determined at different temperatures. The maximum difference between any two diameters of the Australian sphere was found to be 195 nm. The diameter at temperature t °C of the ULE sphere was given by

$$d_t = d_0[1 - 6.408 \times 10^{-8}(t\,°\mathrm{C}^{-1}) + 9.785 \times 10^{-10}(t\,°\mathrm{C}^{-1})^2]. \qquad (1.5)$$

The volume of the sphere can be calculated either by the mean diameter method or by taking the sum of the volumes of the sectors subtending an angle ω_{ij} at the centre of the sphere.

Volume of a sphere by the summation method

A great circle of the sphere is taken as the reference circle and called the equator. The equator is divided at $2n$ equally spaced points. Joining the opposite points gives us n diameters. Draw n great circles through these n diameters. Each great circle is again divided into $2m$ equi-angular points, thus giving us m diameters on each great circle. Every diameter of the equator is measured by the interferometric method. All diameters on other great circles are measured with reference to the equatorial diameter of that great circle. In fact only the difference from the equatorial diameter is measured.

Consider an elementary area bounded by the arcs of two great circles and the arcs of two latitude circles as shown in figure 1.2. Let the points A, B, C and D be, respectively, designated as $(i + 1, j - 1)$, $(i, j - 1)$, (i, j) and $(i + 1, j)$. i will take values 1 to n and j will take values 1 to m. Let O be the centre of the sphere, while P and Q are centres of the two latitude circles. Then

$$OB = OA = r_{i,j-1}$$
$$OC = OD = r_{i,j}$$
$$PA = PB = r_{i,j-1}\sin((j-1)\pi/m)$$
$$QD = QC = r_{i,j}\sin(j\pi/m)$$
$$\text{Arc } AB = r_{i,j-1}\sin((j-1)\pi/m)\pi/n \qquad (1.6)$$
$$\text{Arc } DC = r_{i,j-1}\sin((j-1)\pi/m)\pi/n$$
$$\text{Arc } BC \cong \text{Arc } AD = (r_{i,j-1} + r_{i,j-1})\pi/2m. \qquad (1.7)$$

If the elementary area is sufficiently small then $r_{i,j-1} \cong r_{i,j}$. This gives us

$$\text{Arc } BC = \text{Arc } AD = r_{i,j}\pi/m.$$

Therefore the elementary area ABCD, which is a trapezium of sides AB and DC, is given by

$$\text{Area } ABCD = r_{i,j}[\sin((j-1)\pi/m) + \sin(j\pi/m)]\pi/2nr_{i,j}\pi/m$$
$$= (\pi^2/2mn)(r_{i,j})^2[\sin((j-1)\pi/m) + \sin(j\pi/m)]. \qquad (1.8)$$

If $\omega_{i,j}$ is the solid angle which this elementary area subtends at the centre of the sphere, then

$$\omega_{i,j} = (\pi^2/2mn)(r_{i,j})^2[\sin((j-1)\pi/m) + \sin(j\pi/m)] \qquad (1.9)$$
$$\text{therefore the volume of the sphere} = \tfrac{1}{3}\sum\sum \omega_{i,j}(r_{i,j})^3. \qquad (1.10)$$

The first summation is taken over i and the second over j.

So in the case of the ULE sphere of the National Measurement Laboratory (NML), the diameter of sphere $d_{i,j}$ is equal to $2r_{i,j}$ and is given by the relation

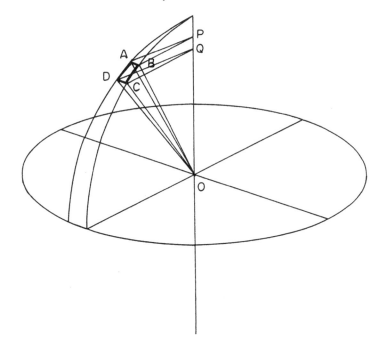

Figure 1.2. Elementary area on the surface of the sphere.

$d_{ij} = d_i + S_{ij}$, where S_{ij} is the difference between the jth diameter of the ith meridian and its equatorial diameter d_{i1}. Hence the volume of the sphere will be

$$V = (1/24) \sum \sum \omega_{ij} (d_i + S_{ij})^3. \tag{1.11}$$

The first summation is over i from 1 to 5 and the second summation is over j from 1 to 18.

The volume of the sphere as a function of temperature is given by

$$V_t = 228.519\,022[1 - 1.922 \times 10^{-7}(t\,°\mathrm{C}^{-1}) + 2.936 \times 10^{-9}(t\,°\mathrm{C}^{-1})^2]\,\mathrm{cm}^3. \tag{1.12}$$

The reported standard uncertainty (one standard deviation) is 0.057 mm³ (0.25 ppm) at 20 °C and 0.25 mm³ (1.1 ppm) at 40 °C.

The sphere is made hollow to weigh less in water so that its apparent mass may be measured with a low capacity balance as this has relatively better discrimination.

NML used two balances—HK 1000 MC and 1 kg Oertling—to measure the mass of the sphere. The true measured mass of the sphere was given as 0.329 618 416 kg with a standard uncertainty of 44 μg (0.13 ppm).

Table 1.3. Particulars of the silicon spheres at NRLM, Japan.

	S4	S5	Uncertainty
Mass	1000.578 93 g	1000.612 19 g	$\pm0.000\,06$ g $\cong 0.06$ ppm
Diameter	93.617 8275 mm	93.618 8621 mm	$\pm0.000\,0007$ mm $\cong 0.08$ ppm
Volume at 22.5 °C	429.609 9173 cm^3	429.624 1607 cm^3	$\pm0.000\,039$ cm^3 $\cong 0.09$ ppm
Density	2.329 040 040 g cm^{-3}	2.329 040 060 g cm^{-3}	$\pm0.000\,000\,024$ g cm^{-3} $\cong 0.1$ ppm

1.4.3   National Research Laboratory of Metrology, Japan

Silicon spheres

The National Research Laboratory of Metrology (NRLM), Japan [7], developed a single-crystal silicon sphere. The floating zone method was used to grow such crystals. The sphere was polished by a technique similar to that used by Leistner and Zosi [8]. Roundness measurements were carried out which showed an asphericity error of not more than 300 nm in any direction. To measure the diameters, Saunders' interferometer was used. To find the order of interference at the centre, the etalon was scanned against the sphere by measuring the changes in intensity at the centre of the modulated fringe pattern. The mass of the sphere was measured against the 1 kg stainless steel standard using the NRLM-II balance [9]. The balance used was the same as that used for comparing mass standards against the prototype kilogram. The density of air was measured *in situ* by using sinkers. The measured parameters and their respective uncertainty are:

Mass $= 1002.734\,69 \pm 0.000\,04$ g $\cong 0.04$ ppm
Diameter $= 93.684\,4370 \pm 0.000\,0086$ mm $\cong 0.09$ ppm
Volume $= 430.527\,578 \pm 0.000\,15$ cm^3 $\cong 0.34$ ppm and
Density $= 2.329\,0835 \pm 0.000\,0008$ g cm^{-3} with a total standard uncertainty
of 3.4×10^{-7}.

Later on, in 1994 [10], two more spheres S4 and S5 were prepared. The asphericity of these spheres was found to be only 70 nm. In addition to diameters, the silicon oxide layer was also measured, which was around 10 nm. The necessary correction was applied. An improved technique was developed to measure the diameters and spherical harmonics were used to calculate the volume. This way, the overall standard uncertainty was improved from 0.34 to 0.1 ppm. The parameters are given in table 1.3.

Quartz spheres

The NRLM [11] also developed three quartz spheres of approximately 85 mm in diameter prepared especially for the Toshiba Co. The quartz has very good optical properties and is capable of acquiring a high degree of polish. These properties

Table 1.4. Quartz spheres at NRLM, Japan.

Sphere no	Diameter (mm)	Coefficient of expansion ($\times 10^{-6}$ K^{-1})	Volume (cm^3)	Uncertainty ($\times 10^{-6}$)
1	84.462 471	1.689	319.996 8013	0.36
2	84.657 252	1.380	317.680 9193	0.36
3	84.519 719	1.320	316.135 1333	0.36

Photograph P1.1. Quartz sphere at NMI, Japan (courtesy NMI).

are highly useful for measurements with an interferometer. The diameter of the sphere was measured at ten points along the meridian. The points were taken with equal angular distance. Three sets of measurements were taken for each sphere. However, the directions of the ten points in each sphere were not kept constant. Measurements on the Talyrond machine showed that the peak-to-peak difference between any two diameters was not more than 400 nm. The coefficients of cubic expansion of quartz for each sphere were not the same. The diameter, coefficient of expansion, volume and uncertainty of each sphere at 16 °C are given in table 1.4.

One such sphere from the National Metrology Institute, Japan is shown in photograph P1.1.

1.4.4 IMGC Italy

Silicon spheres

The Istituto di Metrologia G Colonneti, Italy (IMGC) [12, 13] developed two silicon spheres to measure Avogadro's constant. The volume of each sphere was determined by both dimensional measurement and hydrostatic comparison against their zerodur density standards. The spheres were made from a single silicon crystal and were designated Si1 and Si2. The diameter of the sphere was approximately 93 mm and the mass was 1 kg. The sphere Si2 was polished at the Optical Workshop of the National Metrology Laboratory CSIRO, Australia. The roundness measurements were taken with reference to a single diameter on 18 equi-spaced meridians each passing through that diameter. The peak-to-peak difference between its diameters was within 220 nm. Si1 incurred extensive damage while its roundness was being assessed, so only the volume of Si2 was calculated by dimensional measurement. Later [14] a correction was applied in the measured diameter of the Si2 sphere. The correction became necessary due to a change in the refractive index of air because of the water vapour term in Elden's equation and due to 0.01% more CO_2 being found in the air inside the interferometer. The volume was accordingly corrected. The mass and volume of the sphere were also measured by NRLM Japan [15] under an inter-comparison programme in 1992. The agreement between the values of mass and volume determined by IMGC and NRLM was excellent. Recently [16] another sphere of a single silicon crystal has been acquired from NML, Australia, with much fewer errors in its form and this is designated Si3. For this sphere, the peak-to-peak difference between its diameters is within 28 nm instead of the Si2's 220 nm. The parameters of the spheres are given in table 1.5.

The values of the volume of the spheres, given here, are at 20 °C and 101 325 Pa. The density values are given after taking into account the density changes due to a thin layer of silicon oxide formed outside the sphere and the concentrations of carbon and oxygen atoms.

Table 1.5. Silicon spheres at IMGC, Italy. The standard uncertainty for the mass is given within the parentheses.

Sphere no	Mass (kg)	Volume (cm³)	Density (kg m⁻³)	Uncertainty in volume (ppm)
Si2	0.993 493 9872 (80 μg)	426.560 076	2329.081 21	0.38
Si3	1.000 685 8918 (44 μg)	429.647 784	2329.083 54	0.13

Table 1.6. Parameters of zerodur spheres at IMGC, Italy.

Sphere no	Mass kg	Uncertainty (μg)	Volume (cm^3)	Uncertainty	
				(mm^3)	(ppm)
SP	0.958 567 66	100	377.989 22	0.16	0.42
S1	0.979 351 39	150	386.521 87	0.20	0.52
S2	0.979 682 15	80	386.675 59	0.07	0.18
S3	0.978 886 91	140	386.356 51	0.18	0.46

Zerodur spheres

Prior to the construction of these spheres, three zerodur spheres were developed at IMGC, Italy. Their volumes were measured in 1981–83 [17]. These three zerodur spheres, designated SP, S2 and S4, were not from the same batch of material. These were purchased and polished at different times: SP in 1979, S2 in 1980 and S4 in 1981 and measurements were conducted on them during 1981–83. Zerodur is a glass ceramic with a coefficient of expansion of 10^{-8} K^{-1}. It is inert to air and water. The diameter was measured in two steps: in the first step the number of integral fringes was determined using a fringe counter system and other mechanical machines. Only a fraction of the fringe width has been measured using a Saunders-type interferometer with a single laser source. S1 is the earliest sphere. The parameters of the zerodur spheres are given in table 1.6.

1.4.5 Physikalisch Technische Bundesanstalt (PTB)

PTB [18] uses cubes of zerodur—a ceramic substance of low thermal expansion— as solid-based density standards. The volumes of the cubes are evaluated (a) by determination of the central distances between opposing faces and (b) by taking into account

- the deviation from rectangularity of adjacent faces,
- the unevenness and roughness of the faces and
- the imperfect sharp edges and broken edges due to damage in polishing.

The central distances are determined both optically, using an interference comparator, and mechanically, with an interference displacement comparator equipped with measuring probes. The central distances, the unevenness and the density of the cubes were in agreement for both procedures. The standard uncertainty of the density measurement was 0.7 ppm, while the mean density determined by two different methods agreed well within 0.5 ppm.

The non-rectangularity between adjacent faces is less than 2 seconds of arc; parallelism between opposing faces is better than 2 seconds of arc; the unevenness is less than $\lambda/10$.

Table 1.7. Particulars of the zerodur cubes maintained at PTB, Germany. The mass has been measured with a standard uncertainty of 7×10^{-8}. The total volume, due to non-evenness, rounded off and broken edges, was just equal to 2×10^{-7} of the mean volume.

| | Dimensions (mm) | | | | Mean |
	Faces 1 & 3	Faces 2 & 4	Faces 5 & 6	Volume (mm^3)	volume (mm^3)
Cube no 1					
Optical	73.582 429	73.564 795	72.916 303	394 701.43	
Mechanical	73.582 424	73.564 761	72.916 305	394 701.13	394 701.28
Difference (nm)	5	34	2	0.30	
Uncertainty (nm)	20	20	20	0.8×10^{-6}	
		Mean mass of the cube 0.999 612 40 kg			
Cube no 2					
Optical	72.882 916	73.572 131	73.579 149	394 542.46	
Mechanical	72.882 947	73.572 163	72.579 155	394 542.73	394 542.60
Difference (nm)	5	34	2	0.26	
Uncertainty (nm)	20	20	20	0.8×10^{-6}	
		Mean mass of the cube 0.999 215 46 kg			

The dimensions of the two cubes determined by optical and mechanical measurements, their differences and the volumes of each cube with the uncertainty are given in table 1.7.

The density of the two cubes was also measured at IMGC, Italy. The accuracy of the results can be judged from the closeness of the density values obtained by each laboratory, which are:

Cube no	Density by PTB (kg m^{-3})	Density by IMGC (kg m^{-3})
1	2532.5798	2532.5796
2	2532.5921	2532.5958

1.4.6 National Physical Laboratory, India

The solid density standard at the NPL, India is in the form of a cylinder. The cylinder is shown in figure 1.3. A solid glass cylinder of diameter around 25 mm and length 230 mm is used as a solid transfer standard. The standard is in the form of a cylinder with conical ends and an eye at the top and is made from glass with low thermal expansion. The volume of the cylinder was determined using a hydrostatic method with reference to the density of triple distilled water at 20 °C.

Figure 1.3. Glass cylinder with an eye (NPL, India).

Table 1.8. Mass and volume of quartz spheres maintained at NPL, India.

Sphere	Mass (g)	Volume (cm^3)
S1	590.4961	268.2251
S2	590.2768	268.1117

Each set of measurements was spread over several days and the mean value of the volume obtained was reported. The whole exercise took several years—from 1983 to 1986 [19]. The peak difference and mean value of the volume at 20 °C was found to be 0.000 03 cm^3 and 102.0576 cm^3, respectively. The volume of the cylinder was also compared through hydrostatic weighing against the ULE cylinder of CSIRO, Australia. Xylene was used as the hydrostatic liquid. The volume so determined was found to be 102.0576 cm^3.

Recently NPL, India has acquired two quartz spheres—their mass and volume have been determined by PTB, Germany. These are also used as primary density standards. Their parameters are given in table 1.8.

1.5 International inter-comparisons of density standards

In the early 1960s an international inter-comparison of the measurement of the density of water was conducted by hydrostatic weighing in water of a 1 kg stainless steel weight. Eight major laboratories participated. The results of

Table 1.9. Inter-comparison of solid standards between NIST and IMGC. The figures in parentheses represent the standard uncertainty.

	Volume (cm^3)		Density (g cm^{-3})	
	X1	X2	X1	X2
NBS	343.625 721	343.612 202	2.329 0800	2.329 0799
	(0.000 108)	(0.000 095)	(0.000 007)	(0.000 0006)
IMGC	343.626 24	343.612 70	2.329 0778	2.329 0778
	(0.000 11)	(0.000 11)	(0.000 008)	(0.000 008)

these density measurements [20] were found to differ by as much as 13 ppm emphasizing the need to establish solid-based density standards, which were developed in the due course of time. Since then, various bilateral/international inter-comparisons, as described here, have been conducted.

1.5.1   Between NIST and IMGC

Two samples of silicon crystals, designated X1 and X2, were taken. Their volumes and density were determined independently by IMGC, Italy and NIST, USA [21, 22]. IMGC, Italy used their zerodur spheres S2, S4 and SP, while NIST used their four silicon crystals each of 200 g and the results are given in table 1.9.

The IMGC values of density are systematically lower by 0.9 ppm. Furthermore, the IMGC values of volume are higher by 1.4 ppm. It should be mentioned that the results from the two laboratories, which are independent of mass, volume or density, are the ratio V_{X1}/V_{X2} of the volumes of the two crystals, but depend only on experimental procedures. The results obtained are:

Laboratory	V_{X1}/V_{X2}	Uncertainty
IMGC	1.000 039 39	0.000 000 17
NBS (NIST)	1.000 039 40	0.000 000 14

It should be noted that the agreement between the two values is within 0.01 ppm.

1.5.2   Between NML (Australia) and NIST

Two silicon crystals designated 703 and 806 were circulated and their density was measured against the ULE sphere by NML, CSIRO, Australia and against their 200 g crystals by NIST (the then NBS) [23]. The results are given in table 1.10.

Table 1.10. Inter-comparison of solid standards between NML and NIST.

	Mass (g)		Density (g cm^{-3})
	703	806	
NIST	206.574 59	207.769 14	2.329 0734
	(0.24 ppm)	(0.24 ppm)	(1.1 ppm)
NML	206.574 65	207.769 15	2.329 0719
	(0.28 ppm)	(0.43 ppm)	(1.3 ppm)
Agreement	0.29 ppm	0.05 ppm	0.64 ppm

Table 1.11. Volumes of IMGC's zerodur spheres as measured at various laboratories.

Institute, year	Method ref. Std	S2 (cm^3)	S4 (cm^3)	SP (cm^3)
IMGC, 1982	Dimensional	386.675 92	382.598 13	377.939 59
	measurement	±1.1 ppm	±1.1 ppm	±1.1 ppm
PTB, 1983	Dimensional	386.676 16	382.598 68	
	measurement	±1.1 ppm	±0.8 ppm	
IMGC–PTB, 1983–1984	Hydro. comp.	386.675 07	382.598 54	377.940 07
	Std PTB C1, C2	±1.3 ppm	±1.3 ppm	±1.3 ppm
IMGC–NML, 1983	Hydro. comp.	386.675 42	382.598 63	
	Std NML ULE	±1.5 ppm	±1.5 ppm	
IMGC–NML, 1983–1984	Hydro. comp.	386.675 15	382.597 68	377.939 25
	Std NIST X1 X2	±1.1 ppm	±1.1 ppm	±1.1 ppm
PTB, 1986	Hydro. comp.	386.675 69	—	—
	Std PTB C1, C2	±1.0 ppm		
IMGC–NIST, 1987	Hydro. comp.	—	382.597 50	377.939 07
	Std NIST X1		±1.1 ppm	±1.1 ppm
IMGC–NML, 1988	Hydro. comp.	386.675 30	382.598 09	377.939 71
	Std NML ULE	±0.6 ppm	±0.6 ppm	±0.6 ppm
IMGC, 1988	Hydro. comp.	386.675 04	382.597 83	377.939 43
	Std IMGC Si2	±0.6 ppm	±0.6 ppm	±0.6 ppm
PTB, 1989	Hydro. comp.	386.674 70	—	—
	Std PTB C1, C2	±0.8 ppm		
Maximum difference (cm^3)		0.001 37	0.001 18	0.001 07
Maximum difference (ppm)		3.5	3.1	2.8

1.5.3 Between IMGC, PTB, NML and NIST

A good number of inter-comparisons of IMGC standards have been carried out, twice each with NIST and NML [21, 22], and three times with PTB [14, 24, 25]. The volumes of the zerodur spheres, as given in table 1.11, have been measured

Table 1.12. Inter-comparison of solid standards between NRLM, PTB and IMGC. Calculated/measured density values including the PTB flotation method.

Sphere	Laboratory	Density (kg m^{-3})	1σ uncertainty	Relative difference with PTB value ($\times 10^7$)
S1(PTB)	PTB	2329.081 72	0.001 6	
	IMGC	2329.080 99	0.000 71	−3.1
	NRLM	2329.082 17	0.000 50	2.0
S2(PTB)	PTB	2329.072 40	0.001 6	
	PTB(F)	2329.072 85	0.001 7	1.9
	IMGC	2329.072 84	0.000 51	1.9
	NRLM	2329.072 48	0.000 32	0.2
S1(NRLM)	PTB	2329.083 04	0.001 6	
	PTB(F)	2329.083 30	0.001 7	1.1
	IMGC (92)	2329.082 70	0.001 26	−1.5
	IMGC (94)	2329.080 73	0.000 94	−9.9
	NRLM	2329.082 78	0.000 82	−1.1
S2(NRLM)	PTB	2329.087 72	0.001 6	
	PTB(F)	2329.087 64	0.001 7	−0.3
	NRLM	2329.088 07	0.000 30	1.5

at several laboratories. These spheres have been used to measure the density of several objects through hydrostatic weighing with a standard uncertainty of 0.4 ppm in volume ratio and 1.1 ppm in assigning the absolute density.

1.5.4   Between NRLM, Japan, PTB and IMGC silicon standards

The reliability of density values obtained by different measuring techniques for silicon crystals was checked by density determinations of four silicon spheres at PTB, Germany, IMGC, Italy and NRLM, Japan [26]. All mass measurements were carried out in air. The diameters of the four silicon spheres were measured at IMGC and NRLM and their volumes calculated. While at PTB, their volumes were determined with reference to zerodur cubes using the hydrostatic method. The density of the four spheres was also determined by a flotation method at PTB. The maximum relative difference was found to be 1 ppm, while typical values were 2×10^{-7}. The values for the density obtained at different laboratories with their respective uncertainties are given in table 1.12.

Table 1.13. Shrinkage of zerodur (in ppm per year).

Source	Sample	Years since production	Shrinkage per year (ppm)
Bayer-Helms	N	2.5	0.17
		4	0.11
		6	0.07
		9	0.04
IMGC	S2	2.5	—
		4	0.13
		6	0.10
		8.8	0.06

1.6   Shrinking of zerodur

It should be mentioned that the volume stability of the zerodur standards is doubtful because of the slight shrinking property of zerodur. It has been observed that the shrinkage is greater in the initial years after manufacture of the material and that it reduces as time passes. Therefore, before deciding to construct solid density standards from zerodur, one should ensure that material of at least 10 years of age is obtainable. The rate of shrinking as studied by Peuto and Sacconi [14] is given in table 1.13 for two samples.

References

[1] Bowman H, Schoonover R M and Carroll L C 1974 A density scale based on solid objects *J. Res. Natl Bur. Stand.* A **78** 13–40

[2] Saunders J B 1972 Ball and cylinders interferometer *J. Res. Natl Bur. Stand.* C **76** 11–20

[3] Johnson D 1974 Geometrical considerations in the measurement of the volume of an approximate sphere *J. Res. Natl Bur. Stand.* A **78** 41–8

[4] Bowman H A, Schoonover R M and Jones M W 1967 Procedure for high precision density determination by hydrostatic weighing *J. Res. Natl Bur. Stand.* C **71** 179–98

[5] Bowman H A, Schoonover R M and Caroll C L 1974 The utilization of solid objects as reference standard in density measurement *Metrologia* **10** 117–21

[6] Patterson J B and Morris E C 1994 Measurement of absolute water density 1 °C to 40 °C *Metrologia* **31** 272–88

[7] Fujii K, Tanaka M, Nezu Y, Nakayama K and Masui R 1993 Accurate determination of density of crystals silicon sphere *IEEE Trans. Instrum. Meas.* **42** 395–400

[8] Leistner A and Zosi G 1987 Polishing of a silicon sphere for a density standard *Appl. Opt.* **26** 600–1

[9] Kobayasi K, Nezu Y, Uchikawa K, Ikeda S and Yano H 1986 Prototype kilogram balance II of NRLM *Bull. NRLM* **35** 7–22

[10] Fujii K *et al* 1995 Absolute measurement of densities of silicon crystals in vacuum for determination of the Avogadro's constant *IEEE Trans. Instrum. Meas.* **44** 542–5

[11] Masui R, Fujii K and Takanaka M 1995/96 Determination of the absolute density of water at 16 and 0.101 325 MPa *Metrologia* **32** 333–62

[12] Sacconi A *et al* 1988 Density standards for determination of Avogadro's constant *Measurement* **6** 41–5

[13] Sacconi A, Peuto A, Pasin W, Panciera R, Lenaers G, Valkiers M, van de Berg M and de Bievre P 1989 Towards the Avogadro's constant—preliminary results on the molar volume of silicon *IEEE Trans. Instrum. Meas.* **38** 200–5

[14] Peuto A and Sacconi A 1991 Volume and density measurements for the IMGC Avogadro's constant *IEEE Trans. Instrum. Meas.* **40** 103–7

[15] Peuto A M, Sacconi A, Mosca M, Fujii K, Tanaka M and Nezu Y 1993 Comparison of silicon density standards at NRLM and IMGC *IEEE Trans. Instrum. Meas.* **42** 242–6

[16] Sacooni A, Peuto A, Mosca M, Pnaciera R, Pasin W and Pettorruso S 1995 The IMGC volume–density standards for the Avogadro's constant *IEEE Trans. Instrum. Meas.* **44** 533–7

[17] Peuto A M, Sacconi A, Panciera R, Pasin W and Rassetti M 1984 Precision measurement of solid artifacts for re-determination of density of water *Precision Measurements and Fundamental Constants* vol II, ed B N Taylor and W D Phillips, pp 449–52

[18] Density standards in Zerodur 1987 *PTB Report* MA-7, Physikalisch-Technische Bundesanstalt, Germany (in German)

[19] Nath M, Bhamra S S and Gupta S V 1988 Establishment of density standards in National Physical Laboratory *Report* NPL-88-A.3-003/0156

[20] Bonhour A and Girard G 1963 *CIPM Proc. Verbaux, 2nd Series* **31** 51–3

[21] Peuto A and Davis R S 1985 Comparison between SDS between IMGC and NBS *J. Res. Natl Bur. Stand.* **90** 217–27

[22] Peuto A 1987 CCM—Working Group 5, Project of comparison of determinations on transfer standards *IMGC Report* P 137, pp 1–12

[23] Patterson J B and Davis R S 1985 A density comparison of silicon artifacts between NML (Australia) and NBS (US) *Natl Bur. Stand.* **90** 285–7

[24] Peuto A, Sacconi A, Balforn R and Kochsiek M 1985 Density determination of Zerodur spheres and cubes by measuring the mass and dimensions *BCR Inform. Appl. Metrol.* EUR 10372

[25] Wagenberth H *et al* 1987 Density inter-comparison measurements on solid standards of PTB and IMGC *PTB Report* BCR Contract 3017/1/0/90/85/3, pp 1–23

[26] Bettin H, Glaser M, Spieweck F, Toth H, Sacooni A, Peuto A, Fujii K, Tanaka M and Nezu Y 1997 International inter-comparison of silicon density standards *IEEE Trans. Instrum. Meas.* **46** 556–8

Chapter 2

Special interferometers

Symbols

λ	Wavelength of radiation used
ω	Angular velocity of a radiation
δ, θ	Phase angles
ε	Fraction of a fringe
$\delta\varphi$	Difference in phase angle
μ	Refractive index
f	Focal length of the lens
N_0	Integer number of fringes or fringe order
$\Delta\mu$	Change in refractive index
α	Coefficient of linear thermal expansion
$\delta f / \delta x$	Partial differential coefficient of function f with respect to an independent variable x
Var	Variance, the square of the standard deviation.

2.1 Introduction

Metrologists working in the field of density measurements may not be experts in interferometry and in interference phenomena, so it is prudent to describe the basics of interference phenomena and give details of the interferometers used in determining the diameters of spheres and the dimensions of a cube. It should be noted that these are described purely from the user's point of view and hence are not comprehensive.

2.2 Interference phenomena

Light travels in the form of electromagnetic waves, which are transverse in nature. Each electric or magnetic vector moves in a sinusoidal wave. A point source emits a divergent beam as a bundle of rays. All points equi-distant from it lie on a concave spherical surface with the source as its centre; every point on the surface acts as an independent secondary source with equal amplitude and the same phase. This spherical surface is called a wavefront. If on one wavefront, the amplitude is at a maximum, there will be two wavefronts, one succeeding and another following, on which the amplitude will be at a minimum. The two surfaces will be equi-distant from the central one. The distance between the two adjacent surfaces will be equal to half the wavelength of light. In other words, the propagation of light may be considered as a system of equi-distant surfaces alternatively experiencing maximum and minimum amplitudes, which are respectively called the surfaces of crests and troughs.

Now let us consider the two point sources A and B. Draw from each point a train of equi-distant spherical surfaces with alternate maximum and minimum amplitudes, which in the plane of the paper will be arcs of concentric circles with the source at the centre. Let alternate arcs be made with dotted lines and let these represent the troughs. Arcs from one source will cut arcs from the other source, i.e. the waves from the two sources will interact/ interfere with each other. The intersection points of the arc of troughs from source A with arc of troughs from source B will naturally represent points where the two waves are in the same phase, i.e. here the amplitudes of the two waves will be added. Similarly the intersection points of the crest from source A and the crest from source B will be points where the waves are in the same phase, i.e. here also amplitudes will be added. Hence, at these points, the energy will be at a maximum. Meanwhile the intersection of the arcs of the crest from one source and trough from the other source will represent the point at which the two waves are in opposite phase. Hence at such points the amplitudes of the two waves will be subtracted and the energy will be at a minimum. This interaction of light waves from two sources in space is known as the interference of light. Therefore, interference is a phenomenon in which the intensity of light (energy) in the area where two light beams are interacting is redistributed. The points at which the energy is at a maximum will be brighter if seen on a screen; similarly the points of minimum energy will appear darker if seen on a screen. The intensity of the light will taper from bright to darker points, hence we get a set of bright and dark strips on the screen normal to the plane of the paper. These we call interference fringes.

From figure 2.1, it should also be noted that the locus of the intersection points, either of the crests or troughs, will be on straight lines. Similarly the intersection points of the crest from one source with the trough of the other and vice versa will also lie on a straight line. Furthermore, the two statements are only true with two conditions: The first is that the set of arcs from the two sources are equidistant, i.e. the two waves have the same wavelength. Hence two interfering beams should be monochromatic. Second, the relative positions of the set of arcs do not change, i.e. the two sources are in the same phase. Such sources are said to be coherent sources. Hence the necessary and sufficient condition for the two beams to be able to produce fringes is that the two beams are coherent. Coherent

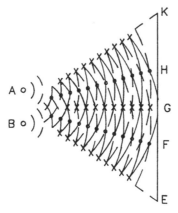

Figure 2.1. Two interfering waves.

beams or sources of light are those which always have the same phase or the phase difference remains unaltered in space as well as in time.

We can observe any effect of light provided it persists for a certain minimum time, say 0.1 s. During this time, millions of waves will be emitted by the source. Therefore to see an interference fringe, all the waves from the two sources must reach a point with a certain specific phase relation and this specific phase relation should remain unaltered for the time required by the eye and mind to perceive the event. Hence no interference effect will be observed unless the two sources continuously emit waves of the same frequency or wavelength. This requirement is called temporal coherence. To obtain interference fringes, the amplitudes of the two waves should be equal or nearly equal. A dark fringe occurs where the two amplitudes are subtracted from each other, so unless two amplitudes are equal or are nearly so, the difference will not be nearly zero. Similarly where the two amplitudes add together the sum may not be very different from their difference, so the eye may not be able to distinguish a difference between the dark and bright fringes. Therefore to obtain a good set of dark and bright fringes the amplitudes of the two waves must be equal or nearly so. Again to see the interference effect at a point, the waves reaching the point from two sources should either have the same phase or a constant phase difference and this phase difference must remain unaltered for the time required by the eye and mind to perceive it. This requirement is known as spatial coherence.

So formally two sources are said to be coherent if they emit light waves of the same frequency, and always bear a constant phase difference relationship, both in space and time. This is possible only when two interfering beams originate from a single source.

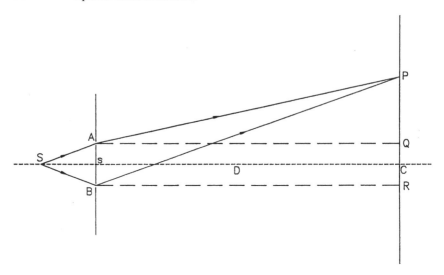

Figure 2.2. Interference from two point sources.

2.3   Fringe width and order

Let us consider a point source S of monochromatic light. The source S illuminates two pinholes A and B equidistant from S. The two pinholes are symmetrically situated with respect to S; i.e. S lies on the perpendicular bisector of AB—this is known as the symmetric axis. The two pinholes act as two coherent sources of light as they are illuminated from a single source. The distance between pinholes A and B is *s*. A screen normal to the symmetric axis is placed at a distance *D* from the midpoint of AB. The distance *s* between the two pinholes is very small in comparison to *D*. This arrangement is shown in figure 2.2.

As the sources at A and B are derived from the same source, these will maintain a constant phase relationship. Let C be the point on the screen, where the axis of symmetry meets the screen. Here the path difference will be zero. Let us consider a point P on the intersection of the planes containing the screen and the paper and *h* units of length from the point C. From figure 2.2, one can see that BP − AP is the path difference at the point P between the light waves from the two sources A and B. Draw the normals to the plane of the screen from A and B which meet the screen at Q and R, respectively.

Let AQ = BR = *D*; PC = *h*; and AB = *s*. Then PQ = *h* − *s*/2 and PR = *h* + *s*/2 so we get

$$AP^2 = AQ^2 + PQ^2 = D^2 + (h - s/2)^2$$
$$BP^2 = AR^2 + PR^2 = D^2 + (h + s/2)^2$$

giving

$$(BP^2 - AP^2) = (h + s/2)^2 - (h - s/2)^2$$
$$(BP - AP)(BP + AP) = 2hs.$$

Furthermore, $h \pm s/2$ is very much smaller than D, so $BP + AP \cong 2D$ giving

$$BP - AP = x_1 = hs/D. \tag{2.1}$$

The amplitude y_1 of a progressive wave of frequency ω from source A at P is given by

$$y_1 = A \sin(\omega t + 2\pi x/\lambda) \tag{2.2}$$

where x is the distance from an arbitrary point and λ is the wavelength of the light. In addition $\omega = 2\pi c/\lambda$, where c is the velocity of light. Similarly the amplitude from the source B at point P is given by

$$y_2 = A \sin(\omega t + 2\pi(x + x_1)/\lambda). \tag{2.3}$$

Putting

$$(\omega t + 2\pi x/\lambda) = \theta \tag{2.4}$$

and

$$2\pi x_1/\lambda = \delta \tag{2.5}$$

and combining the amplitudes at P, we get

$$\begin{aligned} y_1 + y_2 &= A \sin(\theta) + A \sin(\theta + \delta) \\ &= A[\sin(\theta) + \sin(\theta)\cos(\delta) + \cos(\theta)\sin(\delta)] \\ &= A[\{1 + \cos(\delta)\}\sin(\theta) + \cos(\theta)\sin(\delta)] \\ &= R \sin(\theta + \phi). \end{aligned} \tag{2.6}$$

Here

$$A\{1 + \cos(\delta)\} = R\cos(\phi) \tag{2.7}$$

and

$$A \sin(\delta) = R \sin(\phi). \tag{2.8}$$

From (2.7) and (2.8) we have

$$R^2 = 4A^2 \cos^2(\delta/2) \tag{2.9}$$

and

$$\tan(\phi) = \tan(\delta/2) \tag{2.10}$$

giving

$$\phi = \delta/2 = \pi x_1/\lambda. \tag{2.11}$$

The time average of the square of the amplitude at a point is the intensity of light at that point. But the time average of $\sin(\theta + \phi)$, which contains ωt as one of its terms is 1. The reason for this is that the argument of the sine function contains ωt and as ω is very large, the sine function fluctuates between 1 and -1 too rapidly. Therefore the intensity at point P is

$$R^2 = 4A^2 \cos^2(\delta/2). \tag{2.12}$$

As a result of the cosine term, the intensity will fluctuate between zero and 1. The intensity will be zero when $(\delta/2)$ is $\pi/2$ or its odd multiple and will be equal to $4A^2$, when $\delta/2$ is 0, π or its multiple. Hence the intensity at P will be at a maximum when $\delta/2$, in general, is $n\pi$, giving

$$\pi x_1/\lambda = n\pi \qquad \text{or} \qquad x_1 = n\lambda. \tag{2.13}$$

Substituting the value of x_1 from (2.1), we get $hs/D = n\lambda$. Putting $n = N_0$ gives

$$h = N_0\lambda D/s. \tag{2.14}$$

Here N_0 always takes an integer value (including zero) and is called the fringe order.

Let the centres of two consecutive bright fringes be at a distance h_1 and h_2 from C. Then $h_1 - h_2$ is the fringe width and is the sum of the widths of one bright and one dark fringe. If

$$h_1 = N_0\lambda D/s \tag{2.15}$$

then

$$h_2 = (N_0 - 1)\lambda D/s \tag{2.16}$$

giving

$$h_1 - h_2 = \lambda D/s. \tag{2.17}$$

A similar expression would have been derived if we had considered the centres of two consecutive dark fringes. Here we see that the fringe width is independent of N_0, which means all fringes are of equal width.

Now the point C is equidistant from A and B so the path difference is zero, hence it is the centre of the bright fringe. Furthermore at C, h is 0, giving $n = 0$. Therefore C is the centre of a bright fringe of order zero. But $\lambda D/s$ is the sum of the widths of one bright and one dark fringe, and the width of each dark or bright fringe is equal, so the width of the bright or dark fringe will be $\lambda D/2s$.

The width of a bright or dark fringe is proportional to $\lambda/2$.   (2.18)

Moreover λ, the wavelength of light, is very small say about 0.5 μm, therefore D/s should be large enough to make its product equal to a few mm to see the fringes well and perform any measurement. For example, for $D = 2000$ mm, s the distance between the two pinholes must be as small as 0.2 mm, to give a dark or bright fringe width equal to 2.5 mm.

- By analogy with sound waves light waves also suffer a phase change on reflection. For perfectly elastic materials it is π or, equivalently, a path difference of $\lambda/2$. However, the actual phase change would depend on surface quality. The phase change also occurs due to the finite size of the aperture of the receiving beam. The uncertainty in the phase change is one source of type B errors.
- Similar wavefronts from interfering sources produce parallel straight-line fringes. Interfering wavefronts may be either both spherical or both plane.
- Fringes represent a locus of equal path (phase) difference between the two interfering beams.
- Interaction between a plane wavefront and a spherical wavefront will produce circular fringes.
- A parallel beam of light has a plane wavefront.

2.4 Fizeau's interferometer

To measure the linear dimensions of objects, a Fizeau interferometer is generally used, so it will not be out of place to give some details of this type of interferometer.

2.4.1 Construction of Fizeau's interferometer [1]

The optical arrangement in a Fizeau-type interferometer is shown in figure 2.3. S is a monochromatic light source. Normally mercury or cadmium lamps are used for this purpose. For precise and accurate measurements, the frequency, i.e. the wavelength of the source, is calibrated against an iodine-stabilized He–Ne laser. The light coming from the pinhole is focused on another aperture A by a converging lens L1. The aperture is placed in the focal plane of another lens L2, which collimates the beam. The collimated beam is transmitted through a constant deviation prism P. The prism has the property that, depending upon the incidence angle, it will transmit a parallel beam of a specific wavelength. With a monochromatic source, which normally emits light waves with quite a few different wavelengths, by rotating the prism carefully a light of a particular wavelength may be obtained in the desired direction. This parallel beam falls on an optical flat F, which serves two purposes: (1) its lower face works as reference surface for the other surfaces under test; and (2) it acts as a beam splitter. Part of the beam is transmitted through it and falls on the objects and part of the beam is reflected back. Let the reflected beam be denoted by 1. The transmitted beam falls on the surface of the object and also on the base plate, which acts as a datum reference surface. The beam that is reflected back from each surface, namely beams 2 and 3, interferes with beam number 1, which was reflected from the optical flat. Two sets of fringes are therefore formed. It should be noted that when a beam is reflected from the denser medium (the metal) to a rarer medium (air), it suffers a phase change of almost π radians. However, the exact phase change will

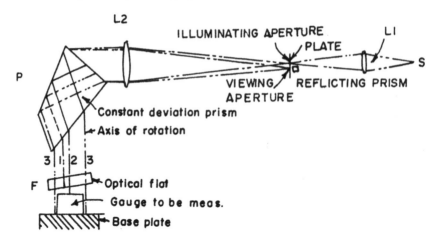

Figure 2.3. Optical arrangement in Fizeau's interferometer.

also depend upon the surface finish of the reflecting surface. To obtain the finite width of the interference fringes, the optical flat is kept slightly inclined to the datum surface. The base plate of the interferometer is highly polished and made optically flat. The base plate can be rotated at will so that one of the many objects can be brought under the light beam to produce interference fringes.

There are two sets of fringes: one obtained with the optical flat and the top surface of the object; the other one, which surrounds the first, is obtained with the optical flat and base plate. The base plate is made about 95% reflecting while the surface of the lower face of the optical flat is made 45% reflecting and 55% of the light is transmitted through it. Hence, the amplitude of the light reflected back from the base plate is about 50% of the amplitude of the original beam, so the amplitudes of the two interfering beams are almost the same. All reflecting beams retrace their path and are diverted by a semi-reflecting mirror in the plane of the aperture A. The interference fringes are examined under a high-power eyepiece with a graticule. From the interferogram only the fractional displacement from the fringes formed with the base plate are estimated. Quite often the source assembly has two lamps, each of which can be brought into action by mechanical means. One source is made from ^{198}Hg and the other is cadmium or the yellow line of krypton ^{86}Kr.

2.4.2 Application

The interferometer is mostly used to calibrate gauge blocks. The gauge blocks are wrung to the base plate along a circular ring. To achieve ringing, the gauge blocks should not only be highly polished and flat but also properly cleaned. These are usually cleaned with ethanol and a chamois leather or a good quality fibre-free cloth. The gauge blocks are wrung to the base plate and left to reach thermal

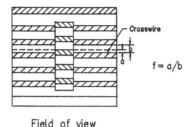

Field of view

Figure 2.4. Fringes formed in Fizeau's interferometer.

equilibrium. The interferometer is installed in a temperature-controlled room or chamber. The temperature of the room should be maintained within $20 \pm 0.5\,°C$. All length measurements are referred to $20\,°C$. Each gauge is brought, in turn, to the centre of the field of view. The displacement of the fringe, situated in the centre of the exposed surface of the gauge from the adjacent fringe from the base plate, is estimated. This displacement a, expressed as a fraction of the fringe separation b, represents the fraction of the half-wavelength by which the gauge length exceeds an integral number of half-wavelengths of the incident light. The fractions are estimated to a hundredth of the fringe width. By rotating the constant deviation prism, different wavelengths of light are brought into action and the fraction of the fringe width is estimated for each wavelength. The estimation is carried out with red, green, blue and violet waves of mercury and cadmium. Knowing the values of these fractions and employing the method of exact fractions, the number of complete fringes (N_0) is calculated for each wavelength. The length of the block is given by

$$L = (N_0 + \varepsilon)\lambda/2$$

where $\varepsilon = a/b$, the fraction of the fringe order. The pressure, temperature and relative humidity of the ambient air in the interferometer are noted. These parameters are required to calculate the refractive index of the ambient air. The refractive index is required for calculating the wavelength of the radiation used in ambient air.

2.4.3 Applicable corrections

To derive the value of the true length of the gauge block at $20\,°C$, corrections will need to be made for the following errors:

(1) errors in the temperature resulting from thermal expansion of the gauge material;
(2) errors in the refractive index of the ambient air;
(3) errors in phase changes at the gauge block and base plate due to differences in their surface finishes; and finally

(4) obliquity errors arising from the inclination of the incident light with the normal to the surfaces of base plate and gauge block.

Temperature correction

If α is the coefficient of linear expansion of the material of the gauge blocks, the temperature correction per unit length will be

$$\alpha((t\,°C^{-1}) - 20).$$

Refractive index correction

The standard values of monochromatic wavelengths in the visible region are quoted for wavelength in air at 760 mm Hg (101 325 Pa) at 20 °C and relative humidity of 50%. If the air is not at the specified temperature and pressure then the necessary correction due to the change in the refractive index is applied. The following relationship is used.

$$\lambda_a/\lambda_s = 1 + [A - B(h_1\,mm^{-1})(1 + \alpha(t\,°C^{-1})) + C(F - 10)] \times 10^{-6} \quad (2.19)$$

where λ_a is the wavelength in ambient conditions of air and λ_s is the wavelength in standard conditions of air.

$$h_1 = h(1 + bh(t\,°C^{-1})).$$

Here $b(t\,°C^{-1}) = (1.049 - 0.0157(t\,°C^{-1})) \times 10^{-6}$ and h is the corrected barometric pressure in mm of Hg. F is the actual vapour pressure of water vapour as against 10 mm of Hg for standard air and A, B and C are constants evaluated for the mean wavelength of the group of radiations used for the measurement.

Although the corrections for reducing the derived length of the gauge to standard conditions are in themselves small they are significant in relation to the accuracy achievable for the determination of length. For example, if the ambient air is at 21.0 °C, pressure 759 mm of Hg and the water vapour present in the air has a partial vapour pressure of 11 mm Hg, then the following corrections are applicable for the 25 mm gauge block.

Cause	Correction (nm)
Temperature (1 °C)	23.2
Pressure (1 mm Hg)	9
Humidity (1 mm Hg)	1.5
Total correction (nm)	33.8

Phase error correction

The phase change on reflection from the base plate and gauge block has the effect that the light appears to be reflected from a surface below or above the mechanical surface and effectively changes the derived length of the gauge.

As a result of the phase change at the base plate, the reflecting planes appear to be p units below the mechanical plane, while the reflecting plane from the top of the gauge block may appear to be g units below the top surface (figure 2.5). If the values of g and p are equal then the optical length as measured by the interferometer will be equal to its mechanical length L. Otherwise the optical length O and mechanical length L will be related as follows

$$O = L + (g - p). \tag{2.20}$$

To estimate the value of $(g - p)$, several gauge blocks from the same set with similar surfaces are taken and their individual optical lengths O_1, O_2, O_3 etc are determined. The second time all the gauge blocks are wrung together and their combined optical length O_c is also determined. Let five such gauge blocks be taken. In the first case

$$O_r = L_r + (g - p)$$

for each gauge. r can take values from one to five, giving us

$$O_1 + O_2 + O_3 + O_4 + O_5 = L_1 + L_2 + L_3 + L_4 + L_5 + 5(g - p). \tag{2.21}$$

In the second case the combined optical length will be related to the mechanical length:

$$O_c = L_1 + L_2 + L_3 + L_4 + L_5 + (g - p) \tag{2.22}$$
$$O_c - (O_1 + O_2 + O_3 + O_4 + O_5) = 4(g - p). \tag{2.23}$$

Obliquity error

There is an additional error in the measurement of the length of the gauge block due to the fact that the incident beam is not normal to the gauge block and base plate. This arises because although the viewing and illuminating apertures are in the same focal plane of the collimating lens L2 they are separated by a distance s. Hence, the incident and reflected beams will be inclined to the optical axis by an angle θ, i.e. the incident and reflected beams will also make an angle θ with the normal to the surface of the base plate and gauge block. Here θ is given by

$$\theta = s/2f \tag{2.24}$$

f being the focal length of the collimating lens. So from the point of view of the incident beam (figure 2.6), the effective length of the gauge block will be X,

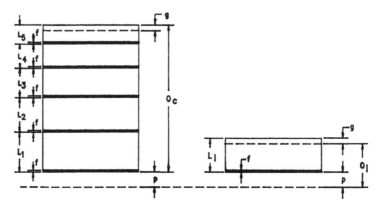

Figure 2.5. Effect of phase change.

while for the reflecting beam it will be Y. From figure 2.6, it can be easily seen that

$$X = L/\cos(\theta) \qquad \text{while } Y = L\cos(2\theta)/\cos(\theta). \tag{2.25}$$

Therefore, the effective optical distance will be

$$(X + Y)/2 = L[1 + \cos(2\theta)]/2\cos(\theta) = L[1 + 2\cos^2(\theta) - 1]/2\cos(\theta)$$
$$= L\cos(\theta). \tag{2.26}$$

Hence, the error $E1$ will be given by

$$E1 = -L[1 - \cos(\theta)]. \tag{2.27}$$

Expanding $\cos(\theta)$ in terms of θ,

$$E1 = -L\theta^2/2 \tag{2.28}$$

or

$$E1 = -L(s/2f)^2/2 = -Ls^2/8f^2. \tag{2.29}$$

Furthermore, there is an error $E2$ due to the finite size of the apertures. For a circular aperture of diameter d,

$$E2 = -L(d/4f)^2 = -Ld^2/16f^2 \tag{2.30}$$

and for a rectangular aperture of length l and width h,

$$E2 = -L(l^2 + h^2)/24f^2. \tag{2.31}$$

Therefore the total error due to the finite size of the apertures and their relative displacement will be given by

$$E = E1 + E2.$$

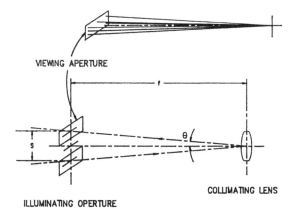

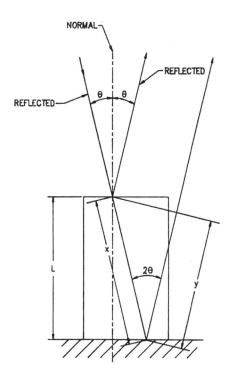

Figure 2.6. Effect of obliquity.

Therefore for a circular aperture

$$E = -L[(s^2/8f^2 + d^2/16f^2] \tag{2.32}$$

and for a rectangular aperture

$$E = -L[(s^2/8f^2 + (l^2 + h^2)/24f^2]. \tag{2.33}$$

2.5   Interferometer to measure the diameter of a sphere

Saunders' interferometer [2], with minor modifications, has been used by most national laboratories to measure the diameters of spherical objects. The interferometer is a combination of two Fizeau interferometers in which the incident coherent beam can approach from either side of the object. The object in this case consists of an etalon and the sphere whose diameter is to be measured. The etalon (figure 2.7) consists of two etalon plates, which are placed parallel to each other at a distance. The outer surface of each plate is slightly inclined with reference to the inner surface to avoid reflections from the outer surface of the plate. The distance between the two plates is a few cm more than the diameter of the sphere to be measured. The inner sides of the two etalon plates are made optically plane and parallel to each other with a high degree of accuracy. Two beams, one reflected from the inner side of the etalon plate and the other from the sphere, interfere. For the plane-wave incident beam, the reflected beam from the plate has a plane-wavefront while the beam reflected from the sphere has a spherical-wavefront and is divergent in nature. So the interference pattern is a set of concentric rings. The centre of the fringe pattern (or interferogram) represents the minimum distance between the etalon plate and the sphere. So if the complete order of the central fringe is known, then the minimum distance d_1 between the sphere and the plate can be found. Similarly the distance d_2 is measured from the other side. The length of the etalon—the distance between the etalon plates L—is also measured by the interferometer. The actual quartz etalon used at NMI, Japan is shown in photograph P2.1.

However, from the interferogram, only the fractional part of the fringe order can be determined. In order to find the integer part of the fringe order various methods are used. One uses two known sources of light followed by the method of exact fractions. Another is to find the integer part by a mechanical measurement using a fringe-counting interferometer. Masui *et al* [3, 4] used the first method. Bell and Patterson [5] and Saconni *et al* [6] used the second method. To measure the fractional order of the central fringe, scanning is carried out along the axis of the beam with photodiodes and electronic devices, as have been used by Fujii *et al* [7, 8] and Succoni *et al* [9]. Some minor modifications have been incorporated to improve the quality of interferogram at NRLM [3].

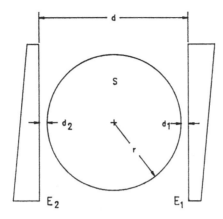

Figure 2.7. Etalon with sphere inside.

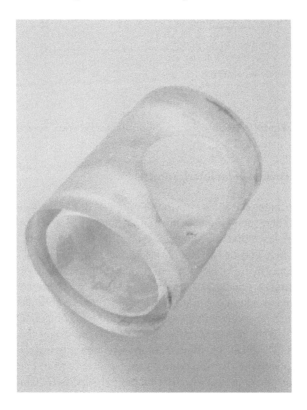

Photograph P2.1. Glass etalon without sphere (courtesy NMI, Japan).

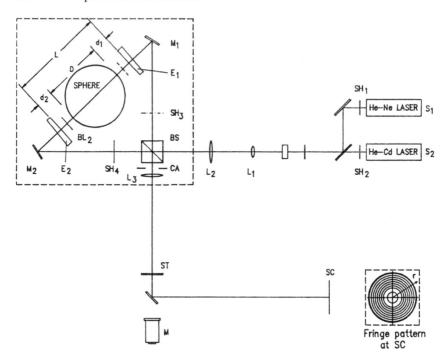

Figure 2.8. Saunders' interferometer for spherical objects.

2.5.1   Construction of an interferometer

The optical arrangement [3] is shown in figure 2.8. It has the following essential components. There are two independent light sources S_1 and S_2. S_1 is a helium–neon (He–Ne) laser and the other, S_2, is a helium–cadmium (He–Cd) laser. Alternatively there may be two He–Ne lasers, one in the red region and the other in the green region. NML [5] used the latter. One of them must have been calibrated against the primary standard of length. The shutters SH_1 and SH_2 are used to allow one source in action at a time. The lenses L_1 and L_2 collimate the light on to the beam splitter BS, which divides the beam into two components; the reflected light beam strikes the mirror M_1 at an angle of 45° and strikes the etalon plate E_1 at right angles. The other transmitted component similarly travels through mirror M_2 to etalon plate E_2. The observations are taken only from one side at a time. For this purpose there are removable shutters SH_3 and SH_4, which can obstruct one of the desired beams.

There are two spacers, which separate etalon plates E_1 and E_2. The inner plane surfaces of E_1 and E_2 are parallel to each other within 2×10^{-5} radian. The distance between the two etalon plates is of the order of the diameter of the sphere to be measured; for example Masui *et al* [3] used 115 mm for a sphere of

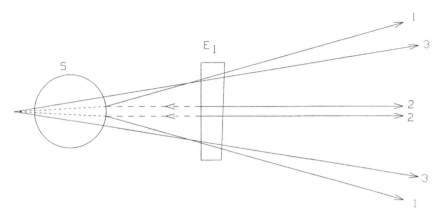

Figure 2.9. Splitting of the incident beam at the inner surface of the plate and sphere.

diameter 85 mm.

Therefore, there are two arms to the interferometer, each of which works as a separate interferometer. The sphere is approached from either side in turn. There are a number of removable stops and apertures to limit the aperture of the beam.

The diameter of the sphere is measured in two steps:

Step I. Measure d_1—the minimum distance between the sphere and the plate E_1—and d_2—the minimum distance between the sphere and plate E_2. The scheme for measuring d_1 and d_2 with two sources will be given in section 2.5.4.

Step II. Measure L, the distance between the plates E_1 and E_2. Then the diameter D of the sphere is given by

$$D = L - d_1 - d_2. \tag{2.34}$$

When the incident beam falls normal to the etalon plate E_1 three things happen to the incident beam (see figure 2.9):

(1) part of the beam is reflected back from the inner surface of the etalon plate;
(2) the transmitted portion of the beam is partly reflected from the sphere surface; and
(3) the remaining part is refracted into the sphere—this part of the beam may ultimately reach the opposite etalon plate. This portion of the beam is due to the transparency of the sphere and does not contain any useful information—it only has a disturbing effect.

Interference takes place between the beam reflected from the inner surface of the etalon plate and the beam reflected from the surface of the sphere, so the useful beams are those reflected from the internal side of the etalon plate and the nearest surface of the sphere, which interfere and produce an interference fringe pattern of concentric rings.

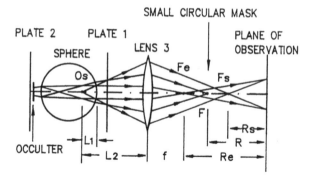

Figure 2.10. Beams converging at different points.

A small removable blade BL_2 (figure 2.8) is placed between the etalon plate E_2 and the sphere to mask the light that leaks to the side of the sphere and would eventually be reflected back from E_2.

The beam reflected from the surface of the sphere is much more divergent so, in order to limit its undesirable reflections from the inside of the tube containing lens L_3, another aperture CA is placed before the collimating lens L_3. Lens L_3 collimates the three beams at three different points on its axis as shown in figure 2.10. A small stop ST is placed at the converging point of the third beam to remove its undesirable effect. The stop is a small circular dot of chromium coating deposited onto a fused quartz plate and is fixed on a small universal stage capable of placing the spot exactly at the point of convergence.

The remaining two beams will interfere and produce fringes in the observation plane SC. As the loci of equal path differences are circles, the interference pattern is in the form of circular concentric rings. The order of interference is a polynomial function of the radius of the ring. As a result of the converging effect of the lens L_3, it is possible to maximize the visibility of the fringes by locating the screen so that the amplitudes of two interfering beams become equal. In this case, referring to figure 2.10 [4], the following relation will hold good.

$$R_s = f^2 L_1 \tau / \{(L_2 - f)(L_2 - f - L_1 \tau)\}$$
$$R_e = f^2 / (L_2 - f - L_1 \tau). \tag{2.35}$$

Here τ is the transmissivity of light at the boundary between the quartz and air.

The etalon, the sphere and the beam-splitting prism are enclosed in a temperature-controlled chamber (shown with dotted lines in figure 2.8). The top lid is removable. The temperature within the etalon cavity is to be kept uniform within 0.02 °C. This difference could not cause any significant errors as both the etalon and sphere are made of fused quartz, which has a very low expansion coefficient.

2.5.2 Analysis of interferograms

The objective of the fringe analysis is to find the fractional part ε of the interference order at the centre of the interferogram. Information on the gap length d_1 or d_2 is contained in the phase difference $\delta\psi$ between two interfering beams at the optical axis. The interference order at the centre of the interferogram is related to $\delta\psi$ by definition:

$$N_0 + \varepsilon = \delta\psi/2\pi. \tag{2.36}$$

N_0 is the integer part of the fringe order and ε is its fractional part. N_0 may be calculated by the method of exact fractions by using two wavelengths. ε is calculated from the interferogram. The fringe order is a polynomial function of the square of the fringe radius and is given by the relation

$$N - N_0 = \varepsilon + Ar^2 + Br^4 + \cdots . \tag{2.37}$$

Normally B is quite small in comparison with A. By assigning values 1, 2 and 3 to $N - N_0$ and measuring the fringe radius starting from the centre we get a set of ordered pairs $(N - N_0, r)$. The constants A and B in this equation are estimated using the least-squares method. Hence, ε is deduced from the interferogram.

2.5.3 Relation between ε and d, the least gap between the etalon plate and sphere

From (2.36) we can write $\delta\psi$ as

$$\delta\psi = 2\pi(\varepsilon + N_0). \tag{2.38}$$

In contrast, the same $\delta\psi$ is related to d, the least distance between the sphere and etalon plate, as

$$\delta\psi = 4\pi d/\lambda + (\varphi_1 - \varphi_2) + (\zeta_1 - \zeta_2) - \pi. \tag{2.39}$$

φ_1 is the phase change at the etalon plate and is equal to π and φ_2 is the phase change due to reflection at the sphere, which is equal to zero. The terms ζ_1 and ζ_2 represent the phase changes which have occurred during propagation and have been calculated by Masui *et al* [11] to be $-\pi$ in each case. An additional $-\pi$ appears as the radii of black fringes rather than white fringes are measured. This gives

$$\delta\psi = 4\pi d/\lambda + (\pi - 0) + (-\pi + \pi) - \pi = 4\pi d/\lambda$$

giving

$$d = (N_0 + \varepsilon)\lambda/2 \tag{2.40}$$

where N_0 is an integer.

2.5.4 Measurement of the length of the etalon

To measure the etalon length, the sphere is removed from the cavity and the collimation system is adjusted so that the incident beam on the etalon plate is slightly divergent. The reflected beams from the two etalon plates will interfere and circular fringes will be formed in the focal plane of lens L3. An interferogram of concentric rings is photographed. In this case the fringe count associated with the rings will decrease as the diameter of the ring increases. So the rings are numbered as $-1, -2, -3$ from the innermost ring. Their diameters are measured and the corresponding radii fitted to a polynomial of degree two and the use of equation (2.37) gives the fractional part of the fringe ε.

2.5.5 Measurement of the diameter of the sphere

The sphere whose diameter is to be measured is first cleaned with methanol and then with a dilute aqueous solution of a neutral detergent. It is then rinsed several times with pure distilled water. The sphere is dried for several hours in a clean chamber maintained at $50\,°C$. The diameter is measured in several uniformly distributed directions, so that the entire solid angle is covered. For each direction a number of photographs of the circular fringes following the scheme given in section 2.5.6 are taken for d_1 and d_2. Measurements are taken alternatively for d_1 and d_2 to avoid errors due to the relative displacement of the sphere. The values of d_1 and d_2 are calculated from equation (2.40). The value of the diameter is obtained from equation (2.33). After one set of measurements is completed, the sphere is rotated through a known angle.

2.5.6 Observation scheme

Let the source S_1 be calibrated against the primary standard of length and let it become operative if SH_1 is open and SH_2 is closed. Similarly source S_2 becomes operative if SH_2 is open and SH_1 is closed. In addition, only one out of the two sources is to be used at a time and observations are only to be taken from one side at a time. In order to reduce the errors due to changes in the environmental conditions, the following scheme of observations should be followed.

Open SH_1	Open SH_3	take observations for d_1
	Close SH_3 and open SH_4	take observations for d_2
	Close SH_4 and open SH_3	take observations for d_1
	Close SH_3 and open SH_4	take observations for d_2
	Close SH_4 and open SH_3	take observations for d_1
Open SH_1	Close SH_3 and open SH_4	take observations for d_2
	Close SH_4 and open SH_3	take observations for d_1
	Close SH_3 and open SH_4	take observations for d_2
	Close SH_4 and open SH_3	take observations for d_1

| | Close SH_3 and open SH_4 | take observations for d_2 |

Close SH_1 open SH_2

	Close SH_4 and open SH_3	take observations for d_1
	Close SH_3 and open SH_4	take observations for d_2
	Close SH_4 and open SH_3	take observations for d_1
	Close SH_3 and open SH_4	take observations for d_2
	Close SH_4 and open SH_3	take observations for d_1

Close SH_2 open SH_1

	Close SH_3 and open SH_4	take observations for d_2
	Close SH_4 and open SH_3	take observations for d_1
	Close SH_3 and open SH_4	take observations for d_2
	Close SH_4 and open SH_3	take observations for d_1
	Close SH_3 and open SH_4	take observations for d_2

Close SH_1 open SH_2

	Close SH_4 and open SH_3	take observations for d_1
	Close SH_3 and open SH_4	take observations for d_2
	Close SH_4 and open SH_3	take observations for d_1
	Close SH_3 and open SH_4	take observations for d_2
	Close SH_4 and open SH_3	take observations for d_1

Close SH_2 open SH_1

	Close SH_3 and open SH_4	take observations for d_2
	Close SH_4 and open SH_3	take observations for d_1
	Close SH_3 and open SH_4	take observations for d_2
	Close SH_4 and open SH_3	take observations for d_1
	Close SH_3 and open SH_4	take observations for d_2

Close SH_2 open SH_1

	Open SH_3	take observations for d_1
	Close SH_3 and open SH_4	take observations for d_2
	Close SH_4 and open SH_3	take observations for d_1
	Close SH_3 and open SH_4	take observations for d_2
	Close SH_4 and open SH_3	take observations for d_1

This way we will have six values of d_1 for an uncalibrated wavelength and 12 for a calibrated one while for d_2 there are four values for an uncalibrated wavelength and 13 for the calibrated wavelength and symmetry is maintained. This observation scheme will cancel or at least minimize the effect of any linear change in temperature and pressure.

2.5.7   Selection of diameters to be measured, and evaluation of their differences and calculation of volume of sphere

The method for selecting and evaluating the variation in diameters was described in section 1.4.2. To determine the volume of the sphere the mean of all the diameters can be taken and the volume calculated by the volume formula for the sphere. However, the method described in the latter part of section 1.4.2 is more accurate and should be used wherever feasible.

2.5.8   Influencing factors and their contribution to uncertainty

The factors which influence the measurement of the diameter of a sphere and their contribution to the uncertainty in the calculation of volume of the sphere are given in table 2.1 [3].

   The combined uncertainty is the square root of the sum of the squares of the component uncertainties (quadrature method) and the relative uncertainty is the combined uncertainty divided by the mean value of the volume of the sphere, which in this case is 0.36 ppm.

2.5.9   Law of propagation of variances

If a quantity y is a function of several variables, then the variance of the quantity y is calculated by the following steps:

(1)   Calculate the partial derivative with respect to each variable and the variance of each variable.
(2)   Square the partial derivative and multiply it by the respective variance of that variable.
(3)   Take the sum for all calculated values in step 2.

This procedure is only valid provided all variables are independent, otherwise their respective co-relations and partial derivatives should also be considered. For example, the diameter d of the sphere is measured from the length L of the etalon and the least air gap between the sphere and two etalon plates. Let, for simplicity, the volume of the sphere be represented as

$$V = 4\pi d^3/3 = 4\pi (L - d_1 - d_2)^3/3. \tag{2.41}$$

There are three parameters, L, d_1 and d_2, which have been measured, and each of them will have a variance. Let their respective variances be denoted by $\text{Var}(L)$, $\text{Var}(d_1)$ and $\text{Var}(d_2)$. Then the combined variance of the volume is calculated by the following steps. The partial derivatives with respect to these variables are:

$$\begin{aligned}
\delta V/\delta L &= 4\pi (L - d_1 - d_2)^2 \\
\delta V/\delta d_1 &= -4\pi (L - d_1 - d_2)^2 \\
\delta V/\delta L d_2 &= -4\pi (L - d_1 - d_2)^2
\end{aligned} \tag{2.42}$$

Table 2.1. Uncertainty budget in the volume measurement of a sphere.

Factors	Uncertainty in each factor	Relative standard uncertainty ($\times 10^6$)
Type B		
1. Calibration of wavelength $\Delta\lambda/\lambda$	8.0×10^{-9}	0.02
2. Refractive index of air $\Delta\mu/\mu$	4.02×10^{-8}	0.12
Air temperature (°C)	0.02	
Atmospheric pressure (Pa)	10	
Water vapour pressure (relative humidity) (Pa)	50	
CO_2 content (mol%)	0.01	
Elden's equation $\Delta\mu/\mu$	1.0×10^{-8}	
3. Sphere temperature	0.02	0.03
4. Phase shift in reflection from etalon in terms of fringes	<0.002	<0.03
5. Phase shift in reflection from the sphere in terms of fringes	<0.01	<0.2
6. Effect of beam stop in terms of fringes	<0.01	<0.2
7. Coefficient of cubical thermal expansion of sphere (°C^{-1})	3.0×10^{-7}	0.03
8. Coefficient of linear thermal expansion of etalon (°C^{-1})	3.0×10^{-9}	0.001
9. Angular alignment of etalon with respect to optical axis in radians	2×10^{-5}	0.002
10. Elastic deformation of sphere due to self-weight		<0.01
Type A		
Standard deviation of mean		0.18

giving variance Var(V) as

$$\text{Var}(V) = (\delta V/\delta L)^2 \, \text{Var}(L) + (\delta V/\delta d_1)^2 \, \text{Var}(d_1) + (\delta V/\delta d_2)^2 \, \text{Var}(d_2) \quad (2.43)$$
$$= 16\pi^2 (L - d_1 - d_2)^4 [\text{Var}(L) + \text{Var}(d_1) + \text{Var}(d_2)]. \quad (2.44)$$

2.5.10 Uncertainty in the volume of the sphere due to uncertainty in the measurement of the diameter

The volume of the sphere V as given by equation (1.34) is

$$V = (1/24) \sum \sum \omega_{ij} (d_i + S_{ij})^3. \quad (2.45)$$

Equation (2.45) gives the volume of the sphere in terms of equatorial diameters and the differences in meridian diameters from their respective

Table 2.2. Variance in measurement of diameter.

	Variance of diameter (nm^2)
Etalon calibration $Var_c(d)$	7.6
Air gap $Var_u(d)$	34
Refractive index $\Delta\mu/\mu\lambda$	12.6
Wavelength of laser used $\Delta\lambda/\lambda$	0.6
Out of roundness $Var_s(d)$	22

equatorial diameter and solid angle ϖ_{ij}. d_i is given by equation (1.34). Therefore the combined variance $Var_{(volume\ at\ 20)}$ will be given by

$$Var_{(volume\ at\ 20)} = (\pi^2 d^4/20)\, Var_u(d) + (\pi^2 d^4/4)\, Var_c(d)$$
$$+ (5d^4/64)\, Var_s(d)\left\{\left(\sum \varpi_{ij}\right)^2 + \sum \varpi_{ij}^2\right\} \quad (2.46)$$

where $Var_u(d)$ is the variance in the air-gap measurements. $Var_c(d)$ is the variance in the measurement of etalon length L and $Var_s(d)$ is the variance in the measurement of the differences S_{ij} between the meridian diameter and the corresponding equatorial diameter. $Var_s(d)$ is the variance in the values of the meridian diameters. Table 2.2 gives the values of these variances as obtained by Patterson and Morris [10].

Substituting the values of these variances in equation (2.45), the variance in the volume of sphere at 20 °C is 0.0026 mm^6.

Further to this, there is a variance due to the arbitrary selection of five meridians. This has been estimated as 0.0007 mm^6. Therefore, the combined variance $Var_{(volume\ at\ 20)}$ from these aforesaid causes is 0.0033 mm^6 and the combined standard uncertainty in volume at 20 °C is 0.06 mm^3. The relative uncertainty in volume at 20 °C is 0.26 ppm.

$Var(V_t)$ is obtained by adding the variance associated with the experimentally determined term $V(t) - V(20)$. This contribution is estimated from the etalon and diameter regressions together with an allowance to account for the spatial temperature variations observed in the interferometer between 30 and 40 °C.

$Var(V_t)$ is found to be 0.061 mm^6 at 1 °C, 0.0033 mm^6 at 20 °C and 0.026 mm^6 at 40 °C.

2.6   New spherical Fizeau interferometer

A new spherical Fizeau interferometer has been developed at PTB, Germany [11]. The optical arrangement is shown in figure 2.11. It consists of two reference

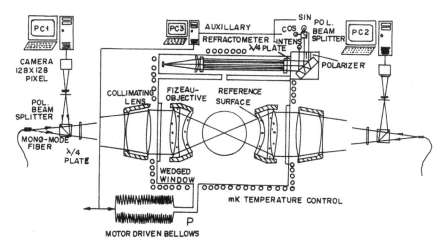

Figure 2.11. Spherical Fizeau interferometer.

spherical surfaces and the sphere whose diameter is to be measured. The spherical waves are generated from plane waves by means of special Fizeau lenses, which also contain the reference spherical surfaces. All spherical surfaces are centred carefully and the measurement is performed on concentric spherical surfaces. The form of the measuring waves is adapted to the surface of the sphere whose diameter is to be measured. As a result, not only one diameter on the optical axis can be evaluated but also many diameters corresponding to the aperture angle of the optical system are obtained. The aperture angle in this case is 60°. Similar to Saunders' interferometer, here also two sets of measurements are needed. The first set is taken with the empty etalon. The sphere is then placed centrally inside the etalon. The second set of measurements is between the surface of the sphere and the adjacent spherical surface of the Fizeau lens. Measurements are possible from both sides but are taken one at a time by interrupting the other beam. The interference pattern is imaged on an electronic camera matrix of 128×128 discrete photodiodes. The data are evaluated with a special phase-stepping algorithm [12, 13]. The 90° phase steps, required for the algorithm, are obtained by changing the effective optical path length. The change in effective path length is obtained by changing the refractive index of air. By changing the air pressure, the refractive index of the air, inside the interferometer, is changed. For this purpose the interferometer is placed inside a pressure-tight housing. Computer-controlled motor-driven bellows P change the air pressure. The refractive index of air is simultaneously measured with the help of a modified Jamin interferometer capable of measuring the refractive index with a relative uncertainty of 5×10^{-9}. The sphere is placed on a three-point support and can be lifted and rotated around two perpendicular axes by means of another motor-driven device, so that it can be positioned in well defined orientations. Because of the electronic camera with 16 000 pixels and wider angular view, the variation in diameter is measured with high angular resolution. The uncertainty in the diameter measurement may be made as small as 1×10^{-8}.

References

[1] Pools S P and Dowell J H 1960 Application of interferometric to block gauge measurement *Optics in Metrology* ed P Mollet (Oxford: Pergamon)
[2] Saunders J B 1972 Ball and cylinders interferometer *J. Res. Natl Bur. Stand.* C **76** 11–20
[3] Masui R, Fujii K and Takenaka M 1995/96 Determination of the absolute density of water at 16 and 0.101 325 MPa *Metrologia* **32** 333–62
[4] Fujii K, Masui R and Sieno S 1990 Volume determination of fused quartz spheres *Metrologia* **27** 25–31
[5] Bell G A and Patterson J B 1984 Density standards, the density and thermal dilation of water *Precision Measurements and Fundamental Constants* vol II, ed B N Taylor and W D Phillips (Gaithersburg, MD: National Bureau of Standards–US Department of Commerce) pp 445–7
[6] Sacconi A, Peuto A, Mosca M, Pnaciera R, Pasin W and Pettorruso S 1995 The IMGC volume–density standards for the Avogadro's constant *IEEE Trans. Instrum. Meas.* **44** 533–7
[7] Fujii K, Tanaka M, Nezu Y, Nakayama K and Masui R 1993 Accurate determination of density of crystals silicon sphere *IEEE Trans. Instrum. Meas.* **42** 395–400
[8] Fujii K *et al* 1995 Absolute measurement of densities of silicon crystals in vacuum for determination of the Avogadro's constant *IEEE Trans. Instrum. Meas.* **44** 542–5
[9] Saconi A, Peuto A, Pasin W, Panciera R, Lenaers G, Valkiers M, van de berg M and de Bievre P 1989 Towards the Avogadro's constant—preliminary results on the molar volume of silicon *IEEE Trans. Instrum. Meas.* **38** 200–5
[10] Patterson J B and Morris E C 1994 Measurement of absolute density of water 1 °C to 40 °C *Metrologia* **31** 272–88
[11] Nicolaus R A and Bonch G 1997 A novel interferometer for dimensional measurement of a silicon sphere *IEEE Trans. Instrum. Meas.* **46** 54–6
[12] Bonch G and Bohme H 1989 Phase determination for Fizeau interferences by phase shifting interferometry *Optik* **82** 161–4
[13] Nicolaus R A 1991 Evaluation of Fizeau interference by phase shifting interferometry *Optik* **87** 23–6

Chapter 3

Standard Mean Ocean Water (SMOW) and the equipment for measuring water density

Symbols

$n(^{16}O)$	Number of atoms, present in given volume, of oxygen of mass number 16
$n(^{18}O)$	Number of atoms, present in given volume, of oxygen of mass number 18
$n(^{17}O)$	Number of atoms, present in given volume, of oxygen of mass number 17
$n(H)$	Number of atoms, present in given volume, of hydrogen of mass number 1
$n(D)$	Number of atoms, present in given volume, of deuteron of mass number 2
R_{18}	$n(^{18}O)/n(^{16}O)$ = abundance ratio of water molecules having oxygen of mass number 18
R_{17}	$n(^{17}O)/n(^{16}O)$ = abundance ratio of water molecules having oxygen of mass number 17
R_D	$n(D)/n(H)$ = abundance ratio of water molecules having heavy hydrogen (deuteron) of mass number 2
δD	Deviation of the ratio of $R_{D(sample)}/R_{D(SMOW)}$ from unity = $[R_{D(sample)}/R_{D(SMOW)} - 1]10^3$
$\delta\,^{18}O$	Deviation of the ratio $R_{18(sample)}/R_{18(SMOW)}$ from unity = $[R_{18(sample)}/R_{18(SMOW)} - 1]10^3$
$\delta\,^{17}O$	$\delta\,^{17}O = 0.5\delta\,^{18}O$
ρ	Density of water
$\rho_{(SMOW)}$	Density of SMOW
$\rho_{(V\text{-}SMOW)}$	Density of V-SMOW. It should be noted that $\rho_{(V\text{-}SMOW)} \equiv \rho_{(SMOW)}$
$\Delta\rho$	Change in density of water

χ	Level of air saturation
W	Mass of body
m	Mass of standard weights
I	Indication by the balance
BC	Buoyancy correction
$\rho_{(p,t)}$	Density of water at pressure p (Pa) and temperature t (°C)
g_h	Acceleration due to gravity at height h from some reference point.

3.1   Introduction

In chapter 1, we discussed solid-based density standards, their uncertainty and international inter-comparisons. It was emphasized that for practical purposes, water and mercury are the most suitable secondary standards of density. Water, apart from in oceanography, is also used as a density standard for the calibration of volumetric glassware using the gravimetric method. Water is available from a large number of sources, all of which may have different densities. The reasons for these differences are twofold. The first is the presence of various chemicals, including air, as impurities both as solute as well as suspended particles. The second is the different isotopic compositions. The impurities, including air in water, can be removed by various processes, but it is difficult to change the isotopic composition of pure water. Hence it is necessary to define water so far as its isotopic composition is concerned. Furthermore, it is also necessary to establish a relationship between the density and isotopic composition. Alternatively we could arrive at a relation which gives the difference in the density of the sample water from that of some standard water and the deviation of its isotopic composition.

It is next to impossible to avoid air dissolution in water when it is used for practical purposes as the density standard, so it is equally important to establish a relationship for its density and the state of air saturation.

In addition, internationally acceptable methods to obtain pure samples of water are explored and finally the method and apparatus used in the most recent determination of the density of water are given.

3.2   Standard Mean Ocean Water (SMOW)

Pure water is composed from four types of water molecule:

(1) The most abundant molecules are those in which one atom of oxygen of atomic mass 16 combines with two hydrogen atoms each of atomic mass 1, $H_2{}^{16}O$.

Other isotopic water molecules are:

(2) one atom of oxygen of atomic mass 18 combined with two hydrogen atoms each of atomic mass 1, $H_2{}^{18}O$;

(3) one atom of oxygen of atomic mass 17 combined with two hydrogen atoms each of atomic mass 1, $H_2 {}^{17}O$; and

(4) one atom of oxygen of atomic mass 16 combined with one atom of hydrogen of atomic mass 1 and with one atom of deuteron of atomic mass 2, $HD {}^{16}O$.

The presence of a particular type of molecule containing a specific isotope of oxygen is defined as the ratio of the number of atoms of that isotopic oxygen to the number of normal oxygen atoms of atomic mass 16 present in a given volume. The case for deuteron is similar. If R_{18}, R_{17} and R_D are the ratios, then these are given by

$$R_{18} = n({}^{18}O)/n({}^{16}O) \tag{3.1}$$
$$R_{17} = n({}^{17}O)/n({}^{16}O) \tag{3.2}$$

and

$$R_D = n(D)/n(H). \tag{3.3}$$

Normally these are expressed as an atomic part per million and the symbol is ppma, i.e. the number of isotopic atoms per million normal atoms.

To arrive at a commonly agreed standard of water, Craig [1] introduced the idea of Standard Mean Ocean Water (SMOW). This idea helped in the comparison of the isotopic composition of different samples of water. The values of R_D and R_{18} for SMOW given by Craig were:

$$R_{18} = (1993.4 \pm 2.5) \times 10^{-6} \tag{3.4}$$
$$R_D = (158 \pm 2) \times 10^{-6}. \tag{3.5}$$

At the insistence of the International Atomic Energy Agency (IAEA), Vienna, several laboratories took up the work of re-determining the isotopic composition of water. The accuracy obtained and the consistency of the results for the same sample analysed by different laboratories have been improved appreciably. Two new reference samples of water have been obtained. These samples allowed the definition of a scale for δ ($\delta {}^{18}O$ and δD) which includes practically all values found in nature.

First reference water: This sample is new SMOW or the Vienna SMOW (V-SMOW), prepared by Craig who obtained water with practically the same isotopic composition as the former reference water. Hagemann *et al* [2] and Baertschi [3] have determined the absolute values of the isotopic ratios. The new values are:

$$R_{18} = (2005.2 \pm 0.45) \times 10^{-6} \tag{3.6}$$
$$R_D = (155.76 \pm 0.05) \times 10^{-6}. \tag{3.7}$$

Second reference water: Standard Light Antarctic Precipitation (SLAP) is the second reference sample. The water is obtained from melted ice from the Antarctic from Plateau Station, which is at an altitude of 3700 m and 79°15′ S and $40°30'$ E. This was collected by Professor E Picciotto. The absolute value of R_{18} and R_D as determined by Hagemann *et al* [2] for this water sample is

$$R_{18} = (1893.9 \pm 0.45) \times 10^{-6} \qquad (3.8)$$

$$R_D(\text{SLAP}) = (89.02 \pm 0.05) \times 10^{-6}. \qquad (3.9)$$

SLAP has the lowest value of R_D found in nature to date.

So the re-determined and internationally agreed values of R_D, R_{18} and R_{17} for V-SMOW, as declared by the Vienna convention, are [4]:

$$R_D = (155.76 \pm 0.05) \times 10^{-6} \qquad (3.10)$$

$$R_{18} = (2005.2 \pm 0.05) \times 10^{-6} \qquad (3.11)$$

$$R_{17} = (371.0) \times 10^{-6}. \qquad (3.12)$$

It should be remembered that there is no difference between SMOW or V-SMOW; either of them represents water with $R_{18} = 2005.2 \times 10^{-6}$ and $R_D = 155.76 \times 10^{-6}$.

Deviations in the isotopic composition of sample water with respect to SMOW

The deviation in the isotopic composition of the sample water is equal to the difference from unity of the ratio of the abundance ratios of the sample and SMOW. The multiplying factor of 1000 is used to avoid smaller fractions. δD, $\delta\,^{17}O$ and $\delta\,^{18}O$ are given as follows:

$$\delta D = [R_D(\text{sample})/R_D(\text{SMOW}) - 1] \times 10^3$$

$$\delta\,^{17}O = [R_{17}(\text{sample})/R_{17}(\text{SMOW}) - 1] \times 10^3 \qquad (3.13)$$

$$\delta\,^{18}O = [R_{18}(\text{sample})/R_{18}(\text{SMOW}) - 1] \times 10^3.$$

The value of $\delta\,^{17}O$ is seldom measured—for natural waters it is related to $\delta\,^{18}O$ by the relation

$$\delta\,^{17}O = 0.4989\delta\,^{18}O = 0.5\delta\,^{18}O. \qquad (3.14)$$

The scale of values for $\delta\,^{18}O$ and δD

Two standards of water, namely SMOW and SLAP, are used to establish the scales for $\delta\,^{18}O$ and δD.

(1) The point of origin is taken to be SMOW. By definition $\delta\,^{18}O = \delta D = 0$.
(2) SLAP: The limit which corresponds to water with the lowest isotopic concentration of ^{18}O and D. The values of $\delta\,^{18}O$ and δD in comparison to SMOW are

$$\delta\,^{18}O(\text{SLAP}) = -55.5 \qquad (3.15)$$

and

$$\delta D = -428. \qquad (3.16)$$

These two values have been universally adopted.

Laboratories interested in the measurement of isotopic ratios may obtain these two samples of water from IAEA, Vienna. They should then determine the isotopic ratios of their water sample, SLAP and SMOW and normalize their results in such a way that the values of $\delta\,^{18}O$ and δD for the SLAP sample are, respectively, -55.5 and -428.

Difference in the density of the sample with known deviations with respect to SMOW

Once the values of $\delta\,^{17}O$, $\delta\,^{18}O$ and δD are known, the density ρ of the sample can be determined from the known value of $\rho_{(SMOW)}$—the density of SMOW or V-SMOW. Initially Menache [5] gave the following relation:

$$\rho - \rho_{(SMOW)} = 0.23 \times 10^{-3}\delta\,^{18}O + 0.018 \times 10^{-3}\delta D \qquad (3.17)$$

where $\rho - \rho_{(SMOW)}$ is in kg m^{-3}. However, Girard and Menache, using their own experimental results [6], later revised this equation to

$$\rho - \rho_{(SMOW)} = 0.211 \times 10^{-3}\delta\,^{18}O + 0.0150 \times 10^{-3}\delta D. \qquad (3.18)$$

Ultimately, Menache *et al* [7] derived the equation theoretically and it is given by

$$\rho - \rho_{(V\text{-}SMOW)} = 0.233 \times 10^{-3}\delta\,^{18}O + 0.0166 \times 10^{-3}\delta D. \qquad (3.19)$$

Girard and Coarasa, at the 1981 *Conference on Precision Measurements and Fundamental Constants II* [8], quoted the same expression. Equation (3.19) is now the internationally accepted relation for the density difference between sample water and SMOW.

Patterson and Morris [9] have used this equation to express their results in terms of V-SMOW. It should be noted that for a water sample with $R_{18} = 1982.4$ and $R_D = 144.8$, the correction applicable for V-SMOW is 0.003 82 kg m^{-3} while it will be only 0.002 33 kg m^{-3} if Craig's value of 1993.4 for R_{18} is taken. This amounts to a difference of 1.6 ppm in the density of water.

All density determinations can now be reduced to the density of Standard Mean Ocean Water (SMOW), which is the same as that of V-SMOW. Menache and Girard [10] have suggested that the density of SMOW should be taken as 999.975 kg m^{-3} at 3.9818 °C—the temperature of its maximum density.

3.3 Preparation of water samples for density measurements

Normally a water sample is either prepared from tap water fit for human consumption or from sea water. Different methods are used for these two water samples.

3.3.1   Water from the tap

The water is filtered through fine-pore filter paper and then kept in the equipment for reverse osmosis. In the second stage the water is de-ionized, and left for carbon absorption. It is finally filtered through filters with a uniform pore size of 0.2 μm. (For this process, commercial equipment is available; one such apparatus is supplied by Millipore Corporation USA.)

    If the water quality is poor, the water should first be distilled preferably in a copper still. The distillate is made to pass through the processes of de-ionization, carbon absorption and filtration through 0.2 μm pores as before.

3.3.2   Water from the sea

Water from the sea should pass through the following processes:

(1)   Add an excess of anhydrous sodium fluoride (NaF), which will change all chlorides of magnesium, calcium and strontium into their respective fluorides. These fluorides have a higher thermal stability. The solution is then distilled.
(2)   In the second stage, small amounts of sulphuric acid (H_2SO_4) and potassium permanganate ($KMnO_4$) are added to the distillate and it is distilled again.
(3)   In the third stage, the distillate is neutralized with a small amount of sodium hydroxide (NaOH) and is distilled again in a fused quartz still. The distillate at this stage should have a resistivity of 0.7 MΩ cm. This stage is repeated several times if necessary to bring the resistivity up to 1 MΩ cm.

Further care should be taken that at least 98% of water by weight, which was taken for treatment, is recovered. Therefore, at each stage, all water should be distilled as far as possible. This will ensure that the isotopic composition of the initial water and that of the final distilled water sample is unchanged. The process is similar to that described by Cox *et al* [11].

3.4   Dissolved air and density of water

3.4.1   De-aeration of sample water

Method 1

One of the easier ways to de-aerate water is to use a water vacuum pump. The method used by Masui *et al* is also described later. The arrangement for the water pump is shown in figure 3.1. About a litre of water is put into a flask of capacity 1.5 litres and connected to a water jet vacuum pump. The water is continually stirred magnetically or ultrasonically. As the pump starts to run, the bubbles in the water start coming out. The water pump should continue until the bubbling ceases. The method ensures that the free oxygen content is less than 1% of its

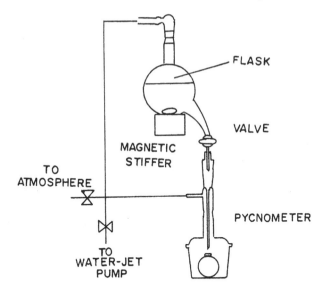

Figure 3.1. De-aeration of water.

saturation value. The effect of the remaining oxygen will be less than 0.03 ppm on the value of the density of the water sample.

Method 2

Another de-aerating method is to put the sample water together with the artefact and its carrier and suspension system in the vessel to be used in the experiment. Raise the temperature of the water to 40 °C and simultaneously evacuate the space above the water sample. Air bubbles should start coming out. Continue the process until the air bubbles stop; 30 min are normally sufficient. Fill the vessel up to its top with boiled water from the same source as the sample was drawn. Insert the thermometer to restrict the area of sample water exposed to air.

Method 3

Yet another procedure is to boil the sample water for half an hour. Isolate the water sample from air until it cools down to 60 °C. Pour the sample water into the sample vessel and add the sphere, its carrier and its suspension arrangement as quickly as possible and isolate the sample vessel. The sample is now ready to be cooled down for measurements.

3.4.2   Effect of dissolved air on the density of water

It is mainly nitrogen and oxygen from air that become dissolved in water. The solubility of oxygen and nitrogen is given by the following expression:

$$\log(X) = A + B/(T\ \mathrm{K}^{-1}) + C \log(T\ \mathrm{K}^{-1})$$

where A, B and C for oxygen and nitrogen are [12]:

	For oxygen	For nitrogen
A	−66.7354	−67.3877
B	87.4755	86.3213
C	24.4526	24.7981

Here T is the temperature in kelvin: $T = 273.15 + t\,°\mathrm{C}^{-1}$.

It can be seen that the values of the constants for oxygen differ only slightly from those for nitrogen. This shows that the saturation levels for nitrogen and oxygen will always be equal. The solubility of the carbon dioxide and argon present in air (only 4% of the total volume of dissolved gases) is much lower. The solubility decreases with increase in temperature, so the effect of the dissolved gases on the density of water also decreases with temperature. It is almost negligible beyond 40 °C.

Marek [13] was perhaps the first to investigate the effect of dissolved air on the density of water. He measured the difference in the density of air-free and air-saturated water between 0 and 14 °C and gave results which can best be represented by a parabola with its maximum at about 5 °C, i.e. the decrease in the density is maximal at 5 °C. However, fully saturated or completely air-free water states are not realizable. Chappuis [14] measured the difference in the density of air-saturated and air-free water from 5 to 8 °C. The difference in density was close to that measured by Marek but it remained practically constant in the range of temperatures which he studied.

Girard and Coarasa [8] measured the difference between the density of air-free water and that of water with a known amount of atmospheric gases at temperatures 4, 10, 16 and 22 °C with the following conclusions:

- The decrease in density is linearly related to the amount of air dissolved.
- The decrease in density between air-free and fully-saturated water is linear with temperature and becomes almost zero around 33 °C and is about 0.0048 kg m^{-3} at 0 °C.

The amount of atmospheric gases dissolved in water is measured on the assumption that the saturation levels of oxygen and nitrogen, in the temperature range under consideration, are equal [15]. Therefore, only the oxygen saturation level represents the water saturation level. To calculate the level of oxygen the Winkler titration method, as improved by Carpenter [16], may be used. The generally accepted relation between change in density and temperature of fully saturated water is:

$$\Delta\rho = -0.004\ 612 + 0.000\ 106t. \tag{3.20}$$

$\Delta\rho$ is in kg m^{-3} and t is the temperature in °C on the IPTS 68 scale.

Alternatively equation (3.20) may be written as

$$\Delta\rho\ (\text{kg m}^{-3})^{-1} = -0.004\ 612 + 0.000\ 106t\ (°\text{C IPTS 68})^{-1}.$$

This form of representation is more correct from the point of view of keeping the dimensions the same for each term. I followed this practice for the previous equations. One can now see that each term in the equation is a pure number. This agrees very well with the previous results except the gradient of this line is a little low, so the effect of the dissolved air becomes zero at about 40 °C. A similar expression has been obtained by Bignell [17] who gave an expression for the uncertainty at the 2σ level:

$$2.56[3.1 \times 10^{-4} + 1.066 \times 10^{-5}(t - 11.52)^2]^{1/2}. \tag{3.21}$$

Combining (3.20) with the linearity of the density differences with respect to the saturation level χ I would like to write the density difference with temperature and saturation level as follows:

$$\Delta\rho = (-0.004\ 612 + 0.000\ 106t)\chi. \tag{3.22}$$

Following the example of (3.20), equation (3.22) may be written as

$$\Delta\rho\ (\text{kg m}^{-3})^{-1} = [-0.004\ 612 + 0.000\ 106t\ °\text{C}^{-1}]\chi.$$

3.5   Hydrostatic method for measuring water density

3.5.1   Principle

Archimedes' principle is used to determine the density of water. When a body of known volume is immersed in a sample of water, the loss in its apparent mass is the mass of water displaced by the body. The mass of water displaced is the product of the volume of the body and the density of water at the measurement temperature. As the volume of the body is known, the density of water at the measurement temperature can be calculated.

The mass of the displaced water can either be calculated by knowing the mass of the body and its apparent mass in water or directly with a little manipulation as will be discussed later.

3.5.2   Mass of the standard body in air

The mass of a body is determined by comparing its mass against standards of known mass and density by weighing in air. Air buoyancy is applied from prior knowledge of the volume of the body and the density of the used weights and air. The air density is calculated by simultaneous measurement of air pressure, temperature and its relative humidity. To calculate air density, an equation, based on IPTS 68, developed by Giacoma in 1981 [20] and later modified by the BIPM in 1990, is used [19]. However, the use of either the 1981 or 1990 equation will cause a difference of about 0.000 001 kg m^{-3} in the density measurement, which is much less than the overall uncertainty expected from such an experiment. Alternatively the air density can be measured *in situ* by measuring the difference in air buoyancy between two objects with the same mass but different volumes.

To eliminate linear time-related variation, a double substitution method with simultaneous determination of the sensitivity of the balance should be used. This method is accomplished as follows.

(1) Place the body on the pan centrally. In the case of a single-pan balance, lift sufficient built-in weights (set of dial weights) in such a way that some standard weights are required to be placed on the balance pan to bring the equilibrium almost to the middle of the scale. In the case of two balance pans, sufficient weights are placed on the other pan in such a way that some standard weights are required to be placed on the balance pan to bring the equilibrium almost to the middle of the scale. Let the indication be I_1.

(2) Lift the body from the pan and substitute the equivalent standard weights so that equilibrium is obtained near to the first rest point. Let the indication be I_2 and the mass of the placed standard weights be S.

(3) Place a small weight, commonly called the sensitivity weight, so that the new equilibrium point is on the extreme end of the scale. Let this be I_3 and the mass of the sensitivity weight be m.

(4) Place the body to be weighed in the pan. Remove the standard weights (not the small sensitivity weight). The equilibrium point will also be closer to I_3. Let the new indication be I_4.

(5) Remove the sensitivity weight and the new indication will be I_5.

Then the mass of the body W is given by

$$W = S + m(I_1 - I_2 + I_4 - I_3)/(I_3 - I_2 + I_4 - I_5) + BC \qquad (3.23)$$

where BC stands for buoyancy correction. An example for recording and calculating for an oscillating two-pan balance is given in table 3.1.

$$W = S + (75.4 - 93.5 + 85.4 - 103.5)$$
$$\times \, 0.1/(85.4 - 75.4 + 103.5 - 93.5) \text{ mg} + BC$$
$$= S + 0.1(-18.1 - 18.1)/(10.0 + 10.0) \text{ mg} + BC.$$

Substituting the value of S,

$$W = 1000.000\,195 \text{ g} + (-36.2)/20.0 \times 0.1 \text{ mg} + BC$$
$$= 1000.000\,195 \text{ g} - 0.181 \text{ mg} + BC$$
$$= 1000.000\,014 \text{ g} + BC.$$

Table 3.1. Recording observations and the calculations for measuring mass in air. The mass of the given standard is 1000.000 195 g. RHP stands for the right-hand pan of a two-pan balance.

				Date:			

			Start	Finish
Time:				
Temperature:			25.3 °C	25.3 °C
Pressure:			752.24 mm Hg	752.36 mm Hg
Relative humidity:			50.0%	50.0%

Sample no	Weight on RHP	Rider position	Scale readings			Mean	Rest point
1	W	5.1	42.2	42.4	42.6	42.4	93.5
			144.7	144.5		144.6	
2	S	5.1	24.3	24.5	24.7	24.5	75.4
			126.4	126.2		126.3	
3	$S + m$	5.0	44.3	44.5	44.7	44.5	85.4
			126.4	126.2		126.3	
4	$W + m$	5.0	52.2	52.4	52.6	52.4	103.5
			154.7	154.5		154.6	
5	W	5.1	42.2	42.4	42.6	42.4	93.5
			144.7	144.5		144.6	

Buoyancy correction

Let the volume of the body be 429.189 cm^3 and that of the standard weight be 119.047 cm^3 and the air density at 25.3 °C and 752.3 mm Hg pressure and 50% relative humidity be 0.001 1676 g cm^{-3}. Hence,

$$BC = 0.001\ 1676(429.189 - 119.047) = 0.362\ 122\ \text{g}$$

and M, the mass of the body, is 1000.362 136 g.

Problems due to static charges

As most of the solid density standards are made from glass or ceramic-like materials, problems due to electrostatic charges collecting on their surfaces may hamper the accuracy in weighing. Care therefore has to be taken to avoid electrostatic charges collecting on the solid standard, which is normally a sphere. For this purpose, the sphere is wrapped in aluminium foil about 25 cm × 25 cm. The two are weighed together by using the double-substitution weighing method. The outer surface is touched occasionally with a well grounded wire, so that no accumulation of electrostatic charge takes place on the sphere. However, a mass difference has been observed between new flat aluminium foil and used foil, which may be due to wrinkling. So foil is wrapped around the body so that whatever wrinkling is to occur is complete. Then unwrap the body and find the mass of the foil separately. The body is wrapped with the foil, taking care that no extra wrinkling takes place. The combination is used in subsequent weighings in air. The mass of the foil is again found after the weighing in air is completed. The mean of the initial and final masses of the aluminium foil is used to find the mass of the body.

3.5.3 Mass of water displaced as the difference between the mass of a body and its apparent mass in water

The apparent mass of a body in water is obtained by carrying out the weighing as follows:

- with the immersed pan empty,
- with the immersed pan loaded with the object of interest,
- with the sensitivity weight added to the balance pan in air,
- with the object removed from the immersed pan but the sensitivity weight still on the balance pan and finally
- with the immersed pan empty and the sensitivity weight removed.

For a single-pan balance, let the apparent mass of the immersed pan be m_i and that of the body in water m_w; also let S_0, S_1, S_2 be the mass of the standard weights placed on the pan. Sufficient built-in weights are removed so that in each case some standard weights have to be placed on the pan to bring the scales to equilibrium. Let the scale indications be I_1, I_2, I_3, I_4 and I_5, then we may write

$$m_i + S_0(1 - \sigma/\Delta) = K I_1 + D$$
$$m_i + m_w + S_1(1 - \sigma/\Delta) = K I_2 + D$$
$$m_i + m_w + m + S_1(1 - \sigma/\Delta) = K I_3 + D$$
$$m_i + m + S_0(1 - \sigma/\Delta) = K I_4 + D$$
$$m_i + S_0(1 - \sigma/\Delta) = K I_5 + D$$

where D is the mass of a dummy and K is a constant to convert scale indications in terms of the unit of mass, giving

$$m_w = (S_0 - S_1)(1 - \sigma/\Delta) + m(I_2 - I_1 + I_3 - I_4)/(I_3 - I_2 + I_4 - I_5) \quad (3.24)$$

where σ and Δ are, respectively, the density of air and the standard weights.

Here the capacity of the balance should be of the order of the apparent mass of the body in water, which is much less than the mass of the body in air. Therefore balances with a smaller capacity but with better readability can be used.

The temperature of the water is measured at every weighing. The whole set is repeated thrice. Before commencing a new set, the water is well stirred so that it acquires a uniform temperature throughout its bulk.

From the mass of the body M, which has already been determined separately, the loss in mass is calculated by subtracting the mass m_w from the mass of the body. But the loss in mass is the product of the volume V of the body and the density ρ of the water; i.e.

$$V\rho = M - m_w.$$

As V, the volume of the solid body, is already known, ρ, the density of water, can be calculated.

3.5.4 Direct determination of the mass of displaced water (NRLM method)

The NRLM found the loss in mass directly using the method described here. A single-pan balance with a capacity of 2 kg with a reproducibility of 20 μg was used. This had three specially built-in weights W_0 equivalent to the mass of the body in air, buoyancy weights S equivalent to the mass of water displaced by the body and some standardized small weights m to bring the equilibrium point almost to the same reading. First, the sphere was placed on the balance pan and weight W_0 lifted from the pan to bring the balance to a certain rest point—small weights were used if necessary. The sphere was then transferred into the water and placed on the lower pan. Now the buoyancy force due to the upthrust of the water acts on the sphere, so some weights are placed in the pan to restore the equilibrium. Buoyancy weight S along with some small weights were therefore loaded onto the pan. The small weights loaded onto the pan were such that the reading obtained when the sphere is on the pan in water remains unchanged when the sphere is removed from the pan and W_0 is loaded. Therefore the mass of the buoyancy weight plus those of the small weights was exactly equal to the mass of water displaced by the sphere. This way, it is not necessary to know the value of W_0—only the mass value of the buoyancy weight plus the small weights needs to be known.

In general, the procedure is as follows.

(1) Put the body in air on the pan of the balance and lift W_0 from the weight hanger, so that equilibrium is obtained at a convenient point on the scale.

(2) Transfer the body into the hydrostatic liquid on the lower pan of the balance. Add standard weights of known mass so that equilibrium is achieved at the same point of the scale.

(3) Remove the body from the lower pan and replace the weight W_0 on the upper pan and add some fractional standard weights so that equilibrium point is the same as in the second weighing.

Writing the equilibrium equations as before, we obtain

$$M + m_i = I_1 + W_0 \tag{3.25}$$

$$m_w + m_i = I_2 + W_0 - S \tag{3.26}$$

$$m_i = I_2 - m \tag{3.27}$$

giving

$$M - m_{\rm w} = \text{Mass of water displaced} = S + m + I_1 - I_2. \qquad (3.28)$$

This method has the disadvantage that a balance with a higher capacity, i.e. equal to the mass of the body, is necessary. However, as the mass of the sphere used is of the order of 1 kg, the balance capacity is not a problem. The most sophisticated balances are only available at the 1 kg level.

3.5.5   Minimizing the varying effect of surface tension on the suspension wire

Method 1

To reduce the varying effect of surface tension on the suspension wire, we at NPL used a thin copper wire, which was pre-oxidized by passing a small current for a few minutes in air.

Method 2

NRLM, Japan uses a wire which has been coated with platinum black by the following procedure. The wire is first heated in an alcohol flame and polished with fine emery paper. The wire is then electroplated by dipping it into 3%wt platinum chloride and 0.6%wt lead acetate for 2–3 min. The current through the wire was kept between 30 and 40 mA. Immediately after plating, the wire is baked in an alcohol flame until the platinum layer turns grey. This process stabilizes the platinum layer, which keeps a constant angle of contact with the water. The surface tension varies with contamination, which in this case is caused by micro dust particles. To avoid this, a guard ring 3 cm in diameter around the wire is provided, so that the water surface near the meniscus is isolated from the rest of the water surface. A Teflon circular shield is provided at the immersion pan, lest flakes of the platinum coating fall on the sphere.

Method 3

Another method to circumvent this problem is the procedure developed by the NML, Australia. A very thin stainless steel wire of diameter 0.08 mm is used. A windshield is provided for the suspension wire. A drop of detergent is added to the water surface through a side tube set at an angle. It has been verified that the added detergent drop did not make a difference to the density of water.

3.5.6   Correction due to the vertical gradient of gravity

As in most of the cases, the weights and body to be weighed are in different horizontal planes, a correction due to a variation in gravity is applied. The correction is about 300 μg on 1 kg for a vertical difference of 1 m.

If a liquid replaces the water then the density of the liquid can be similarly determined.

3.6   Thermostatic baths

3.6.1   Bath at NPL, New Delhi

As all precision measurements of density refer to a standard temperature, these are carried out at the specified temperature. For this purpose, a bath with a high degree of temperature stability is required. A bath designed and fabricated at NPL, India is used to determine the density of liquids including that of water, solids and other liquids using the solid density standards. A thermostatic bath essentially consists of three major items. These are (1) a cell, which contains a hydrostatic liquid, (2) various arrangements for placing and suspending a standard solid body from the balance and (3) temperature-measuring equipment. The cell is surrounded with a system for circulating water at the desired temperature and the arrangement to produce it. One such bath, constructed at NPL, India [21], is described in the following paragraphs. Improvements and additions carried out in the thermostatic baths are also given with appropriate references.

Construction

A vessel of rectangular section is connected to a cylindrical one through two tubes B and C. Water at the desired temperature is produced in the cylindrical vessel. The ensemble is placed in a bigger rectangular open box and is thermally insulated by covering it all around with an insulating material about 100 mm thick and leaving an air-gap of 10 mm to make the insulation more effective. The outer wall of the open box is made from a conducting layer of an aluminium sheet. The air between the insulation material helps to homogenize the temperature by convection, while the conductivity of the aluminium helps to reduce the effect of any localized heating or cooling occurring outside the bath. Water is used as the thermostatic liquid. The bath has a double-walled window of plate glass. Plate glass is used to avoid any distortion of the image. The bath is schematically shown in figure 3.2.

The cylindrical vessel is fitted with (1) a propeller P, (2) a cooling coil E, (3) a controlled heater H_1 and (4) a booster heater H_2. The thermostatic water is maintained at the required temperature by balancing the heat supplied by the controlled heater against the heat loss in cooling coils. The coils housed in the upper half of the cylindrical vessel completely surround the two heaters. The propeller P fitted just below the coil not only mixes the hot and cold water but also circulates it all around the bath through tubes B and C like a pump. The heater H_1 is controlled through a toluene regulator F and a set of electronic relays, which in turn regulate the temperature of the bath.

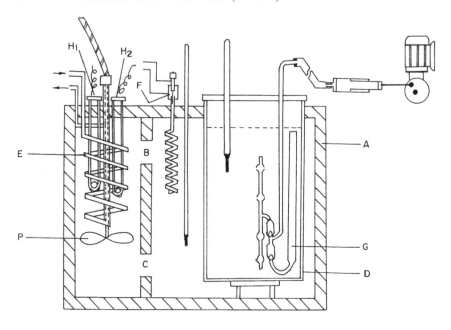

Figure 3.2. Thermostatic bath (NPL, India).

Figure 3.3. Toluene regulator.

Toluene regulator

The toluene regulator is made from low-expansion glass and is shown in figure 3.3. The coiled portion contains a liquid with a high thermal coefficient of expansion. Toluene has been used in this case. The straight portion terminating in a fine capillary at the other end contains mercury as the contact liquid. There is a cup surrounding the capillary. One small piece of tungsten wire is fused through the cup in the lower portion of the capillary tube. The lower end of the wire just touches the mercury in the capillary. A proportioning head H is fitted at the upper end of the cup, surrounding the capillary. The head H has a fixed wire, which dips in the mercury contained in the cup, and another wire whose height can be adjusted by turning the screw knob. The proportioning head is connected through relays to the controlled heater H_1. The head H is so adjusted that as soon as the desired temperature is reached, contact is made between its two terminals through the mercury in the capillary and the cup. So the heater is tripped. When the temperature falls, the mercury in the capillary recedes, due to contraction of the toluene and mercury in the regulator so the contact breaks and heater H_1 starts working again. The current in the heater is adjusted through the variable transformer so that the heat produced by the heater is compensated by heat loss in the surroundings. Once a proper setting of the head H is made the temperature can be maintained at the desired value. An electronic device [22] may be used instead of the toluene regulator. If the ambient temperature of the room is higher than the desired temperature, the temperature inside the water cell is attained by circulating pre-cooled water and balancing the heat produced by the controlled heater and heat required by the cold water. It has been observed that if the temperature of the pre-cooled water is 2 °C below the required temperature, then the heat balance is perfect to maintain the temperature within a few milli-kelvins. The hydrostatic liquid is constantly stirred except when weighing is carried out through a special glass pump G.

Glass pump

The glass pump consists of three limbs as shown in figure 3.4. Limb A is connected through a rubber tube to a cylinder provided with a reciprocating piston run by a small electric motor. Limb B is a suction tube and limb C has four or five equally spaced perforations. Limbs A and B are connected to limb C through valves V_1 and V_2 as shown in figure 3.4. The pump is placed in a liquid to be stirred in a vertical position such that limb B is completely immersed in the liquid. On the outward stroke of the piston (the piston is moving right), valve V_1 opens and a small amount of liquid is sucked from limb B. During the backward stroke, valve V_2 opens and valve V_1 closes so that the liquid is pushed out into the entire liquid to be stirred, through the perforations of limb C. Thus the pump keeps the liquid thoroughly stirred without dissolving any air or causing turbulence. This way, the surface of the liquid continues to be renewed, which eliminates any

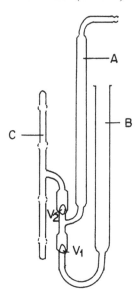

Figure 3.4. Glass pump.

change in surface tension due to contamination. The valves are in the form of two small glass cylinders with conical ends, which fit into the cross section of the respective tubes. The bath is suitable for all density measurement and calibrations of reference-grade hydrometers. The bath along with its balance and suspension device etc is shown in figure 3.5.

Discussions about stirring and contact of water with air

At NPL, India the aforesaid pump is used for stirring the sample water before and after each weighing creating mechanical vibrations in the sample water, which may take a long time to die out. In contrast, NRLM, Japan took special care to avoid mechanical vibration in the sample water. They used a special copper shield between the circulating water and the cell containing the sample water. All hydrostatic weighings are carried out in the cell to determine the density of the water sample. But this method has the danger of having a vertical temperature gradient.

Similarly special care has been taken to avoid contact between the sample water and air. NRLM used special quartz circular blocks to avoid contact with the sample water. Only holes for the thermometer, suspension wire etc were the sources of contact with air. We took no such care in the aforesaid bath.

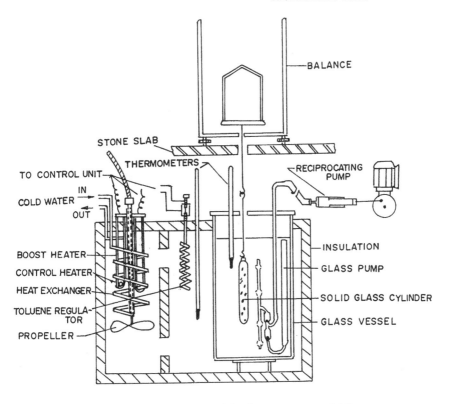

Figure 3.5. Arrangement of density measurement (old).

3.6.2 Computer-controlled density measurement system at NPL, India

NPL, India has recently acquired a complete system including TEMSEN electronic water baths, a special vertical hanger capable of taking three samples of solid objects and a new hydrostatic liquid to determine the volumes/density of solids against the known volume of standard quartz spheres. The density measurement is fully automatic and computerized [22].

Hydrostatic liquid

Fluninert FC-40 is used as the hydrostatic liquid which has the following advantages over water:

- a low surface tension of 16 mN m^{-1} as opposed to the 75 mN m^{-1} of water,
- a high density of 1981 kg m^{-3} as opposed to the 1000 kg m^{-3} of water,
- a compressibility of 1.3×10^{-4} bar^{-1} as opposed to that of 0.176×10^{-4} bar^{-1} for water and

- a volume expansion coefficient of 1.2×10^{-3} K^{-1} as opposed to that of 0.2×10^{-3} K^{-1} for water.

Temperature control

The bath temperature is maintained through a three-stage heat-exchanging system. The first stage has a compressor to bring the temperature of the circulating water down to 14 °C within ±0.1 °C. The circulating water is heated to 18 °C and the temperature is controlled within ±0.01 °C in the second stage. At the third stage the water is heated further and the temperature is maintained at 20 °C within ±0.005 °C. The TEMSEN bath pre-set at 20 °C is used for density measurements. The bath is placed on a hydraulic lift, which is capable of moving in all three directions.

Vertical sample holder

Instead of a horizontal object changer, a vertical one is used. There are three lifting devices, one for each object, and each one is moved by its own stepper motor. In a horizontal object changer, the system carrying all the standards and sample has to move about 90°, disturbing the liquid. The disturbances so produced take a long time to die down. In a vertical object changer, however, the vertical movement is restricted to about 1.5 mm. In addition, only one object together with its lifting device moves by the small distance of 1.5 mm [23], so the hydrostatic liquid remains almost undisturbed. Consequently, the residual density drift becomes a linear function of time. For the purpose of centring, three point supports are provided so that the spheres resting on these points always sit centrally on the hanging pan of the balance. For flat-based objects, three equally inclined rods are provided so that the object is placed on the hanging pan of the balance centrally. Standard solid bodies of known volume are normally placed on the first and third pans and the body whose volume, and hence density, is to be determined is kept on the middle pan.

Balance and other equipment

A Mettler balance, model AT 201, of capacity 200 g and resolution 10 μg, with a hanging pan is used for measuring the apparent weight of the body in the hydrostatic liquid. The temperature of the hydrostatic liquid is measured with a Guildline platinum resistance thermometer connected to the computer through an IEEE-488 interface. The relative humidity and air temperature in the weighing chamber is measured with a VAISALA HMI-36 hydrometer and the pressure is measured with a SETRA-370 pressure transducer. The vertical object changer, the balance and instruments measuring other parameters like pressure, temperature and relative humidity are connected with the appropriate interfaces. The object is lifted and placed onto the pan balance automatically using computer programming. All data are acquired and stored through a multiplexer and RS 232C interface.

Sequence of weighing

The weighing of the empty hanger and objects is carried out in the following sequence:

- W_0, the weight of the empty hanger,
- W_1, the weight of the hanger with solid standard S_1 in the top pan,
- W_x, the weight of the hanger with object U in the middle pan,
- W_2, the weight of the hanger with solid standard S_2 in the bottom pan, and finally
- W_0, the weight of the empty hanger.

To eliminate time-dependent variations the weighing is also undertaken in the reverse sequence.

Weighing and storing of data

The computer first records the apparent weight of the hanger together with the temperature of the hydrostatic liquid, pressure, temperature and relative humidity of air in the weighing chamber. The object in the highest position is picked up and its apparent weight after a 5 min waiting period is calculated. For one weighing, the computer takes about 35 observations and calculates the mean and standard deviation. This constitutes one weighing observation; four such weighing results are obtained and the mean of these four weighings, after the appropriate buoyancy corrections have been made, is stored in the memory. The computer-aided balance performs the weighing in the aforesaid sequence for different objects. The computer may be programmed in such a way that additional weights may be placed for simultaneous sensitivity calculations in one specific weighing or in every weighing.

Calculations

From W_1 and W_0 one can find the density of the liquid at level h_1 by the following equation:

$$\rho_{l1} = \rho - (W_1 - W_0)(1 - \sigma/\Delta)g_{h_1}/(g V_1). \tag{3.29}$$

A similar equation is obtained for the density of the hydrostatic liquid ρ_{l2} in the plane of S_2:

$$\rho_{l2} = \rho - (W_2 - W_0)(1 - \sigma/\Delta)g_{h_2}/(g V_2). \tag{3.30}$$

For calculating the density ρ_{l1}, the first value of W_0 is taken while for ρ_{l2} the last value of W_0 is taken. Here g_{h_1} and g_{h_2} are the acceleration values due to gravity at

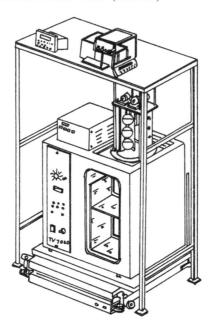

Figure 3.6. Bath, balance and hanger with three spheres outside the bath, NPL, India.

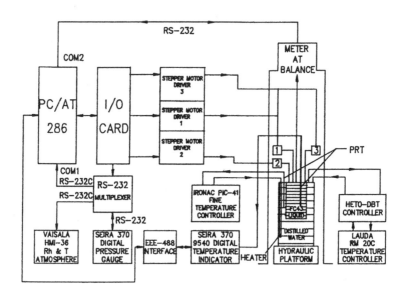

Figure 3.7. Block diagram of the computerized hydrostatic measuring system.

Photograph P3.1. Computer-aided density measurement system at NPL, India (courtesy National Physical Laboratory, India).

heights h_1 and h_2. As the density of the liquid will vary due to its own hydrostatic pressure, the density of the liquid at the level of the object is given by

$$\rho_{lx} = \rho_{l2} + (\rho_{l2} - \rho_{l1})(h_2 - h_x)/(h_2 - h_1). \tag{3.31}$$

In order to make use of substitution weighing with separate mass standards of density $\Delta 1$, an additional weight of mass m_{ad} is placed in the weighing pan. The mass of m_{ad} must be such that it is not necessary to change any built-in weights when the standard objects of known volumes are placed on the hanging pan in the liquid. Only the scale reading may change slightly. In this case V_x, the volume of the unknown body, is given by

$$V_x = [m_x - (W_x - W_{0m})(1 - \sigma/\Delta)g_{h_x}/g + m_{ad}(1 - \sigma/\Delta 1)g_{h_x}/g]/\rho_{lx} \tag{3.32}$$

where W_{0m} is the mean of the two values of apparent weights of the empty hanger and m_x is the mass of the object.

Computing

All calculations and weighings are carried out through a computer; the number of such complete sets of measurements (normally ten) to be taken is fed into the computer. This process takes about 4.5 h but no personal intervention is required at any stage. The overall uncertainty is about 2 ppm. The computer-controlled bath and balance are shown in figure 3.6.

The block diagram in figure 3.7 depicts the computerized set-up with three stepper motors (one for each object) and various interface devices.

The system for density measurement with three spheres in the position used at NPL, India, is shown in photograph P3.1.

3.6.3 Apparatus at NRLM, Japan [18]

The balance is housed in a box, 60 cm × 75 cm × 85 cm, three sides of which are made from glass to permit observation from outside. The box is supported on four legs, each 110 cm high. This allows other equipment used for hydrostatic weighing to be kept under it. Figure 3.8 shows the main water cell schematically. The sample cell is a cylindrical vessel of fused quartz with a diameter of 20 cm, height 50 cm and capacity 16 litres. The immersed pan is also made from fused quartz and annular in shape. The inner and outer diameters of the pan are 8 and 12 cm while it is 8 mm thick. The sphere is placed on this pan. The immersed pan is suspended by an aluminium chain attached to the bottom of the pan at one end and to a 0.2 mm stainless steel wire at its other end. The wire crosses the water surface and is attached to a rectangular frame and the pan is suspended from the frame. The loading and unloading device for the sphere is in the shape of a fused quartz cylinder 7 cm in diameter and height 8 cm with a vertical travel of 3 cm. A glass rod projecting from the bottom of the cell guides this. The loading device is connected to a control panel by a system of tie rods, a pipe and cam mechanism allowing the operator to move the device up and down by 3 cm. The pipe around the suspension wire also acts as its windshield.

To reduce air dissolution, three fused quartz discs are placed just below the water surface. The three discs were held apart by 2 cm long spacers and are horizontally placed covering most of the cross section of the cell. A small hole for the suspension wire and three other holes, one for the platinum resistance thermometer and another two for the tie rods, are provided. To facilitate their placement, the disc assembly was cut into two semi-circular parts, which could be placed conveniently around the suspension wire.

To obtain a stable environment, the balance and other apparatus are installed in a 300 m long underground (12 m below the ground) tunnel. The tunnel is not air conditioned except for normal ventilation, but the temperature variation in a day is not more than 0.05 °C with an annual variation of only 2 °C. To keep the concentration of carbon dioxide unchanged despite the respiration of the operator the balance is kept open for a few hours with no operator inside the laboratory, so that the air in the balance acquires the same composition as the air inside the tunnel. It is assumed that the air inside the tunnel has a natural composition because of the large volume (3000 m^3) of the tunnel, there being only moderate ventilation to remove excess moisture.

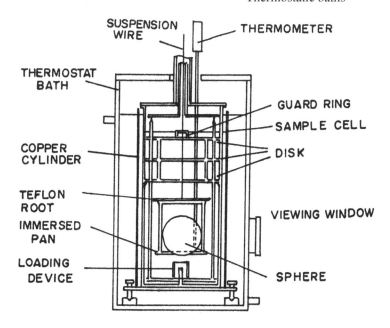

Figure 3.8. Main water cell with sphere and hanging arrangement at NRLM, Japan.

Temperature control

The cell is secured on a levelling plate and placed in a thermostatic bath of circular section of diameter of 44 cm and height 64 cm. A 12 mm thick layer of foam insulates the outside of the bath. Constant temperature water is circulated from outside the bath. To shield the cell from mechanical vibrations due to the circulating water, a copper cylindrical shell is placed around the cell. The temperature drift between most sessions is not expected to be more than 1 mK. One session normally lasts about 1.5 h. The temperature of the water is measured with a calibrated platinum resistance thermometer with a resolution of 1 mK.

Assembling the sphere and suspension device

The water of the cell is heated with an immersed 500 W water heater for 1 h. The heater is kept enclosed in a fused quartz sheath. When the water temperature decreases to about 50 °C, the cell is moved to the drained (without water) bath. The loading device, the pan with the sphere on it and the disc assembly are then submerged in the water in the cell. The sphere and other components are checked for any visible air bubbles on their surfaces—if bubbles are present the process of submersion has to be repeated. Finally the bath is filled with tap water and the temperature control is started. The cart carrying the bath is rolled under the balance and the suspension wire and loading device are connected to the

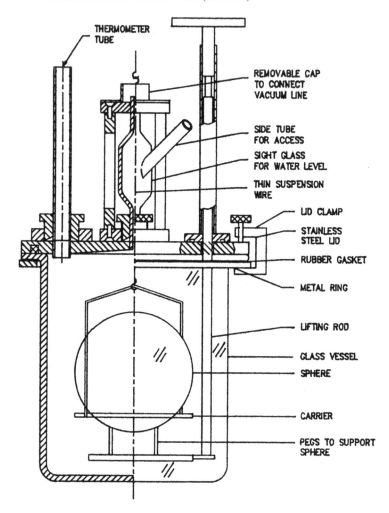

THERMOMETER
TUBE

REMOVABLE CAP
TO CONNECT
VACUUM LINE

SIDE TUBE
FOR ACCESS

SIGHT GLASS
FOR WATER LEVEL

THIN SUSPENSION
WIRE

LID CLAMP

STAINLESS
STEEL LID

RUBBER GASKET

METAL RING

LIFTING ROD

GLASS VESSEL

SPHERE

CARRIER

PEGS TO SUPPORT
SPHERE

Figure 3.9. Diagram of the water cell and sphere at NML, Australia.

balance. The measurements are normally started a few days later, when the water temperature becomes sufficiently stable. It has been noted that there are more fluctuations in the hydrostatic weighing if the temperature of the bath is greater than the ambient temperature, so the ambient temperature is kept between 2 to 4 °C above the cell temperature.

3.6.4   Water cell at NML, Australia [9]

The water cell has essentially the same design features as that of NMRL, Japan. However, two very important points are stressed: the vessel was designed in such a way that (1) the contact between the sample water and air is minimal; and (2) the level of the water, at the intersection of suspension wire and water surface, is kept constant.

The glass sight tube above the lid is sealed on the outside to prevent water entering from the bath and it has a constricted neck of only a few millimetres internal diameter. The two tubes through the lids, one for holding the thermometer and the other for the lifting rod, are also sealed. The water level is observed and is always kept at the cross-wire of the telescope fixed with the system. The water cell is shown in figure 3.9.

The temperature of the bath can be set to any temperature between 1 and 40 °C. The spatial and temporal accuracy of the bath is such that the temperature variation is within 0.005 °C.

3.7 Results of recent determinations of the density of water

3.7.1 NRLM, Japan

To determine the density of water at 16 °C [18], all measurements were taken in air, so that air dissolution was natural and unavoidable. Before using a particular sample of water, some portion of it was taken out and termed the reference sample; the water sample put in the water cell was called the working sample. The difference in density between the reference and working samples of the water was measured after one series of measurements with one sphere. Due account of this difference in density was taken in subsequent calculations. Special pyknometers were used for this purpose [24]. The air dissolution may decrease the density of water from 2.5 to 3 ppm.

The correction due to deformation of the sphere and immersed pan was calculated but found to be negligible. The buoyancy weights were 120 cm above the centre of the upthrust of the water so the necessary correction to the vertical gradient in the acceleration due to gravity was applied.

The temperature of the sample water was maintained very close to 16 °C. A correction in the density due to these small temperature deviations was applied and the values of the density of SMOW have been reported at 16 °C.

The measurements were carried out in the following format: Once the sphere and sample water had been placed in the cell, five separate weighing sessions were taken over several days. A weighing session consists of the following weighing sequence:

(1) the sphere plus hanger on the pan,
(2) the hanger only on the pan and the sphere lifted,
(3) the sphere plus hanger on the pan,
(4) the hanger only on the pan and the sphere lifted,
(5) the sphere plus hanger on the pan,
(6) the hanger only on the pan and the sphere lifted and

(7) the sphere plus hanger on the pan.

The water temperature was measured before and after each weighing session while the air temperature, pressure and relative humidity were measured for each weighing of the session. To minimize the effect of disturbance in the water while loading or unloading, a 5 min waiting period was allowed between loading or unloading and the weighing process. The sensitivity of the balance was measured at the end of each session. Five weighing sessions formed one weighing series. Six series were carried out for each sphere.

For each series a fresh sample of water was inserted into the sample cell and termed the work sample. Corresponding to each work sample there was a reference de-aerated sample which was not used in the hydrostatic weighing. It was necessary to maintain reference samples to find the air dissolution in the work sample by finding the difference in density between the reference and work samples. The density difference was measured with the help of two equal pyknometers [24]. Three spheres of known volume were used. Thus 18 samples of water were used in the experiment.

The temperature of the water at the time of measurement was not exactly 16 °C, so each time the volume of the sphere was calculated by using the appropriate coefficient of cubic expansion for the sphere. From the loss in mass measured by hydrostatic weighing, the density of the sample water at a given temperature and pressure was obtained. To reduce each measurement to 16 °C and to standard atmospheric pressure, namely 0.101 325 MPa, the following relation due to Kell [25] was used:

$$\rho_{std} = \rho_{(p,t)}\{1 - 0.000\,465(P - 0.101\,325)\}/\{1 - 0.000\,1625(t - 16)\} \quad (3.33)$$

where P is in MPa and t in °C on the ITS-90 scale.

Equation (3.33) may be written in a dimensionally more correct form as

$$\rho_{std} = \rho_{(p,t)}\{1 - 0.000\,465((P\,\mathrm{MPa}^{-1}) - 0.101\,325)\}$$
$$\times \{1 - 0.000\,1625((t\,°\mathrm{C}^{-1}) - 16)\}^{-1}.$$

The correction to the density value from each measurement session due to air dissolution has also been applied to give the density of SMOW at 16 °C and 0.101 325 MPa.

Result and discussions

For each sample of water, five values for the density for SMOW corresponding to each session were obtained and the mean taken so there were 18 values for the density, each corresponding to SMOW at the same temperature and pressure. The peak-to-peak difference between these 18 values was 3.2×10^{-3} kg m^{-3}. The standard uncertainty, which is equal to the standard deviation of the mean of these 18 values, was found to be 0.6×10^{-3} kg m^{-3}.

Details of factors influencing the uncertainty and their respective contributions to the uncertainty are given in table 3.2.

The mean density of SMOW at 16 °C and 0.101 325 MPa was given as 998.9468 kg m^{-3} with a standard uncertainty of 0.0006 kg m^{-3}.

It was observed that the variation in the values within a session was smaller than the variation in the density between the 18 mean values of SMOW each reduced to the same conditions. Since the standard uncertainty in the volume measurement was only 0.18 × 10^{-6} and corrections for impurity and isotopic effects were applied within 0.1 to 0.2 × 10^{-6}, most variations must have occurred during the hydrostatic weighing.

The maximum density at 4 °C, from the results of the measurements at 16 °C and with previous measurements of thermal expansion of water, was deduced to be 999.975 ± 0.0008 kg m^{-3}.

3.7.2 NML, Australia

In earlier work, a modified H51 Mettler balance was used. The movement of the suspension was very small (0.03 mm). However, the intersection point of the suspension wire and the water surface could have been different. So in later measurements, a Sartorius balance—a mass comparator with null reading device—was used, so that the position of the suspension system was the same in all the readings. The Sartorius balance was adjusted to a resolution of 0.01 mg.

The apparent mass of the sphere in water was determined by comparing it with standard masses placed in the pan of the balance and reading the mass difference on the balance readout system. To manipulate the standard masses from outside a special stainless steel mass of 100 g in the form of a pan was made, so that it was lifted from outside with the help of an external lever. Any additional weights required to counterpoise the apparent mass, within 10 mg, were placed on it.

The temperature of the bath was maintained at different temperatures and the apparent mass of the spheres was measured for each temperature. However, the variation in temperature outside the cell was found to be within 0.005 °C and no significant temperature difference was observed within the water sample in the cell.

From knowing the volume of the sphere at that temperature together with the mass of the sphere, the density of the water at the temperature and pressure of the measurement was determined.

Ten water samples, numbered 1–10, were taken. Measurement of the density of water took about 12 years to complete. To reduce density values at integral values of temperature and at a pressure 0.101 325 MPa, Kell's data [25] for the thermal expansion and compressibility of water were used. Finally the value of the density of V-SMOW [2] was deduced. The isotopic composition of all but sample 1 was determined. For each water sample, the density at 4 °C was determined and these are given in table 3.3—the third column with heading *N*

Table 3.2. Uncertainty budget for the density of SMOW at 16 °C.

Factors	Uncertainty in each factor	Relative standard uncertainty ($\times 10^6$)
I Uncertainty in measurement of volume (table 2.1)		0.36
II Uncertainty in weighing		
Type B		
1. Standards of mass used in weighing the sphere in water (μg)	64	0.2
2. Standards of mass used in the sphere in water (μg)	100	0.0
3. Mass of aluminium foil used for weighing in air (μg)	50	0.17
4. Density of air	0.0003	0.26
Air temperature (°C)	0.01	
Atmospheric pressure (Pa)	10	
Water vapours (Pa)	50	
CO_2 content (mol%)	0.01	
Use of BIPM-81 equation for density calculation	0.01	
5. Temperature of water (mK)	1	0.16
6. Isotopic composition of water		
$\delta\,^{18}O$		0.021
δD		0.015
7. Use of the difference in density relation due to isotopic composition (%)	10	0.15
Type A		
Standard deviation of mean density of SMOW		0.2
Combined relative uncertainty in density measurement		0.57

gives the number of measurements carried out on the date given in the second column.

All other density values after due correction of the isotopic composition to V-SMOW at different temperatures have been divided by the density of the sample at 4 °C. These data have been fitted to a Theisen-type formula:

$$\rho/\rho_0 = 1 - A(t - t_0)^2(t + B)/(t + C) \tag{3.34}$$

where t_0 is the temperature at which the water attains its maximum density ρ_0. A, B and C are parameters determined by the method of least squares. Since t_0 is very close to 4 °C, ρ_0 differs from $\rho_{(4)}$ by less than four parts in 10^9. So for

Table 3.3. Density of SMOW at 4 °C (NML) [9]. The peak-to-peak variation is 2.86×10^{-3} kg m^{-3}, the mean of all values is 999.973 64 kg m^{-3} and the mean SD is 0.000 72 kg m^{-3}.

Sample no	Date	N	Mean density of SMOW/ V-SMOW
2	20/05/80	12	999.973 27
3	07/0780	12	999.973 49
4	06/08/80	12	999.972 86
5	22/10/80	9	999.974 33
6	03/06/80	9	999.974 44
7	05/09/90	11	999.973 44
	05/09/90	11	999.973 45
	06/09/90	12	999.973 37
	07/09/90	11	999.973 36
	10/09/90	10	999.973 47
	18/09/90	9	999.973 75
	18/09/90	6	999.974 28
8	20/09/90	11	999.972 09
	27/09/90	12	999.972 90
	03/10/90	6	999.973 34
9	26/06/92	18	999.974 95
10	30/06/92	21	999.974 31

this work $\rho_{(4)}$ and ρ_0 have been taken to be the same, so the dilatation data $\rho/\rho_{(4)}$ have been fitted to equation (2.34). To obtain the value of t_0, data below 10 °C were taken and all the data then used in a second fit to obtain the values for the other parameters. The values obtained for A, B, C and t_0 are:

$$A = 1.858 \times 10^{-6} \, °C^{-2}$$
$$B = 316.338\,08 \, °C \tag{3.35}$$
$$C = 70.699\,73 \, °C$$
$$t_0 = 3.9818 \, °C.$$

The same dilatation data ρ/ρ_0 have been fitted to a fifth-degree polynomial with $(t - t_0)$ as the variable:

$$\rho/\rho_0 = 1 - \{A(t-t_0) + B(t-t_0)^2 + C(t-t_0)^3 + D(t-t_0)^4 + E(t-t_0)^5\}. \tag{3.36}$$

The following values for A, B, C, D and E have been obtained:

$$t_0 = 3.9818 \, °C$$
$$A = 7.0134 \times 10^{-8} \, °C^{-1}$$

Table 3.4. Variances due to various causes and their contribution towards the density of water.

Temperature (°C)	Volume	Mass in air	Apparent mass in water	Temperature	Small contribution	Total
1	1.17	0.031	0.626	0.016	0.008	1.90
3	0.858	0.031	0.058	0.002	0.008	0.96
4	0.732	0.031	0.029	0.000	0.008	0.80
7	0.448	0.031	0.073	0.013	0.008	0.58
10	0.274	0.031	0.119	0.048	0.008	0.48
15	0.103	0.031	0.121	0.143	0.008	0.41
20	0.063	0.031	0.099	0.269	0.008	0.47
25	0.063	0.031	0.140	0.414	0.008	0.66
30	0.134	0.031	0.200	0.572	0.008	0.95
35	0.268	0.031	0.542	0.743	0.008	1.6
40	0.502	0.031	0.302	0.928	0.008	1.8

The header row spans: Variance of density of water ($\times 10^{12}$)

$$B = 7.926\,504 \times 10^{-6}\,°\text{C}^{-2} \tag{3.37}$$
$$C = -7.575\,677 \times 10^{-8}\,°\text{C}^{-3}$$
$$D = 7.314\,894 \times 10^{-10}\,°\text{C}^{-4}$$
$$E = -3.596\,458 \times 10^{-12}\,°\text{C}^{-5}.$$

Theisen's formula and the polynomial function appear to represent the data equally well and give very similar figures for the sum of squares of residuals. However, the polynomial fit was chosen to represent the results of this work.

Uncertainty budget in water density at NML

Variances have been calculated for

- the volume of the sphere,
- the mass of the sphere in air,
- the apparent mass of the sphere in water,
- the temperature measurement,
- the combined contributions due to air density,
- the gravity ratio,
- the density of the mass standard used,
- the water pressure and
- the isotopic composition of the water samples for different temperatures.

Their contributions to the variance of density are given in table 3.4.

References

[1] Craig H 1961 *Science* **133** 1833–4

[2] Hagemann R, Nief G and Roth E 1970 *Tellus* **22** 712

[3] Baertschi P 1976 *Earth Planet Sci. Lett.* **31** 341

[4] Marsh K N (ed) 1987 *Recommended Reference Materials for the Realisation of Physico Chemical Properties* (Oxford: Blackwell Scientific) pp 14–15

[5] Menache M 1967 Du problem de la masse volumique de l'eau *Metrologia* **3** 58–63

[6] Girard G and Menache M 1971 Variation de la masse volumique de l'eau en fonction de sa composition isotopique *Metrologia* **7** 83

[7] Menache M, Beaverger C and Girard G 1978 *Ann. Hydrographiques 5th Series* **6** 37–75

[8] Girard G and Coarasa M J 1984 Effect of dissolved air on the density of water *Precision Measurements and Fundamental Constants* vol II, ed B N Taylor and W D Phillips (Gaithersburg, MD: National Bureau of Standards–US Department of Commerce) pp 453–7

[9] Patterson J B and Morris E C 1994 Measurement of absolute water density 1 °C to 40 °C *Metrologia* **31** 272–88

[10] Menache M and Girard G 1973 Concerning the different tables for thermal expansion of water between 0 and 40 °C *Metrologia* **9** 662–8

[11] Cox R, McCartney M J and Culkin F 1968 *Deep-Sea Res.* **15** 319–25

[12] Lida D R (ed) 1998 *CRC Handbook of Chemistry and Physics* 77th edn (London: Chemical Rubber Company) pp 6–13

[13] Marek W J 1891 *Ann. Phys. Chem.* **44** 171

[14] Chappuis P 1910 *Trav. Mem. Bur. Int. Poids Mes.* **14** D1

[15] Menache M, Beaverger C and Girard G 1978 *Ann. Hydrographiques* **6** 37

[16] Carpenter J H 1965 *Limnol. Oceanogr.* **10** 135

[17] Bignell N 1983 The effect of dissolved air on the density of water *Metrologia* **19** 57–9

[18] Masui R, Fujii K and Takenaka M 1995/96 Determination of the absolute density of water at 16 and 0.101 325 MPa *Metrologia* **32** 333–62

[19] Davis R S 1992 Equation for determination of density of moist air (1981–1991) *Metrologia* **29** 67–70

[20] Giacoma P 1982 Equation for determination of density of moist air (1981) *Metrologia* **18** 33–40

[21] Mohindernath, Bhamra S S and Gupta S V 1988 Establishment of density standards in National Physical Laboratory *Report* No NPL-88-A.3-003/0156

[22] Saxena T K, Sharma D C and Sinha S 1997 An automatic set up for solid density standard *Indian J. Eng. Mater. Sci.* **4** 102–5

[23] Spieweck F, Kozdon A, Wagenbreth H, Toth H and Hoburg D 1990 A computer controlled solid-density measuring apparatus *PTB Mitteillungen* **100** 169–73

[24] Masui R, Watanabe H and Iuzuka K 1978 Precision hydrostatic pyknometer *Japan. J. Appl. Phys.* **17** 755–6

[25] Kell G S 1977 *J. Phys. Chem. Ref. Data* **6** 1109–31

Chapter 4

Dilatation of water and water density tables

Symbols

M	Mass of mercury
β	Relative increase in density with increase in pressure (Pa)
α	Coefficient of linear thermal expansion
$\Delta L / L$	Fractional change in length
ρ_m	Density of mercury
ΔP	Difference in pressure from 101 325 Pa
ΔU	Change in volume of water
ρ_w	Density of water
a_i	Coefficients of various powers of temperature t in water density and temperature relations
b_i	Coefficients of various powers of temperature t in water density and temperature relations
d	Relative density of water
h	Depth or height of water column
P, p	Pressure in Pa
U	Volume of water
V	Capacity of dilatometer
W	Mass of water.

4.1 Introduction

The dilatation of water or the density of water and temperature relationship has been a discussion point since the beginning of the 20th century. In the meantime a large number of relations have been formulated by various authors, which have added to the confusion. Some authors have derived density relationships from their own experimental data, while others have applied statistical considerations and used existing experimental data. In general the density of water has been expressed by two types of mathematical relation, namely polynomials and rational

82

functions of temperature. It should be noted that all the density measurements which we are going to discuss in the following pages are at normal atmospheric pressure, i.e. 101 325 Pa.

4.2   Earlier work on the density and dilatation of water

4.2.1   Chappuis' work

Chappuis, working at the International Bureau of Weights and Measures (BIPM), Paris measured the expansion of water from 0 to 41 °C. He used three dilatometers, each of capacity 1 litre, two made from glass and the third from platinum. The first series of measurements was carried out in 1891 and the second series with the platinum dilatometer in 1897. Up to 20 °C, his results [1] were consistent within 1 ppm, but beyond 20 °C, the agreement was only within 5 ppm. To construct water density tables, he expressed the ratio of the relative density at 0 °C to that of water at temperature t °C as a third-degree polynomial in temperature t. In order to obtain the values of the coefficients of different powers of t, he partitioned his results into three temperature ranges. The expressions with their temperature ranges are:

Temperature range 0–10.3 °C:

$$(0.999\,8681/d - 1) \times 10^6 = -67.464\,645(t\,°C^{-1}) + 8.934\,223(t\,°C^{-1})^2$$
$$- 0.078\,919\,46(t\,°C^{-1})^3. \tag{4.1}$$

Temperature range 10.3–13 °C:

$$(0.999\,8681/d - 1) \times 10^6 = -54.7835 - 55.242\,760(t\,°C^{-1})$$
$$+ 7.945\,055(t\,°C^{-1})^2 - 0.048\,001\,50(t\,°C^{-1})^3. \tag{4.2}$$

Temperature range 13–41 °C:

$$(0.999\,8681/d - 1) \times 10^6 = -114.5565 - 42.940\,141(t\,°C^{-1})$$
$$+ 7.106\,115(t\,°C^{-1})^2 - 0.029\,057\,59(t\,°C^{-1})^3. \tag{4.3}$$

The relative density of water at 0 °C is 0.999 8681. Here d is the relative density at $t/4$ °C.

4.2.2   Thiesen's work

Almost in the same period (1896), Thiesen, Scheel and Diesselhorst of the Physikalisch Technische Reichsanstalt (PTR), Germany, also measured the density of water by balancing two columns of pure water, each 2 m in height. The columns were maintained at different temperatures. The results were published in 1900 [2]. Their results agreed very well with those of Chappuis at lower temperatures but differed by 6 ppm at 25 °C and 9 ppm at 40 °C. The final density temperature table was calculated by fitting the observation data to the following expression:

$$(1 - d) \times 10^3 = \{(t - A)^2(t + C)\}/\{B(t + D)\}. \tag{4.4}$$

The values of A, B, C and D were obtained by using the method of least squares as the following:

$$A = 3.98 \,°C$$
$$B = 503.570 \,°C^2$$
$$C = 283 \,°C \tag{4.5}$$
$$D = 67.26 \,°C.$$

The temperature t was measured in °C as maintained by each laboratory almost independently.

4.2.3 Mendeleev's table

Mendeleev [3], using Thiesen's results, suggested the following formula for the much wider range of $-10\,°C$ to $200\,°C$:

$$(1 - d) = (t \,°C^{-1} - 4)^2/[118\,932 + 13\,666.75(t \,°C^{-1}) - 4.13(t \,°C^{-1})^2]. \tag{4.6}$$

The values of the density obtained by this formula agreed within 1 ppm with those of Thiesen. Water outside the temperature range 0–100 °C is considered to be in a meta-stable state.

4.2.4 Steckel and Szapiro's work

Steckel and Szapiro in 1962 carried out relative density measurements on water of known isotopic composition and heavy water. Dilatometers of volume 4–5 cm^3 were used over an extended range 1.4–77.8 °C [4]. All water samples were purified in exactly the same way, so that the isotopic composition did not vary from sample to sample. They constructed a water density table using Thiesen's expression (equation (4.4)) but with the following different values of A, B, C and D.

$$A = 3.986 \,°C$$
$$B = 407.507 \,°C^2$$
$$C = 196.577 \,°C \tag{4.7}$$
$$D = 57.002 \,°C.$$

Between 0 and 40 °C, the tabulated values agreed better with Thiesen's rather than Chappuis' results. The maximum difference was 3.7 ppm at 40 °C, otherwise all other values agreed to within 2–3 ppm.

4.2.5 Stott and Bigg's table

Stott and Bigg [5] constructed a density table in steps of 0.1 °C in the range 0–40 °C for the International Critical Tables, by adopting the mean of Thiesen's and Chappuis' values for the water density at each temperature.

4.2.6 Tilton and Taylor's table

Tilton and Taylor [6] found that Chappuis' data fitted Thiesen's expression better than Chappuis' own three cubic equations, so they published a water density table derived on this basis. The expression used was

$$(1 - d) \times 10^3 = ((t\,°C^{-1}) - 3.9863)^2((t\,°C^{-1}) + 288.9414)$$
$$\times \{508.9292((t\,°C^{-1}) + 68.129\,63)\}^{-1}. \qquad (4.8)$$

4.2.7 Bigg's table

Bigg [7], while preparing the water density tables in SI units, assigned weight factors to Chappuis' and Thiesen *et al*'s results. He assigned the following weight factors to various results: two to Chappuis' results obtained from glass dilatometer 1; three to those obtained with the help of glass dilatometer 2; and five each to the results obtained by Chappuis using the platinum dilatometer and to those obtained by Theisen *et al*.

After taking into account the difference in the litre as defined in 1901 and the decimetric cube, he fitted all the results to a fifth-degree polynomial and published water density tables in kg m^{-3} in 1967.

Almost on a similar line by assigning different weight factors to three series of Chappuis' results and those of Thiesen, Alexandrov and Trakhtengerts [8] published tables for the temperature range 0–100 °C. Wagenbreth and Blanke [9] also published water density tables for the temperature range 0–40 °C. These authors were of the view that the standard scale of the hydrogen thermometer used by Chappuis and Thiesen coincided with IPTS-48 and accordingly applied corrections to bring the results into line with IPTS-68. However, Menache and Girard [10] were of the opinion that the hydrogen scale coincides better with IPTS-68.

4.2.8 Kell's formulation

Kell [11] used a fifth-degree polynomial in temperature t divided by a linear factor in t and expressed density d as

$$d = \{a_0 + a_1t + a_2t^2 + a_3t^3 + a_4t^4 + a_5t^5\}/\{0.999\,972(1 + b_1t)\} \qquad (4.9)$$

and ascribed the following values to the different constants:

$$a_0 = 0.999\,8396 \qquad\qquad a_1 = 18.224\,944 \times 10^{-3}\,{}^\circ\mathrm{C}^{-1}$$
$$a_2 = -7.922\,210 \times 10^{-6}\,{}^\circ\mathrm{C}^{-2} \qquad a_3 = -5.544\,846 \times 10^{-8}\,{}^\circ\mathrm{C}^{-3}$$
$$a_4 = 1.497\,562 \times 10^{-10}\,{}^\circ\mathrm{C}^{-4} \qquad a_5 = -3.932\,952 \times 10^{-13}\,{}^\circ\mathrm{C}^{-5} \qquad (4.10)$$
$$b_1 = 18.159\,725 \times 10^{-3}\,{}^\circ\mathrm{C}^{-1}.$$

Later in 1977, Kell [12] expressed the density of SMOW in kg m^{-3} as a ratio of two polynomials. The polynomial in the numerator contained only odd powers of temperature t, while the polynomial in the denominator had only even powers of t. The expression was:

$$\rho_{(\mathrm{SMOW})} = \left(\sum A_0 + A_i t^{2i-1}\right) \bigg/ \left(1 + \sum B_j t^{2j}\right)$$

with the following values for A_i and B_j.

$$A_0 = 999.8427 \qquad\qquad B_1 = 9.090\,169 \times 10^{-6}\,{}^\circ\mathrm{C}^{-2}$$
$$A_1 = 67.8782 \times 10^{-3}\,{}^\circ\mathrm{C}^{-1} \qquad B_2 = 1.451\,1976 \times 10^{-9}\,{}^\circ\mathrm{C}^{-4}$$
$$A_2 = 103.1412 \times 10^{-6}\,{}^\circ\mathrm{C}^{-3} \qquad B_3 = 134.848\,63 \times 10^{-15}\,{}^\circ\mathrm{C}^{-6} \qquad (4.11)$$
$$A_3 = 15.958\,35 \times 10^{-9}\,{}^\circ\mathrm{C}^{-5} \qquad B_4 = 2.008\,615 \times 10^{-18}\,{}^\circ\mathrm{C}^{-8}$$
$$A_4 = 636.8907 \times 10^{-15}\,{}^\circ\mathrm{C}^{-7}.$$

4.2.9   International recommendation for the determination of the density of water

In 1971, the International Union of Geodesy and Geophysics (IUGG) adopted a recommendation from the Association Internationale pour les Sciences Physiques de l'Ocean and the International Union of Pure and Applied Chemistry. In 1973 they called for a determination of the following properties:

- the absolute density of well characterized water at a few well defined temperatures in between 0 and 40 °C,
- the relative density of water in the range 0–40 °C,
- the effect of isotopic composition and air dissolution on the density of water and
- the maximum value of the density of water and the temperature at which water acquires it.

Several national laboratories took up the work and carried out density measurements using standard solid bodies of known volume and mass. Recent available results are from Bell and Clarke [13], Bell and Patterson [14] and Patterson and Morris [15] all from NML, CSIRO, Australia and Takenaka and Masui [16] and Watanabe [17] from NRLM, Japan.

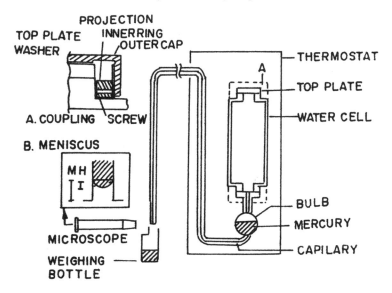

Figure 4.1. Dilatometer used by Takenaka and Masui: (A) details of the coupling and (B) view of the meniscus through the microscope.

4.3 Thermal dilatation of water by the mercury displacement method

In 1990, Takenaka and Masui [16] measured the thermal expansion of pure water with a natural isotopic abundance using the dilatometer method in the temperature range 0–85 °C.

4.3.1 Experimental set-up

Dilatometer

The dilatometer is shown in figure 4.1. It has four parts: a water cell, a top plate, a capillary with a bulb and a weighing bottle. All parts are made from fused quartz. The cylindrical water cell has an approximate volume of 100 cm^3, is about 10 cm in height and has an outer diameter of 4 cm. The cell has a projection on each end; a stainless steel ring with fine threads is fitted on the two projections. A metal cup holds the top plate. Details of the coupling are shown in figure 4.1(A). The flanged end of the capillary above the bulb is similarly held with another cup. The coupling device is designed to minimize the change in the capacity of the water cell due to elastic deformation when the end projections are squeezed together with screws for mounting the rings. The capillary tube, with an internal diameter of 0.7 mm, ends in a bulb and a flange to be secured with a metal ring.

The water cell is kept in a thermostatic bath and the capillary connected to it

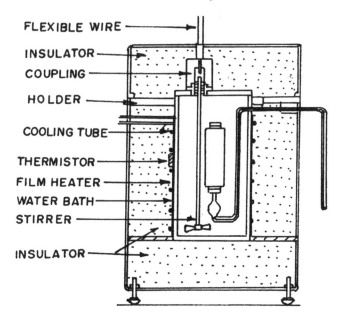

FLEXIBLE WIRE

INSULATOR

COUPLING

HOLDER

COOLING TUBE

THERMISTOR

FILM HEATER

WATER BATH

STIRRER

INSULATOR

Figure 4.2. Dilatometer in its thermostat.

penetrates the thermostat wall and opens to the atmosphere outside the thermostat. The water bath comprises an aluminium cylinder, outer diameter 16 cm with wall thickness 1 cm. A film heater and a copper cooling coil are wound uniformly all over the surface of the cylinder. A thermistor, which controls the temperature of the water bath, is embedded in the wall of the cylinder. An AC bridge is used to detect the temperature and send a corresponding signal to control the supply voltage to the heater. The water cylinder is surrounded by urethane—a thermal insulator with a thickness of about 10 cm. A propeller, with a flexible shaft, stirs the water in the bath. The flexible shaft is used to avoid excessive vibrations. The dilatometer and its bath are shown in figure 4.2.

The thermal expansion of water rises sharply with a rise in temperature. At around 80 °C, a change of 1 mK in temperature corresponds to a 0.7 ppm change in density, so a thermostat with a temperature stability of 0.2 mK over the entire range is used for this purpose.

Assembling and filling the dilatometer

All quartz parts of the dilatometer are cleaned, by keeping them dipped in chromic acid for several days and then drying them at 70 °C in a chamber. All parts of the dilatometer are assembled except the top plate. Some mercury (5 cm^3) is poured into the dilatometer and pure bi-distilled water added under vacuum and the top plate is fixed. To avoid any leakage, a trace of grease might be used if necessary.

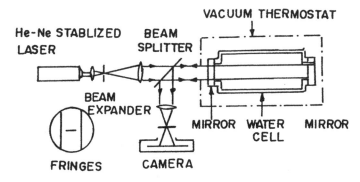

Figure 4.3. Optical arrangement for measurement of linear expansion.

The end of the capillary outside the bath is cut, so that its tip is at the level of centre of the bulb of the dilatometer. This ensures that the mercury level in the capillary remains constant.

Roughly speaking, when the temperature of the water cell is raised, the increase in the volume of the water will expel the mercury which, if weighed and divided by its density, will give the expansion of water. However, in this case we have not taken into account the expansion of the cell. We may take the value of the quartz expansion coefficient from the literature and use this but this will entail an inaccuracy of a few ppm, as the coefficient of quartz varies from sample to sample by this order. In order to achieve an accuracy of better than 1 ppm, it is advisable to measure the coefficient of linear expansion of the cell used in the experiment.

4.3.2   Measurement of the linear thermal expansion of quartz

As the cell can be separated from the rest of the dilatometer and its ends are well ground and polished, two semi-reflecting mirrors (reflectivity 0.8) are attached at each end and the system is used as a spacer in a Fizeau interferometer. The two mirrors may be inclined to each other at 7×10^{-5} radian, so that two to three fringes are always visible. A stabilized He–Ne laser is used as the light source. The light beam is expanded to 10 mm in diameter by a convex lens. A beam splitter reflects the beams coming back after reflection. After being condensed to a point, all spurious reflections are removed by a pinhole placed at the focus of the condensing lens. The fringes are then photographed. Two triangles are marked on one of the mirrors for reference purposes to locate the relative positions of the fringes in different photographs. The optical arrangement is shown in figure 4.3.

The interferometer is set in vacuum to avoid errors due to the refractive index of air. To evacuate the cell a small hole is provided in each mirror. The residual pressure should not be more than a few pascals. The vacuum chamber is placed in a thermostat whose temperature may be varied and measured accurately. Three copper–constantan thermocouples are used to monitor the temperature of the cell and any gradients along it. The difference in temperature between the two faces of the cell should not be more than 0.2 mK.

The measurements are taken at at least ten temperatures, which are uniformly distributed over the temperature range. The change in the length of the cell is given by

$$\Delta L = (N + \varepsilon)\lambda/2. \tag{4.12}$$

N, the integer part of the fringe, is counted by observing the fringe movement with respect to the reference mark in the mirror. ε, the fractional part of the fringe, is estimated by enlarging the photographs and examining them on the screen of a projector.

Thus $\Delta L/L$ is obtained as a function of temperature. The volume expansion is calculated from the linear expansion assuming the material of the cell is isotropic, i.e. by multiplying it by three.

4.3.3   The principle of measurement

The mass of mercury in the dilatometer and weighing bottle as well as the mass of water in the water cell remain constant. The loss in the mass of mercury due to the expansion of water from the dilatometer is the gain in the mass of mercury in the weighing bottle.

Let the mass and density of the mercury and water be respectively denoted by M, ρ_m, W and ρ_w. V is the volume of the dilatometer and U is the volume of water. Furthermore, subscripts 0, t and max are used to indicate the variables at the reference temperature, any temperature and the temperature of maximum density. Therefore at the reference temperature, the mass of mercury in the dilatometer is given by

$$(V_0 - U_0)\rho_{m0}. \tag{4.13}$$

Let the temperature of the dilatometer be raised to t °C. If the mercury expelled by the expansion of water is ΔU, the volume of the dilatometer becomes V_t. Therefore the mass of mercury left in the dilatometer is given by

$$(V_t - U_0 - \Delta U)\rho_{mt}. \tag{4.14}$$

M, the mass of mercury gained in the weighing bottle, is, therefore, given by

$$M = (V_0 - U_0)\rho_{m0} - (V_t - U_0 - \Delta U)\rho_{mt}. \tag{4.15}$$

Therefore,

$$M/\rho_{mt} = (V_0 - U_0)\rho_{m0}/\rho_{mt} - (V_t - U_0) + \Delta U$$

giving

$$\Delta U = M/\rho_{mt} - (V_0 - U_0)\rho_{m0}/\rho_{mt} + (V_t - U_0). \tag{4.16}$$

Furthermore, if U_t and U_0 are the volume of the water at $t\,^\circ$C and at the reference temperature then

$$U_t = U_0 + \Delta U. \tag{4.17}$$

If W and ρ_{wt} are the mass and density of water at $t\,^\circ$C, then equation (4.17) can be written as

$$W/\rho_{wt} = W/\rho_{w0} + \Delta U \qquad \Delta U = W[1/\rho_{wt} - 1/\rho_{w0}]. \tag{4.18}$$

Multiplying by ρ_{m0}, we get

$$\Delta U \rho_{w0} = W(\rho_{w0}/\rho_{wt}) - W$$
$$W + \Delta U \rho_{w0} = W(\rho_{w0}/\rho_{wt})$$

or

$$\rho_{wt}/\rho_{w0} = W/\{W + \Delta U \rho_{w0}\}. \tag{4.19}$$

Dividing both sides of (4.19) by ρ_{max}/ρ_{w0}, where ρ_{max} is the maximum density of water, we get

$$\rho_{wt}/\rho_{max} = W/[W\rho_{max}/\rho_{w0} + \Delta U \rho_{max}]. \tag{4.20}$$

ΔU can be calculated from equation (4.16).

U_0, the volume of water in the cell at the reference temperature, is determined by weighing the cell filled with water only and the volume of the capillary plus bulb is determined by weighing it when it is filled with mercury. The sum of the volumes occupied by the water and mercury at the reference temperature gives V_0. V_t is determined by measuring the coefficient of linear expansion of the cell using an interferometer. Here the reference temperature is taken as $4\,^\circ$C, the difference in density at $4\,^\circ$C and the maximum density is of the order of 10^{-8}, so for all practical purposes ρ_{max}/ρ_{w0} is 1.

The density of the mercury may be taken from the mercury density–temperature table, so the ratio ρ_{wt}/ρ_{w0} or ρ_{wt}/ρ_{max} may be determined at various temperatures.

4.4 Measurement of the expansion of water by the upthrust method

The upthrust acting on a solid (sinker) immersed in a liquid is the product of the volume of the solid and the density of the liquid in which it is immersed. This principle has been used by Watanabe [17] for measuring the ratio of the density of water at any temperature and its maximum density.

4.4.1   Principle of measurement

The upthrust acting on the sinker is normally taken as the product of the volume of the sinker and the density of the water sample. For accurate work, there are several more effects which have to be taken into account.

(1) The compressibility of water, as the density increases with increase in pressure, so if ρ is the density of water at $t\,°C$ at normal pressure, i.e. at 101 325 Pa, and β is the coefficient of the increase in density with pressure, then the effective value of ρ will be

$$\rho[1 + \beta(\Delta P + \rho g h)] \tag{4.21}$$

where ΔP is the difference in atmospheric pressure from 101 325 Pa. The term $\rho g h$ appears due to the hydrostatic pressure of water.

(2) The effective volume of the sinker will be a function of the temperature and the coefficients of expansion of the material of the sinker. So instead of V it should be multiplied by a factor F given by

$$F = (1 + G(t)). \tag{4.22}$$

(3) $G(t)$ may not be a linear function of t, the temperature. In addition, the pressure inside and outside the sinker will not be the same, so its volume will be decreased by a factor

$$\{1 - \alpha(\Delta P + \rho g h)\}. \tag{4.23}$$

Therefore the effective volume of the sinker will be

$$V\{1 + G(t)\}\{1 - \alpha(\Delta P + \rho g h)\}. \tag{4.24}$$

Hence if M is the mass of the sinker and m is its mass equivalent in water at temperature $t\,°C$, then the equilibrium equation in terms of a vertically downward force and an upthrust will be given by

$$(M - m)g = V\rho[1 + \beta(\Delta P + \rho g h)](1 + G(t))\{1 - \alpha(\Delta P + \rho g h)\}. \tag{4.25}$$

Similarly, the equilibrium equation at the temperature at which the density of water is a maximum, denoted by ρ_{max}, will be

$$(M - m_0)g = V\rho_{max}[1 + (\beta - \alpha)(\Delta P + \rho_{max}g h)](1 + G(t_0)). \tag{4.26}$$

Dividing (4.25) by (4.26), we get

$$\rho/\rho_{max} = [(M - m)/(M - m_0)]f\{\alpha, \beta, G(t), \Delta P\}. \tag{4.27}$$

The function $G(t)$ is determined experimentally by finding $\Delta L/L$ using a proper interferometer as described later in section 4.4.2. The combined effect of the compressibility coefficient β of water and α of the sinker is determined separately by using a piezometer. Other quantities are measured during the experiment; ρ and ρ_{max} in the function f may be taken to be equal.

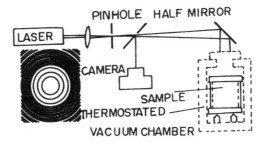

Figure 4.4. Optical arrangement for measuring the thermal expansion of sinkers.

4.4.2 Measurement of the thermal expansion of sinker

For $G(t)$, the thermal expansion of the sinker is measured with a Fabry–Perot interferometer using a He–Ne laser. The etalon is constructed using a cylinder cut from one of the used sinkers. The optical arrangement for measuring the thermal expansion of these sinkers is shown in figure 4.4.

The pinhole is 0.1 mm in diameter, the wedge angle of the two reflectors, each 12 mm thick, is 1 minute of arc. One of the reflectors has a 2 mm diameter hole to evacuate the air from the cylindrical specimen. The experiment is repeated in various orientations with respect to the axis of the cylinder to eliminate any systematic error from a lack of parallelism between the ends and from undulations on the reflecting surfaces. The diameters of the circular fringes are measured by photographing the fringes and then examining the photograph on a projection microscope with 1 μm readability and analysed by measuring the diameters of eight rings and using the method described in section 2.5.3. The regression equation of the observed expansion against temperature was derived for a series of measurements. The standard deviation of the sum of residual errors was found to be one-hundredth of the fringe width (3 nm). This corresponds to an uncertainty of 0.16 ppm. No thermal hysteresis was detected. The mean results of the expansion are:

$$(\Delta L/L) = 0.4377 \times 10^{-6}(t\,^\circ\mathrm{C}^{-1})$$
$$+ 1.187 \times 10^{-9}(t\,^\circ\mathrm{C}^{-1})^2 \quad \text{for larger sinkers}$$
$$(\Delta L/L) = 0.3299 \times 10^{-6}(t\,^\circ\mathrm{C}^{-1})$$
$$+ 1.594 \times 10^{-9}(t\,^\circ\mathrm{C}^{-1})^2 \quad \text{for smaller sinkers.}$$

4.4.3 Measurement of the compressibility of the sinker

The combined effect of α, the deformation of air-filled sinkers due to pressure, and the isothermal compressibility of water, is measured by the piezometer shown in figure 4.5. The sinker floats in the piezometer P, which has two parts. The lower part is a cylinder with a flange and the upper part has a capillary with a

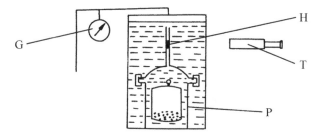

Figure 4.5. Line diagram of a piezometer.

small mercury thread H and a flange to tightly fit onto the flange of the lower cylindrical portion. The two portions of this flask are sealed. The inner diameter of the capillary is determined by finding the mass of a given length of a thread of mercury. Dividing the mass per unit length by the density of mercury gives the area of the cross section and hence the diameter. The pressure in the system is changed and measured with a pressure gauge G. The movement of the thread of mercury, which is seen through cathetometer T, will enable us to note any effective change in volume due to deformation of the hollow sinker and the isothermal compressibility of water. The effective change in volume due to these two opposing causes can be measured as we have already determined the area of the cross section of the capillary tube, and the linear movement of the thread is measured with the cathetometer. The product of these two gives the effective change in volume.

The volumes of the hollow sinkers used were in the neighbourhood of 350 cm^3. The change in the upthrust of water on the cylinder was 3.3 g when the temperature changed from 4 to 44 °C. As the amount of water required in weighing was large and the thermal dilation of water does not depend on the isotopic composition of water, only double-distilled tap water was used instead of SMOW.

Any arrangement for hydrostatic weighing may be used and one such apparatus is shown in figure 4.6.

4.5   Recent determinations of the density and dilatation of water at NML, Australia

4.5.1   Work by Bell and Clarke

Bell and Clarke [13] determined the density of water at seven temperatures, namely 3.98, 10.4, 15.4, 20.0, 30.0, 34.8, 40.1 °C, and expressed the results as a ratio of the density at temperature t and the density at 3.98 °C. They also fitted these values to a Thiesen-type formula:

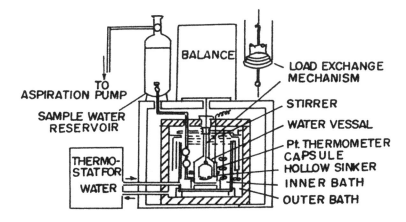

Figure 4.6. Apparatus for hydrostatic weighing.

For water in equilibrium with the ambient air:

$$(1 - d) \times 10^3 = ((t\,^\circ\mathrm{C}^{-1}) - 3.98)^2((t\,^\circ\mathrm{C}^{-1}) + 299.666)$$
$$\times \{519.985((t\,^\circ\mathrm{C}^{-1}) + 69.223)\}^{-1}. \qquad (4.28)$$

For air-free water

$$(1 - d) \times 10^3 = ((t\,^\circ\mathrm{C}^{-1}) - 3.98)^2((t\,^\circ\mathrm{C}^{-1}) + 283.263)$$
$$\times \{503.375((t\,^\circ\mathrm{C}^{-1}) + 67.335)\}^{-1}. \qquad (4.29)$$

The fit was found to be excellent, all calculated and observed values agreeing within ±0.8 ppm. Indeed, most of the values were within ±0.3 ppm, however the expressions suffer from fewer degrees of freedom.

4.5.2 Work by Bell and Patterson

Bell and Patterson [14] from the same laboratory took more measurements and reported a modified formula, in 1981, again a Thiesen-type one:

$$(1 - d) \times 10^3 = ((t\,^\circ\mathrm{C}^{-1}) - 3.989)^2((t\,^\circ\mathrm{C}^{-1}) + 335.1596)$$
$$\times \{558.7808((t\,^\circ\mathrm{C}^{-1}) + 72.2553)\}^{-1}. \qquad (4.30)$$

4.5.3 Work by Patterson and Morris

Patterson and Morris [15] in 1994 determined the relative density between 0 and 40 °C and expressed their results as follows.

$$(1 - d) \times 10^3 = ((t\,^\circ\mathrm{C}^{-1}) - 3.9818)^2((t\,^\circ\mathrm{C}^{-1}) + 316.338\,08)$$
$$\times \{538.211((t\,^\circ\mathrm{C}^{-1}) + 70.699\,73)\}^{-1}. \qquad (4.31)$$

The data were also fitted to a fifth-degree polynomial in terms of $(t - t_0)$:

$$\rho/\rho_{max} = 1 - \{A(t-t_0) + B(t-t_0)^2 + C(t-t_0)^3 + D(t-t_0)^4 + E(t-t_0)^5\}. \quad (4.32)$$

The values obtained for A, B, C, D and E were:

$$\begin{aligned}
t_0 &= 3.9818\,°C \\
A &= 7.0134 \times 10^{-8}\,°C^{-1} \\
B &= 7.926\,504 \times 10^{-6}\,°C^{-2} \\
C &= -7.575\,677 \times 10^{-8}\,°C^{-3} \\
D &= 7.314\,894 \times 10^{-10}\,°C^{-4} \\
E &= -3.596\,458 \times 10^{-12}\,°C^{-5}.
\end{aligned} \quad (4.33)$$

Here it should be emphasized that although all these results are from the same laboratory they vary too much in regard to the values of the various constants.

4.6 Recent determinations of the density and dilatation of water at NRLM, Japan

4.6.1 Work by Takenaka and Masui

Using the method described in section 4.3, four sets of data were recorded comprising 72 independent ordered pairs of the relative density ρ/ρ_{max} and temperature. The data were fitted to the following Thiesen-type expression:

$$\begin{aligned}
(1-d) \times 10^3 &= ((t\,°C^{-1}) - 3.981\,52)^2((t\,°C^{-1}) + 396.185\,34) \\
&\quad \times ((t\,°C^{-1}) + 32.288\,53) \\
&\quad \times \{609.6286((t\,°C^{-1}) + 83.123\,33) \\
&\quad \times ((t\,°C^{-1}) + 30.244\,455)\}^{-1}.
\end{aligned} \quad (4.34)$$

It should be noted that there is an extra factor of $(t+32.288\,53)/(t+30.244\,455)$. This is based on an expression Thiesen used in his later work [18]. He also fitted the same data to a Kell-type relation [12]:

$$\rho/\rho_{max} = \left(A_0 + \sum A_i t^{2i-1}\right) \bigg/ \left(1 + \sum B_j t^{2j}\right). \quad (4.35)$$

Here i and j take values from 1 to 4. The values of the constants are:

$$\begin{aligned}
A_0 &= 9.998\,6784 \times 10^{-1} & A_1 &= 6.782\,6308 \times 10^{-5}\,°C^{-1} \\
A_2 &= 1.036\,5704 \times 10^{-7}\,°C^{-3} & A_3 &= 1.748\,5485 \times 10^{-11}\,°C^{-5} \\
A_4 &= 8.415\,2542 \times 10^{-16}\,°C^{-7} \\
B_1 &= 9.088\,7089 \times 10^{-6}\,°C^{-2} & B_2 &= 1.497\,4442 \times 10^{-9}\,°C^{-4} \\
B_3 &= 1.600\,6519 \times 10^{-13}\,°C^{-6} & B_4 &= 2.810\,6977 \times 10^{-18}\,°C^{-8}.
\end{aligned} \quad (4.36)$$

Table 4.1. Uncertainty budget of Takenaka's results.

Factors contributing to the uncertainty	Amount	Uncertainty in ρ/ρ_{max} (ppm)		
		At 20 °C	At 40 °C	At 85 °C
Type B				
I Thermal expansion of water				
Mass of sample water	1 mg	0.01	0.01	0.01
Density of water at reference temp.	< 3 ppm	<0.01	<0.01	<0.01
Volume of dilatometer	1 mm^3	0.03	0.07	0.15
Calibration of balance for Hg bottle	0.1 mg	0.07	0.07	0.07
Density of mercury in dilatometer	1.5 ppm	0.04	0.1	0.1
Mean temperature of the part of capillary having temperature gradient	3 K	0.3	0.3	0.3
Calibration of PRT	0–0.3 mK	0.02	0.08	0.2
Temperature distribution in the thermostat	0.2–1 mK	0.04	0.12	0.65
Instability in volume of dilatometer	$0.7(t-4)/81$ ppm	0.2	0.37	0.7
II Thermal expansion of water cell				
Laser wavelength	$< 3 \times 10^{-8}$	<0.1	<0.1	<0.1
Initial length of cell	0.1 mm	0.03	0.05	0.05
Misalignment of cell	2×10^{-5} rad	0.05	0.05	0.05
Residual gas pressure	4 Pa	0.03	0.03	0.03
Cell temperature	0–0.2 K	0.0	0.1	0.3
Type A				
In measurement of mass of mercury		0.2	0.2	0.2
Due to curve fitting		0.2	0.2	0.2
Combined standard uncertainty		0.5	0.6	1.1

Furthermore he fitted the data to an eighth-degree polynomial of the form

$$\rho/\rho_{max} = \sum A_i t^{i-1}.$$ (4.37)

The obtained coefficients were:

$A_1 = 9.998\,6785 \times 10^{-1}$

$A_2 = 6.781\,9907 \times 10^{-5}\,°C^{-1}$

$A_3 = -9.085\,8952 \times 10^{-6}\,°C^{-2}$

$A_4 = 1.028\,8239 \times 10^{-7}\,°C^{-3}$

$A_5 = -1.407\,7910 \times 10^{-9}\,°C^{-4}$

$A_6 = 1.635\,5966 \times 10^{-11}\,°C^{-5}$ (4.38)

$A_7 = -1.368\,8193 \times 10^{-13}\,°C^{-6}$

$A_8 = 6.969\,9179 \times 10^{-16}\,°C^{-7}$

$A_9 = -1.591\,4816 \times 10^{-18}\,°C^{-8}$.

Table 4.2. Dimensions of the sinkers.

Inner condition	Designation	A (mm)	B (mm)	C (mm)	Volume (cm^3)	Mass (g)
Vacuum	LV1	80	74.5	3	351.7	352.9
	LV2	80	74.5	3	351.9	354.9
	SV1	75	70.8	3	296.3	301.1
	SV2	75	70.8	3	303.3	307.4
Air filled	LA	80	74.5	3	351.8	354.9
	SA	75	70.8	3	303.4	307.6

The standard deviation from the fit was 0.2 ppm in each case.

Factors influencing measurement uncertainty and their contributions are given in table 4.1

4.6.2 Work by Watanabe

In 1991, Watanabe [17] used the method described in section 4.4 to measure the thermal expansion of water in the temperature range 0–44 °C by finding the apparent mass of a hollow sinker in the form of a cylinder (artefact). In fact six such sinkers were used. Two pairs of sinkers were used and the members of each pair were made with great care so that the thermal properties of each member were the same as far as possible. The larger sinkers were labelled L and the smaller ones S. The dimensions and approximate mass and volume of each sinker are given in table 4.2. All sinkers were initially evacuated and denoted LV and SV. After the measurements were concluded, one sinker of each pair was filled with air and denoted LA and SA. The air was filled at atmospheric pressure to evaluate the thermal effects on their elastic deformation. All sinkers were made of fused quartz. The apparent density of each hollow sinker was made a little bigger than that of water by loading it with small lead shots so that these would work as good sinkers. One such sinker is shown in figure 4.7.

All the results were pooled and the final results were expressed as a sixth-degree polynomial as follows:

$$\rho/\rho_{max} = \sum A_i t^{i-1} \tag{4.39}$$

$A_1 = 9.998\,6775 \times 10^{-1}$ $\qquad A_2 = 6.786\,687\,54 \times 10^{-5}\,°C^{-1}$

$A_3 = -9.090\,991\,73 \times 10^{-6}\,°C^{-2}$ $\qquad A_4 = 1.025\,981\,51 \times 10^{-7}\,°C^{-3}$

$A_5 = -1.350\,290\,42 \times 10^{-9}\,°C^{-4}$ $\qquad A_6 = 1.326\,743\,92 \times 10^{-11}\,°C^{-5}$

$A_7 = -6.461\,418 \times 10^{-14}\,°C^{-6}.$

$$\tag{4.40}$$

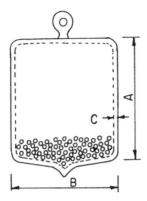

Figure 4.7. Hollow sinker.

Table 4.3. Uncertainty budget of Watanabe's results.

| Factors contributing uncertainty | Standard uncertainty in ρ/ρ_{max} (ppm) Temperature (°C) | | | | | |
	0	10	20	30	40	44
Type A						
Mass in air			Negligible			
Apparent mass in water	0.1	0.1	0.1	0.1	0.1	0.1
Temperature variation	(0.2)	(0.2)	(0.2)	(0.35)	(0.50)	(0.55)
	0.01	0.02	0.05	0.11	0.19	0.22
Temperature distribution	(0.2)	(0.2)	(0.2)	(0.35)	(0.50)	(0.55)
	0.01	0.02	0.05	0.11	0.19	0.22
Type B						
Isotopic fluctuations			Negligible			
Dissolved air	0.2	0.2	0.2	0.2	0.2	0.2
Coefficient of expansion of sinker	0.03	0.05	0.09	0.11	0.11	0.12
Thermostatic effect of sinker	0.03	0.05	0.13	0.22	0.30	0.33
Combined standard uncertainty in ρ/ρ_{max} (quadrature method)	0.23	0.24	0.28	0.37	0.47	0.52

The uncertainty budget of water density measurement by Watanabe is indicated in table 4.3.

As we can see, the complexity due to the large number of expressions adds to the confusion. Fortunately the values of ρ/ρ_{max} obtained from any of the formulae or expressions given by Patterson [15], Takenaka [16] or Watanabe [17] agree very well within, say, 0.2 ppm. It is again emphasized that all results for the density of water refer to an ambient pressure of 101 325 Pa.

4.7   Harmonized formula for water density by the author

We can see from the previous discussions that a variety of formulae have been used to express the density–temperature relationships for water. Some are based on the actual measurement data and others are merely derived from statistical considerations. The situation has become even more complex, when one type of expression, for example a Theisen-type one, has been used with many different values for the constants. Hence, it has become necessary to look for a harmonized expression, which can represent recent measurements by various scientists equally well. In order to harmonize the various formulae, the present author [19] obtained values for ρ/ρ_{max} from the formulae derived from the actual measurement data by various researchers [14–17]. If a researcher has given Theisen-type expressions as well as polynomial forms, the values of ρ/ρ_{max} have been calculated using both expressions. In this way, I obtained six sets of values for the density of water for each temperature. Equal weights were assigned to each datum and the mean value of ρ/ρ_{max} was obtained for each °C from 0–41 °C. A polynomial function for ρ/ρ_{max}, in the temperature range 0–41 °C, was fitted to the data, so obtained, by using the least-squares method. The temperature range was chosen by considering the range in which most metrologists are interested. The range of temperatures in which water is used as a density standard for all practical purposes would be between 10 and 35 °C. The fit is so good that the square root of the average sum of residual errors is only 0.1 ppm. No calculated value differs from the corresponding mean value by more than 0.17 ppm.

I obtained the following expression:

$$(1 - \rho/\rho_{max}) \times 10^6 = A_1 T + A_2 T^2 + A_3 T^3 + A_4 T^4 + A_5 T^5 \qquad (4.41)$$

where $T = t - t_0$, t is in °C, $t_0 = 3.983\,035$ °C and

$$
\begin{aligned}
A_1 &= -2.381\,848 \times 10^{-2}\,°\mathrm{C}^{-1} \\
A_2 &= 7.969\,992\,983\,°\mathrm{C}^{-2} \\
A_3 &= -7.999\,081 \times 10^{-2}\,°\mathrm{C}^{-3} \\
A_4 &= 8.842\,680 \times 10^{-4}\,°\mathrm{C}^{-4} \\
A_5 &= -5.446\,145 \times 10^{-6}\,°\mathrm{C}^{-5}.
\end{aligned}
\qquad (4.42)
$$

The value of ρ_{max} has been taken to be

$$999.974\,950 \pm 0.000\,84\ \mathrm{kg\ m}^{-3}. \qquad (4.43)$$

It should be noted that the second term on the right-hand side contributes the most, so its coefficient has been determined to a higher number of decimal points. The values of ρ_{max} and temperature at which water attains its maximum density have been taken according to the recommendation by the BIPM. The density values for SMOW in steps of 0.1 °C in the range 0–41 °C at a pressure of 101 325 Pa

Table 4.4. The density of SMOW (Standard Mean Ocean Water) in kg m^{-3} on ITS-90 at 101 325 Pa. *Note*: Wherever an asterisk (*) appears, the integer value to be taken, thereafter in the row, will be one less than the figure written in the second column of the row. The value of ρ_{max} taken is 999.974 950 ± 0.000 84 kg m^{-3}. The uncertainty given is 2SD. To take full advantage of the table, find the values of $\delta\,^{18}$O and δD of the water sample, taking $R_D = (155.76 \pm 0.05) \times 10^{-6}$ and $R_{18} = (2005.2 \pm 0.05) \times 10^{-6}$ for SMOW and use $\rho - \rho_{(V-SMOW)} = 0.233 \times 10^{-3}\delta\,^{18}$O $+ 0.0166 \times 10^{-3}\delta$D. If the water is partially saturated with air, a further correction is applied using the relation $(\Delta\rho/\text{kg m}^{-3}) = (-0.004\,612 + 0.000\,106t\,°\text{C}^{-1})\chi$. χ is the level of saturation.

Temp.		0.0	0.1	0.2	0.3	0.4	0.5	0.6	0.7	0.8	0.9
0	999	.8431	.8598	.8563	.8626	.8687	.8747	.8804	.8860	.8915	.8967
1	999	.9018	.9067	.9114	.9159	.9203	.9245	.9285	.9324	.9361	.9396
2	999	.9429	.9461	.9491	.9519	.9546	.9571	.9595	.9616	.9636	.9655
3	999	.9671	.9687	.9700	.9712	.9722	.9731	.9738	.9743	.9747	.9749
4	999	.9749	.9748	.9746	.9742	.9736	.9728	.9719	.9709	.9697	.9683
5	999	.9668	.9651	.9633	.9613	.9592	.9569	.9545	.9519	.9492	.9463
6	999	.9432	.9400	.9367	.9332	.9296	.9258	.9218	.9177	.9135	.9091
7	999	.9046	.8999	.8951	.8902	.8851	.8798	.8744	.8689	.8632	.8574
8	999	.8514	.8453	.8391	.8327	.8261	.8195	.8127	.8057	.7986	.7914
9	999	.7840	.7765	.7689	.7611	.7532	.7451	.7370	.7286	.7202	.7116
10	999	.7029	.6940	.6850	.6759	.6666	.6572	.6477	.6380	.6283	.6183
11	999	.6083	.5981	.5878	.5774	.5668	.5561	.5452	.5343	.5232	.5120
12	999	.5007	.4892	.4776	.4659	.4540	.4420	.4299	.4177	.4054	.3929
13	999	.3803	.3676	.3547	.3418	.3287	.3154	.3021	.2887	.2751	.2614
14	999	.2475	.2336	.2195	.2053	.1910	.1766	.1621	.1474	.1326	.1177
15	999	.1027	.0876	.0723	.0569	.0414	.0258	.0101	*.9943	.9783	.9623
16	998	.9461	.9298	.9133	.8968	.8802	.8634	.8465	.8296	.8125	.7952
17	998	.7779	.7605	.7429	.7253	.7075	.6896	.6716	.6535	.6353	.6170
18	998	.5985	.5800	.5613	.5425	.5237	.5047	.4856	.4664	.4471	.4276
19	998	.4081	.3885	.3687	.3489	.3289	.3089	.2887	.2684	.2480	.2275
20	998	.2069	.1863	.1654	.1445	.1235	.1024	.0812	.0599	.0384	.0169
21	997	.9953	.9735	.9517	.9297	.9077	.8855	.8633	.8409	.8185	.7959
22	997	.7733	.7505	.7276	.7047	.6816	.6585	.6352	.6118	.5884	.5648
23	997	.5412	.5174	.4936	.4696	.4455	.4214	.3971	.3728	.3483	.3238
24	997	.2992	.2744	.2496	.2247	.1996	.1745	.1493	.1240	.0986	.0731
25	997	.0475	.0218	*.9960	.9701	.9441	.9180	.8918	.8656	.8392	.8128
26	996	.7862	.7596	.7328	.7060	.6791	.6521	.6250	.5978	.5705	.5431
27	996	.5156	.4881	.4604	.4326	.4048	.3769	.3488	.3207	.2925	.2642
28	996	.2358	.2074	.1788	.1501	.1214	.0926	.0636	.0346	.0055	*.9763
29	995	.9470	.9177	.8882	.8587	.8290	.7993	.7695	.7396	.7096	.6795
30	995	.6494	.6191	.5888	.5583	.5278	.4972	.4666	.4358	.4049	.3740
31	995	.3430	.3118	.2806	.2494	.2180	.1865	.1550	.1234	.0917	.0599
32	995	.0280	*.9960	.9640	.9319	.8996	.8673	.8350	.8025	.7700	.7373
33	994	.7046	.6718	.6389	.6060	.5729	.5398	.5066	.4733	.4399	.4065
34	994	.3729	.3393	.3056	.2718	.2380	.2040	.1700	.1359	.1017	.0675
35	994	.0331	*.9987	.9642	.9296	.8949	.8602	.8254	.7905	.7555	.7204
36	993	.6853	.6501	.6148	.5794	.5439	.5084	.4728	.4371	.4013	.3655
37	993	.3296	.2936	.2575	.2213	.1851	.1488	.1124	.0760	.0394	.0028
38	992	.9661	.9294	.8925	.8556	.8186	.7815	.7444	.7072	.6699	.6325
39	992	.5951	.5576	.5200	.4823	.4446	.4067	.3688	.3309	.2928	.2547
40	992	.2166	.1783	.1400	.1016	.0631	.0245	*.9859	.9472	.9085	.8696
41	991	.8307									

have been given in table 4.4. This table supersedes the one given by Jones and Harris [20]. The density values refer to SMOW or V-SMOW with R_D, R_{17} and R_{18} respectively equal to

$$R_D = (155.76 \pm 0.05) \times 10^{-6}$$
$$R_{18} = (2005.2 \pm 0.05) \times 10^{-6} \tag{4.44}$$
$$R_{17} = (371.0) \times 10^{-6}.$$

The density values given in table 4.4 agree very well with those given by the task group formed by the Working Group on Density, Consultative Committee for Mass and Related Quantities (CCM) [21]. The CCM is the committee constituted by the International Committee of Weights and Measures, an executive organ of the General Conference of Weights and Measures (CGPM) under the Convention du Metre. The difference is never more than a few parts in 10 million.

For the density of a sample with an isotopic composition which differs from that of SMOW, the following equation should be used:

$$\rho - \rho_{(V\text{-SMOW})} = 0.233 \times 10^{-3} \delta\,^{18}O + 0.0166 \times 10^{-3} \delta D \tag{4.45}$$

where $\rho - \rho_{(V\text{-SMOW})}$ is in kg m^{-3}.

Similarly to find the density differences $\Delta\rho$ between air-free and partially air-saturated water the following equation should be used:

$$(\Delta\rho/\text{kg m}^{-3}) = (-0.004\,612 + 0.000\,106(t\,^{\circ}C^{-1}))\chi. \tag{4.46}$$

The dilatation data of water, which should, in my view, be the ratio of the volume at any temperature to the minimum volume $V/V_{\min}$, have also been expressed as a fifth-degree polynomial function:

$$(V/V_{\min} - 1) \times 10^6 = B_1T + B_2T^2 + B_3T^3 + B_4T^4 + B_5T^5 \tag{4.47}$$

where $T = t - t_0$, t is in $^{\circ}C$, $t_0 = 3.983\,035\,^{\circ}C$ and

$$B_1 = -1.314\,238 \times 10^{-2}\,^{\circ}C^{-1}$$
$$B_2 = 7.954\,235\,386\,^{\circ}C^{-2}$$
$$B_3 = -7.815\,959 \times 10^{-2}\,^{\circ}C^{-3} \tag{4.48}$$
$$B_4 = 8.654\,438 \times 10^{-4}\,^{\circ}C^{-4}$$
$$B_5 = -5.008\,041 \times 10^{-6}\,^{\circ}C^{-5}.$$

The square root of the average residual error is only 1.3 ppm.

References

[1] Chappuis P 1907 *Trav. Mem. Bur. Poids. Mes.* **3** D1–40

Chappuis P 1910 *Trav. Mem. Bur. Poids. Mes.* **14** D63

[2] Theisen Mm Scheel K and Diesselhorst H 1990 *Wiss. Abh. Phys-tech. Reichsanst.* **3** 1–17

[3] Mendeleev D I 1897 *Vremenik Glavnoj Palati Mer I Vesov* **3rd Part** 133

[4] Steckel F and Szapiro S 1963 *Trans. Faraday Soc.* **59** 331

[5] Stott V and Bigg P H 1928 *International Critical Tables* vol III, pp 24–6

[6] Tilton L W and Taylor J K 1937 *J. Res. Natl Bur. Stand. Wash.* **18** 205–14

[7] Bigg P H 1967 Density of water in SI units over the range 0–40 °C 1967 *Br. J. Appl. Phys.* **18** 521–5

[8] Aleksandrov A A and Trakhtengrets M S 1970 *Therm. Eng.* **17** 122

[9] Wagenbreth H and Blanke W 1971 *PTB Mitteilungen* **6** 412

[10] 1976 *International Union of Pure and Applied Chemistry* **45** 1–9
Menache M and Girard G 1973 Thermal expansion of water between 0 and 40 °C *Metrologia* **9** 2–68

[11] Kell G S 1967 *J. Chem. Eng. Data* **12** 66

[12] Kell G S 1977 Density of liquid water *J. Phys. Chem. Ref. Data* **6** 1109–31

[13] Bell G A and Clarke A L 1976 A determination of dilatation of pure water of known isotopic composition *Atomic Masses & Fundamental Constants* vol 5, ed J H Sanders and A H Wapstra (New York: Plenum) pp 615–21

[14] Bell G A and Patterson J B 1984 Density standards, the density and thermal dilation of water *Precision Measurements and Fundamental Constants* vol II, ed B N Taylor and W D Phillips (Gaithersburg, MD: National Bureau of Standards–US Department of Commerce) pp 445–7

[15] Patterson J B and Morris E C 1994 Measurement of absolute water density 1 °C to 40 °C *Metrologia* **31** 272–88

[16] Takenaka M and Masui R 1990 Measurement of the thermal expansion of pure water in the temperature range 0–85 °C *Metrologia* **27** 165–71

[17] Watanabe H 1991 Thermal dilation of water between 0–44 °C *Metrologia* **28** 33–43

[18] Thiesen M 1904 *Phys. Tech. Reichs. Wiss. Abbandl.* **4** 1–32

[19] Gupta S V 2001 New water density at ITS 90 *Indian J. Phys.* **October**

[20] Jones F E and Harris G L 1992 ITS-90 density of water formulation for volumetric standards calibration *J. Res. Natl Inst. Stand. Technol.* **97** 335–40

[21] Tanaka M, Girard G, Davis R, Peuto A and Bignell N 2001 Recommended table for the density of water between 0 °C and 40 °C based on recent experimental reports *Metrologia* **38** 301–9

Chapter 5

Mercury density measurement

Symbols

β	Coefficient of isothermal compressibility of mercury as a function of pressure
β_0	Coefficient of isothermal compressibility independent of pressure
ρ_0	Density of mercury at $0\,°C$
ρ	Density of mercury at $t\,°C$
P or p	Pressure
a_i	Coefficients of temperature in density relation
b_i	Coefficients of pressure in compressibility relation.

5.1 Introduction

The density of mercury is important as apart from being used for calibrating smaller capacity volumetric glassware, it is also used as the hydrostatic liquid in primary barometers for measuring atmospheric pressure. Mercury is used as a reference standard of density for determining the capacity of small volumetric measures. In addition the accuracy of a primary standard barometer at the 1 atmosphere level depends directly on the density of mercury. The requirement for accuracy at the 1 ppm level makes it necessary that the density of mercury is also known at least with an uncertainty of better than 1 ppm. Mercury is used in thermometers as it is a linearly expanding liquid and as a confining liquid in investigations of the pressure–volume–temperature relations of other fluids. Cook and Stone [1] and Cook [2] carried out the most accurate and basic determination of its density. There are two methods for determining the density of mercury, namely the displacement method and the content method. However, before discussing these methods, let us consider its purification.

Table 5.1. Impurities present in mercury which can affect its density at the 1 ppm level.

Metal	Quantity (ppm)	Metal	Quantity (ppm)
Platinum	2.7	Tin	1.1
Gold	3.4	Iron	1.4
Zinc	1.5	Sodium	0.13
Copper	1.9	Calcium	0.13
Lead	4.3	Aluminium	0.25

5.2 Purification of mercury

Normally the density of the impurities found in mercury is much less than that of mercury, so impurities comprising only a few parts in 10 million by weight can alter the density of mercury by more than 1 ppm. Hence the mercury samples have to be purified very carefully. All equipment which comes into contact with the mercury should also be kept scrupulously clean. Furthermore, the isotopic abundance ratio is also to be determined, which is really a big problem for mercury.

5.2.1 Level of impurities that might change the density of mercury by 1 ppm

The amounts of impurities that might change the density of mercury by 1 ppm are given in table 5.1. In addition, silver forms an amalgam with mercury which has the same density as that of mercury even if silver is present up to 30% by weight.

5.2.2 Pre-cleaning procedures

(1) Wash once or twice with 10% KOH, followed by rinsing with distilled water and 10% HNO_3 and then rinse thoroughly with distilled water before triple distillation in reduced air pressure.
(2) In the case of recycled mercury, the steps are:
   - Filter through a filter paper with a pore size of 15–40 μm.
   - Wash for 5 h with acetone to degrease it.
   - Wash for 15 h with a mixture of nitric acid, hydrogen peroxide and water mixed by volume in the ratio of 1:1:3 in order to remove base metals and sulphides.
   - Wash for 6 h with distilled water and dry at 80 °C and at a pressure of 2.7 kPa.

5.2.3   Distillation and cleaning of mercury

The mercury is shaken well with dilute nitric acid for a few minutes and then distilled three times under a reduced pressure of, say, 20 mm Hg. Hulett and Minchin [3] studied the distillation of mercury. They have shown that quite large amounts of zinc and cadmium rise to the surface when mercury is distilled in vacuum. However, if it is distilled in air at reduced pressure of about 20 mm Hg (2.7 kPa), the following happens:

- the amounts of silver, gold and platinum are reduced to a few parts in 10^8 or less after two distillations;
- metals with high boiling points remain in the still; and
- oxidizable metals are oxidized in air and are deposited in the upper part of the still.

5.3   Displacement method for determining the density of mercury

A solid, the density of which is a little more than that of mercury, is chosen. Its volume is determined by dimensional measurements. The loss in apparent mass when the solid is immersed in mercury is measured. If the volume of the solid is known then the density of mercury is calculated from the loss in apparent mass. The method used by Cook and Stone [1] is briefly described in the following sections.

5.3.1   Solid standard

A solid in the form of a cube is taken for this purpose. As the mercury has a density of about 13 545 kg m^{-3}, there is limited choice in finding a material heavier than this, so normally tungsten carbide mixed with 15%wt cobalt is used. The cube is formed by sintering it in a carbon mould. The size of the tungsten carbide particles is kept below 5 μm. In order to bring it to the true geometrical shape of a cube it is lapped with diamond dust 0.5–3 μm in size. The dust may be mixed with paraffin oil. To acquire a good surface finish, the final polishing is carried out with diamond dust mixed in paraffin of particle size less than 2 μm.

5.3.2   Polishing defects in a cube

A solid cube made by lapping and polishing may suffer from the following defects:

- the faces may not be perfectly flat;
- adjacent faces may not be at right angles;
- the faces may have rough surfaces; and
- the edges may be broken while it is being made or used.

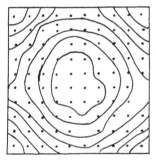

Figure 5.1. A typical fringe pattern from the face of the cube.

Flatness test

To test the flatness of the faces, two experiments are carried out:

(1) to see whether the central portion of the face is depressed; and
(2) to see whether all the four corners of a face lie in the same plane, a
     plane passing through any three corners is considered and the depression
     or elevation of the fourth corner is measured.

For (1), an optical flat of known flatness is used and a fringe pattern showing
contours with various depths is obtained. Alternatively the cube is placed on a
plane surface and a small amount of liquid paraffin is poured onto the face of the
cube, so that a film about 3 mm thick is formed. If the face is flat then the thickness
of the film will be uniform; otherwise it will form a lens. The upper surface of
the liquid film will be planar while the lower surface of the liquid film will be
the exact replica of the form of the face of the cube. When a collimated beam
of monochromatic light illuminates the face then interference fringes are formed
by two light beams, the first reflected from the top surface of the liquid paraffin
and the other from the surface of the cube. The fringe pattern will be similar to
Newton's rings. The fringes in this case show contours of equal thickness from
the upper surface of the film, which will act as a reference surface. One such
pattern obtained by Cook is shown in figure 5.1.

For (2), use an autocollimator and a moveable mirror with a flat base and
follow the normal method of flatness measurement, which will give the position
of the fourth corner of each face of the tungsten carbide cube. The mean depth of
the fourth corner for each face with standard deviations for Cook's cube is given
in table 5.2.

The cube with depressions at the corners would appear as shown in
figure 5.2.

Table 5.2. Dimensions and standard deviation of the tungsten carbide cube.

Face no	Fourth face	Depth (nm)	SD (nm)
1	(2, 4)	21.5	11
2	(3, 6)	15.3	11
3	(5, 6)	6.1	9
4	(1, 2)	13.8	18
5	(3, 6)	23.0	14
6	(3, 5)	−21.5	4

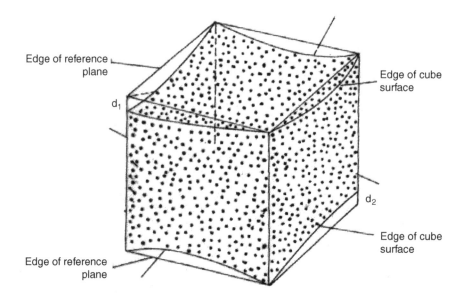

Figure 5.2. Tungsten carbide cube (the dotted portion shows the material).

5.3.3   Dimensional measurement and volume

The edge lengths were measured using a Fizeau interferometer as described in section 2.4. Corrections due to faces not being square to each other are applied. Cook also investigated a correction due to elastic deformation of the cube, under different methods of support, but found this to be negligible. The edges as measured by Cook for his cube were 88.868 266 mm, 88.865 388 mm and 88.865 752 mm.

Uncertainty budget in volume

The volume of the cube was measured with a relative standard deviation of 1.6×10^{-7}. The uncertainty due to different causes can be broken up as follows:

Surface roughness	3×10^{-8}
Form of faces	1.0×10^{-7}
Length of edges	1.1×10^{-7}
Correction due to broken edges	1×10^{-8}

5.3.4   Weighing a solid standard (a block of tungsten carbide) in air

A solid standard (block) of known volume is weighed in air. Such blocks have a tendency to be slightly ferromagnetic, therefore the weighing is carried out six times so that every face is used as a top face and the mean of these weighings is taken. The appropriate air buoyancy corrections are applied using the latest BIPM equation for air density given in [4].

5.3.5   Weighing a solid standard (a cube of tungsten carbide) in mercury

For suspending the cube, a highly polished and flat plate smaller than the cube can be used. The cube is attached to the plate by wringing them together. A very fine tungsten wire of diameter 0.05 mm is attached to the plate. The other end of the wire is attached to a stainless steel rod. The rod can be transferred to a double V hook attached to the pan of the balance. The arrangement is shown in figure 5.3.

Immersing the cube in mercury

The mercury is poured around the cube at a highly reduced pressure from a reservoir with a stainless steel tube at its bottom, so that mercury from the bottom is drained into the container with the cube.

Container for holding mercury with the cube inside it

The container shown in figure 5.4 is made of five glass plates cemented together to make an open container. The sixth plate, which is used as a cover, is in two parts so that the container is covered without disturbing the suspended cube. The plate has proper holes for the glass tubes for temperature-measuring devices. The glass container is reinforced with an aluminium box, which may be surrounded with an insulating material. The whole box is placed on a brass plate, which can be covered with a large bell jar so that the desired low pressure may be created. When creating low pressure, the insulating material, being fibrous in nature, is removed. The top of the box has a stand with double V, on which the stainless steel rod attached to the suspension wire carrying the plate is placed

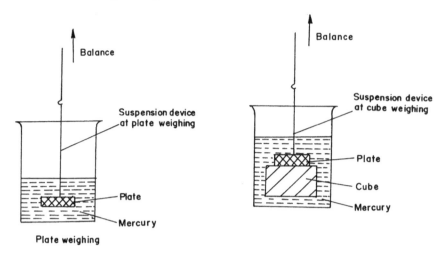

Figure 5.3. Tungsten carbide cube hooked onto a highly polished plate for weighing in mercury.

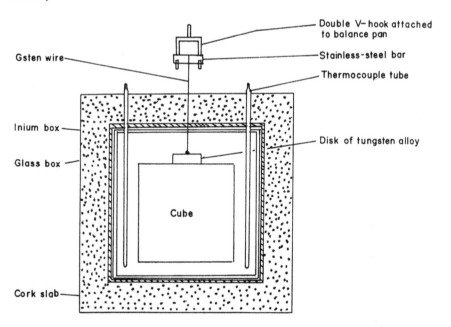

Figure 5.4. Container for holding mercury with the cube inside it.

when it is not connected to the balance via another, double-V shaped hook (figure 5.5(*a*)). The whole arrangement can be moved in the *x*, *y* and *z* directions. By lowering the arrangement the rod carrying the hanging plate with or without

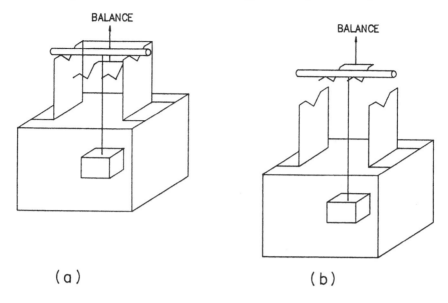

Figure 5.5. (*a*) Cube resting on its supports. (*b*) Cube attached to the balance.

the cube can be loaded to the double-V shaped hook attached to the balance (figure 5.5(*b*)). Whenever the cube is to be weighed in mercury, it is wrung to the plate. Substitution weighing is used to find the apparent mass of the cube and the plate in mercury. Special attention is given to ensure the surface tension effect on the wire from which the cube is suspended in mercury is properly taken into account. Cook monitored the varying nature of the angle of contact and applied the appropriate correction.

A similar experiment is carried out for weighing the suspension arrangement in mercury without the cube. Care is taken that in each case the suspension is immersed exactly up to the same level and the appropriate correction applied due to the buoyancy of mercury on the wire, if necessary.

Uncertainty budget of the upthrust

Uncertainty components in various measurements and due to surface tension are given in table 5.3 The quadrature method has been used to obtain the standard uncertainty.

5.3.6 Density of mercury by displacement method

Cook measured the density of four, well-documented, samples. The mean density at 20 °C according to ITS-48 at one atmospheric pressure is given as 13.545 8924 g cm^{-3}.

Table 5.3. Uncertainty in upthrust measurement.

Mass of cube	1×10^{-7}
Weight of cube in mercury at 20 °C	1.5×10^{-7}
Surface tension of wire	1.0×10^{-7}
Temperature measurement (maximum)	2.0×10^{-7}
Standard uncertainty	2.8×10^{-7}

Using the expansion formula developed by Beattie *et al* [5], the density of mercury at 0 °C was given as $13.595\,0861$ g cm^{-3}.

5.4   Content method for determining the density of mercury

This method essentially consists of creating a vessel of known capacity, filling it with mercury and finding the mass of the mercury contained in it. The density is obtained by dividing the mass of the mercury by its volume.

5.4.1   Content measure (vessel)

Six highly polished and flat thick plates of quartz are joined together to form a cubical content measure. The plates should be so polished that they can be wrung together. For example Cook [2] took two flat square plates (A, B) $12.5 \times 12.5 \times 2.5$ cm^3 in size, two plates (C, D) $12.5 \times 7.3 \times 2.5$ cm^3 and two plates (E, F) $7.3 \times 7.3 \times 2.5$ cm^3. The plates E and F are wrung to plate C on each edge of it and are made parallel to each other. Plate D is finally wrung to the plates E and F. These four plates form the vertical walls of a box of size $7.3 \times 7.3 \times 7.3$ cm^3 approximately. These four plates are attached to the base plate A to form an open box as shown in figure 5.6.

Isopropyl ether helps in attaching and adjusting the plates parallel to each other. Plate B serves as the lid of the box. A capillary tube of uniform diameter is fitted at the centre of plate B and is flushed by lapping with the inner face of the plate. The other end of the capillary tube is fitted with a lapped stainless steel flange to serve as the datum for the depth of the mercury thread in the capillary tube and also as a joint to the vacuum system. The cross section of the capillary tube is calibrated by weighing different lengths of the mercury thread in the tube. An equation is obtained, connecting the mass of the mercury contained in the tube to its length from the inner surface of the lid-plate. The plate is finally attached to the walls of the box to form a closed box. Opposite plates should be made parallel to each other, as far as possible so that interference fringes are obtained from the two beams reflected from the inner surfaces of the opposite walls. For this purpose, a Fizeau interferometer, as described in section 2.4, is used.

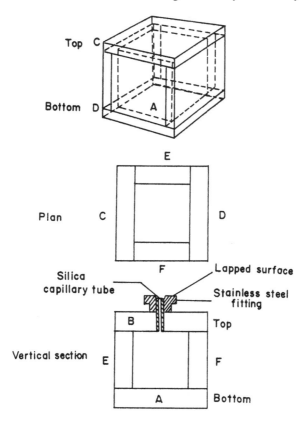

Top C

Bottom D   A

E

Plan   C   D

F   Lapped surface

Silica
capillary tube   Stainless steel
fitting

B   Top

Vertical section   E   F

A   Bottom

Figure 5.6. Content measure (showing the position of plates of different sizes).

Measuring the volume of the content measure (vessel)

To measure the internal dimensions of the content measure (vessel) so formed, the method described in section 2.5.4 for the measurement of the etalon length may be easily employed. Interference fringes are obtained with a laser to give the fractional part of the fringe order. By using two lasers and employing the method of exact fractions, the integer part of the number of fringes is also obtained. One of the lasers should be calibrated against the primary standard of length. To obtain the value of the mean separation between opposite walls, each face is divided into 100 equal squares and the distance between the centres of corresponding squares is measured and the mean value taken as the separation between opposite plates. To eliminate undesired reflections, the outer surfaces of all the plates are matted and the outer surface of each plate is inclined to about a few minutes of arc with its corresponding inner surface.

If the linear dimensions are measured in vacuum, then a correction for

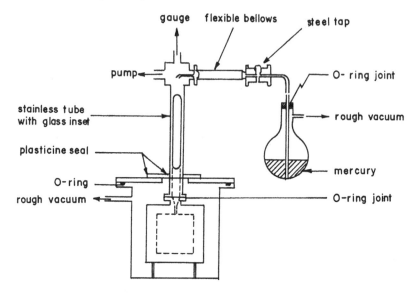

Figure 5.7. Outline of apparatus used for filling the content measure with mercury.

the compressibility of quartz plates is also applied. The volume of mercury is measured under a pressure of one atmosphere plus the hydrostatic pressure due to mercury at the centre of the vessel. To obtain the density of mercury at a pressure of 101 305 Pa, the compressibility of mercury also has to be taken into account.

The finite width of the edges of the plates and any chipped portion should be properly estimated. Edges of width 10 μm may produce an error of one part in 10^6. Cook found that the aggregate volume of such chipped and rounded-off edges was 1.310 mm^3, which was 3×10^{-6} of the volume of his cube. The major contribution to the uncertainty in the measurement came from the largest chip, for which the two observers estimated values of 0.624, 0.694 and 0.639 mm^3, a range of 0.07 mm^3, which amounted to 1×10^{-7}. In addition, the total uncertainty, due to ignorance of the exact form of the chipped portion, might perhaps be overestimated by up to 10% of the chipped volume.

Another source of error is due to the difference in the capacity of the vessel and the actual volume of mercury filling it. The angle of contact of surface tension force prevents the mercury from filling the edges, corners and chipped portions of the vessel. The radius of curvature of the cylindrical surface of the mercury is about 5 μm and the total empty space between the cube and the mercury along the edges under normal air pressure will therefore be 4×10^{-6} cm^3 or 1×10^{-8} of the cube volume. The empty volume at the corners and chips is even less. The empty volume is difficult to detect as, on looking through the sides of the vessel, the mercury always appears to fill the edges, corners and chips.

5.4.2   Filling the content measure (vessel) with mercury

The vessel is to be filled with mercury under a very low pressure of 0.1 Pa. An outline of the apparatus used by Cook [2] is shown in figure 5.7. The content measure is placed in a steel box, which can be roughly evacuated, so that there is no great difference in the pressure inside and outside the vessel. The stainless steel flange fitted to the capillary tube is clamped through an O-ring to a vertical filling tube. The vertical tube consists of a glass liner cemented with Araldite into an outer stainless steel tube partly cut away so that the level of mercury in the glass tube is visible. A small diffusion pump is connected to create a low pressure of 0.1 Pa in the vessel and the filling tube. The mercury is contained in a glass flask and it can be drawn into the filling tube through a tube reaching the bottom of the flask so that mercury from the bottom is sucked into the vessel, leaving the contaminated mercury surface undisturbed. The pressure above the mercury is not reduced until the vessel is ready to be filled otherwise the mercury may flood into the filling tube before the desired low pressure is attained in the vessel. The pressure of 0.1 Pa is maintained for a few hours so that it becomes certain that, despite the small conductance of the capillary tube, the pressures in the vessel and the filling tube are equal. As soon as the desired low pressure of 0.1 Pa is achieved, the pressure over the mercury surface is slowly increased so that the mercury fills the tube and starts filling the vessel. The mercury flow is controlled through a stainless steel tap. When the vessel is nearly full, the filling tube is allowed to drain completely into the vessel. Air is then allowed into the apparatus and the vessel removed. The excess mercury ensures that no air enters the vessel. The vessel is now ready to be placed inside the thermostatic bath.

After the mass of the mercury in the vessel is determined, the vessel is emptied through a hypodermic needle introduced through the capillary tube.

5.4.3   Thermostatic bath

The density of mercury changes by 1 ppm for a change of 5 mK. To maintain the temperature within 1 mK and the gradients within a tighter limit, a suitable thermostat is chosen. The problem is all the more difficult, as it is not possible to insert any temperature-measuring device inside the vessel. Furthermore, to keep the pressure outside and inside the vessel more or less equal, the vessel is immersed in a jacket of mercury. The mercury level in the jacket should be almost up to the upper surface of the lid. A diagram of the complete apparatus with the vessel, its bath, probe, glass container and insulating arrangement is shown in figure 5.8.

5.4.4   Gauging and volume determination

After ensuring that the temperature of the mercury inside the vessel has attained the temperature of the bath, the excess mercury is removed until the meniscus in the capillary is near the middle of the lid. The height of the meniscus from

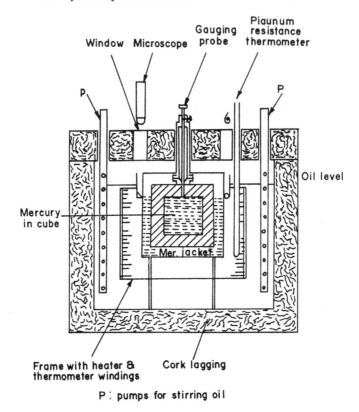

Figure 5.8. Thermostatic bath used by Cook.

the inside surface of the lid is measured using some suitable means. Cook [2] measured it through a probe as shown in figure 5.9. The probe is held in a cylinder in which one end is ground and lapped flat to rest on the upper lapped surface of the flange fitted on the lid of the vessel. The contact between the probe and the meniscus of the mercury is seen through the side of the lid of the vessel and the passage in the insulated lid of the thermostatic bath. A similar arrangement is made to illuminate the meniscus from behind the vessel. The probe with a large cylinder is properly insulated, to prevent it affecting the temperature inside the vessel. The probe is pushed down until it touches the mercury surface and it is then clamped. The pressure of the atmospheric air and temperature of the bath are measured. The gauge assembly is then removed, care being taken that no mercury is removed with it. The projection of the gauge beyond the lapped surface of the gauge assembly is measured by supporting it on two equal slip gauges and another slip gauge touching the probe as shown in figure 5.9(b). The difference in the lengths of the two slip gauges gives the depth of mercury from the lapped surface of the flange. The distance between this lapped surface and the inner surface of

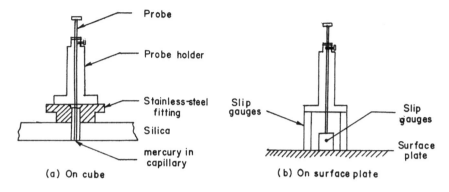

Figure 5.9. Gauging of mercury height (*a*) on cube and (*b*) on surface plate.

the lid is measured; subtracting the depth of the mercury surface from it gives the length of the mercury thread in the capillary and hence the mass of mercury in the capillary.

Quartz or fused silica has a tendency to absorb water from the atmosphere and accumulate static charges, which creates problems in determining the mass of mercury in the vessel. Use of a similar vessel of the same material as counterpoise eliminates the water absorption problem to a great extent. Irradiating the air inside the balance with a strong beam of β-rays removes any static charge on the vessel. Proper earthing of the balance and occasionally touching the vessel with a conducting wire is another solution. Corrections are applied for the pressure acting at the centre of the vessel and for air buoyancy on the weights. The mass of mercury is obtained from the simultaneous determination of the mass of the vessel filled with mercury and when emptied. Care is taken that no mercury traces are left in the vessel. The weights used must also be calibrated almost immediately after the weighing.

Example for a calculation of the volume and mass of mercury

An example of calculation of volume and mass of mercury as given by Cook is reproduced here:

Atmospheric pressure	756.97 mm Hg
External pressure at the level of centre of the vessel	793.37 mm Hg
Internal pressure at the level of centre of the vessel	800.75 mm Hg
Mean	797.06 mm Hg
Mean temperature	21.0317 °C
Volume of cube at 21 °C in vacuum	393.700 142 cm^3
Expansion to 21.0317 °C	+0.000 019 cm^3
Contraction of mercury under 797.06 mm of Hg	−0.001 115 cm^3

Volume under actual conditions   393.699 046 cm^3

Mass of mercury in vessel and capillary   5332.0824 g
Subtracted mass of mercury in the capillary   0.0781 g
Mass of mercury in the vessel   5332.0043 g

Density of mercury under conditions of measurement
          5332.0043/393.699 046 =   13.543 351 g cm^{-3}

Correction for meniscus and air drawn into the vessel   −0.000 003 g cm^{-3}
Correction to 20 °C   +0.002 530 g cm^{-3}
Correction to 101 305 Pa pressure   +0.000 003 g cm^{-3}

Density at 20 °C   13.545 881 g cm^{-3}

Coefficient of thermal expansion of the vessel 1.56×10^{-6} °C^{-1}
Compressibility of vessel 2.70×10^{-5} MPa^{-1}
Thermal expansion of mercury as per Beattie's formula
$\alpha = (18\,144.01 + 0.7016(t\,°C^{-1}) + 2.8625 \times 10^{-3}(t\,°C^{-1})^2$
$\quad +2.617 \times 10^{-6}(t\,°C^{-1})^3) \times 10^{-8}$
Compressibility of mercury 4.07×10^{-5} MPa^{-1}

Uncertainty budget in measurement of density of mercury

Uncertainties at one standard deviation level as quoted by Cook [2] are as follows:

Source of uncertainty	Parts per 10 million
Wavelength of light source used	0.6
Volume defects of internal edges	1.0
Changes of volume	1.4
Gauging of volume of mercury contained	1.0
Mass of mercury determined	2.5
Mass standard used	0.4
Temperature measurement	2.0

Using the quadrature method, the overall standard uncertainty is found to be 0.35 ppm.

5.5   Thermal expansion of mercury

The most recent accurate measurement of the thermal expansion of mercury made was by Beattie *et al* [5]. Sommer and Poziemski [6] critically examined the data of Beattie *et al* and that of all other workers. The measuring process is based upon the gravimetric determination of the mass of the mercury emerging from a fused quartz capillary due to thermal expansion of a known volume of mercury. The temperature range is 0–300 °C. Beattie's formula has been modified by the introduction of the new temperature scale ITS-90. The ratio of density ρ at $t\,°C$ to that of ρ_0 at $0\,°C$ can be expressed as follows:

$$\rho = \rho_0[1 + a_0(t\,°C^{-1}) + a_1(t\,°C^{-1})^2 + a_2(t\,°C^{-1})^3 + a_3(t\,°C^{-1})^4)]^{-1} \quad (5.1)$$

$$a_0 = 1.815\,868 \times 10^{-4} \qquad a_1 = 5.458\,43 \times 10^{-9}$$
$$a_2 = 3.4980 \times 10^{-11} \qquad a_3 = 1.5558 \times 10^{-14}. \qquad (5.2)$$

Taking the value of the density of mercury at 20 °C as 13 545.848 kg m^{-3} and using this relation the value of ρ_0 has been derived as 13 595.076 kg m^{-3}.

5.6   Isothermal compressibility

Sommer and Poziemski [6] have also reviewed the work done between 1949 and 1971 in regard to compressibility. The values of isothermal compressibility β at 20 °C and 50 °C at pressures of 0.1, 100 and 800 MPa have been measured by various researchers, a summary of which is given here:

Worker(s)	Year	$t\,°C$	β (10^{12} Pa) at			Reference
			0.1 MPa	100 MPa	800 MPa	
Kleppa	1949	50	41.00	—		[16]
Bett *et al*	1953	20	41.15	38.64	31.70	
		50	41.60	40.20	32.50	[7–10]
Pena and	1959	20	40.70	—	—	[11]
McGlashan		50	42.00	—	—	
Seemann	1965	20	40.13	—	—	[12]
and Klein		50	41.58	—	—	
Davis and	1966	20	40.17	38.74	31.50	[13]
Gordan		50	41.50	40.05	32.30	
Hayward	1970	20	39.92	38.68	31.80	[14]
Grindley	1971	20	40.59	39.18	32.00	[15]
and Lind		50	41.97	40.44	32.70	

The average value of β at 20 °C is 40.44 × 10^{-12} Pa. Using these values, an approximate relation for the compressibility of mercury with respect to pressure

Table 5.4. Results of density measurements of last century.

Researcher	Sample	Year	ρ (kg m^{-3}) at 20 °C	Uncertainty at 2σ level
Cook	A	1959	13 545.856	1×10^{-6}
Cook	B	1959	13 545.851	1×10^{-6}
Cook	C	1959	13 545.851	1×10^{-6}
Cook	D/E	1959	13 545.868	1×10^{-6}
Cook	NSL	1959	13 545.859	1×10^{-6}
Cook	NBS	1959	13 545.852	1×10^{-6}
Furtig	1	1971	13 545.830	1×10^{-6}
Furtig	1	1977	13 545.833	1.5×10^{-6}
Adametz	1	1990	13 545.839	1×10^{-6}
Sommer *et al*	1	1991	13 545.836	1×10^{-6}
Furtig	2	1976	13 545.848	1.5×10^{-6}
Adametz	2	1979	13 545.860	2×10^{-6}
Adametz	2	1981	13 545.867	2×10^{-6}
Patterson	AV	1984	13 545.869	2×10^{-6}
and Prowse	VDW	1984	13 545.853	1×10^{-6}
Adametz	VNIITTRI	1989	13 545.832	1×10^{-6}
Adametz	NIST	1990	13 545.840	1×10^{-6}

p in pascals, at 20 °C has been given as

$$\beta = \beta_0(1 + b_1 p + b_2 p^2 + b_3 p^3) \tag{5.3}$$

where

$$\beta_0 = 40.25 \times 10^{-12} \text{ Pa} \qquad b_1 = -3.730\,11 \times 10^{-10} \text{ Pa}^{-1}$$
$$b_2 = 1.938\,77 \times 10^{-19} \text{ Pa}^{-2} \qquad b_3 = -7.299\,26 \times 10^{-29} \text{ Pa}^{-3}. \tag{5.4}$$

5.7 Results of density measurements

Since Cook carried out his density measurements, which are still regarded as being very accurate, several other researchers have measured the density of mercury. However, in the long period of 40 years, the temperature scale has been revised twice. Cook and Stone's [1,2] results were in terms of ITS-48, while those of Furtig [17], Adametz [18, 19], Patterson and Prowse [20] were in terms of IPTS-68. All these results when adopted to ITS-90 would read as given in table 5.4.

The mean value of density at 20 °C, giving equal weight to all results, is 13 545.850 kg m^{-3}.

Table 5.5. Density of mercury in kg m^{-3} on ITS-90 at 101 325 Pa.

Temp.	0.0	0.1	0.2	0.3	0.4	0.5	0.6	0.7	0.8	0.9
0	13 595.076	13 594.829	13 594.583	13 594.336	13 594.089	13 593.842	13 593.595	13 593.348	13 593.102	13 592.855
1	13 592.608	13 592.361	13 592.114	13 591.868	13 591.621	13 591.374	13 591.127	13 590.881	13 590.634	13 590.387
2	13 590.140	13 589.894	13 589.647	13 589.400	13 589.154	13 588.907	13 588.660	13 588.414	13 588.167	13 587.920
3	13 587.674	13 587.427	13 587.180	13 586.934	13 586.687	13 586.441	13 586.194	13 585.947	13 585.701	13 585.454
4	13 585.208	13 584.961	13 584.714	13 584.468	13 584.221	13 583.975	13 583.728	13 583.482	13 583.235	13 582.989
5	13 582.742	13 582.496	13 582.249	13 582.003	13 581.756	13 581.510	13 581.263	13 581.017	13 580.770	13 580.524
6	13 580.278	13 580.031	13 579.785	13 579.538	13 579.292	13 579.046	13 578.799	13 578.553	13 578.306	13 578.060
7	13 577.814	13 577.567	13 577.321	13 577.075	13 576.828	13 576.582	13 576.336	13 576.089	13 575.843	13 575.597
8	13 575.351	13 575.104	13 574.858	13 574.612	13 574.365	13 574.119	13 573.873	13 573.627	13 573.381	13 573.134
9	13 572.888	13 572.642	13 572.396	13 572.149	13 571.903	13 571.657	13 571.411	13 571.165	13 570.919	13 570.672
10	13 570.426	13 570.180	13 569.934	13 569.688	13 569.442	13 569.196	13 568.950	13 568.704	13 568.457	13 568.211
11	13 567.965	13 567.719	13 567.473	13 567.227	13 566.981	13 566.735	13 566.489	13 566.243	13 565.997	13 565.751
12	13 565.505	13 565.259	13 565.013	13 564.767	13 564.521	13 564.275	13 564.029	13 563.783	13 563.537	13 563.291
13	13 563.045	13 562.800	13 562.554	13 562.308	13 562.062	13 561.816	13 561.570	13 561.324	13 561.078	13 560.832
14	13 560.587	13 560.341	13 560.095	13 559.849	13 559.603	13 559.357	13 559.112	13 558.866	13 558.620	13 558.374
15	13 558.128	13 557.883	13 557.637	13 557.391	13 557.145	13 556.900	13 556.654	13 556.408	13 556.162	13 555.917
16	13 555.671	13 555.425	13 555.179	13 554.934	13 554.688	13 554.442	13 554.197	13 553.951	13 553.705	13 553.460
17	13 553.214	13 552.968	13 552.723	13 552.477	13 552.232	13 551.986	13 551.740	13 551.495	13 551.249	13 551.004
18	13 550.758	13 550.512	13 550.267	13 550.021	13 549.776	13 549.530	13 549.285	13 549.039	13 548.794	13 548.548
19	13 548.303	13 548.057	13 547.812	13 547.566	13 547.321	13 547.075	13 546.830	13 546.584	13 546.339	13 546.093
20	13 545.848	13 545.602	13 545.357	13 545.112	13 544.866	13 544.621	13 544.375	13 544.130	13 543.885	13 543.639
21	13 543.394	13 543.149	13 542.903	13 542.658	13 542.412	13 542.167	13 541.922	13 541.676	13 541.431	13 541.186
22	13 540.941	13 540.695	13 540.450	13 540.205	13 539.959	13 539.714	13 539.469	13 539.224	13 538.978	13 538.733
23	13 538.488	13 538.243	13 537.997	13 537.752	13 537.507	13 537.262	13 537.017	13 536.771	13 536.526	13 536.281
24	13 536.036	13 535.791	13 535.546	13 535.300	13 535.055	13 534.810	13 534.565	13 534.320	13 534.075	13 533.830
25	13 533.585	13 533.339	13 533.094	13 532.849	13 532.604	13 532.359	13 532.114	13 531.869	13 531.624	13 531.379
26	13 531.134	13 530.889	13 530.644	13 530.399	13 530.154	13 529.909	13 529.664	13 529.419	13 529.174	13 528.929
27	13 528.684	13 528.439	13 528.194	13 527.949	13 527.704	13 527.459	13 527.214	13 526.969	13 526.724	13 526.479
28	13 526.235	13 525.990	13 525.745	13 525.500	13 525.255	13 525.010	13 524.765	13 524.520	13 524.276	13 524.031
29	13 523.786	13 523.541	13 523.296	13 523.051	13 522.807	13 522.562	13 522.317	13 522.072	13 521.827	13 521.583

Temp.	0.0	0.1	0.2	0.3	0.4	0.5	0.6	0.7	0.8	0.9
30	13 521.338	13 521.093	13 520.848	13 520.604	13 520.359	13 520.114	13 519.869	13 519.625	13 519.380	13 519.135
31	13 518.891	13 518.646	13 518.401	13 518.156	13 517.912	13 517.667	13 517.422	13 517.178	13 516.933	13 516.688
32	13 516.444	13 516.199	13 515.955	13 515.710	13 515.465	13 515.221	13 514.976	13 514.732	13 514.487	13 514.242
33	13 513.998	13 513.753	13 513.509	13 513.264	13 513.020	13 512.775	13 512.531	13 512.286	13 512.041	13 511.797
34	13 511.552	13 511.308	13 511.063	13 510.819	13 510.574	13 510.330	13 510.086	13 509.841	13 509.597	13 509.352
35	13 509.108	13 508.863	13 508.619	13 508.374	13 508.130	13 507.886	13 507.641	13 507.397	13 507.152	13 506.908
36	13 506.664	13 506.419	13 506.175	13 505.930	13 505.686	13 505.442	13 505.197	13 504.953	13 504.709	13 504.464
37	13 504.220	13 503.976	13 503.732	13 503.487	13 503.243	13 502.999	13 502.754	13 502.510	13 502.266	13 502.022
38	13 501.777	13 501.533	13 501.289	13 501.045	13 500.800	13 500.556	13 500.312	13 500.068	13 499.823	13 499.579
39	13 499.335	13 499.091	13 498.847	13 498.603	13 498.358	13 498.114	13 497.870	13 497.626	13 497.382	13 497.138
40	13 496.893	13 496.649	13 496.405	13 496.161	13 495.917	13 495.673	13 495.429	13 495.185	13 494.941	13 494.697
41	13 494.453									

5.8 Mercury density tables

Based on the aforesaid mean value of the density at $20\,°C$ and using the Beattie formula given in section 5.5, tables giving the density of mercury have been generated. Apart from thermometry, mercury is used as a reference standard for determining the volume of small capacity measures and to maintain the primary standard barometer. In either case, the temperature range of interest is 0–$41\,°C$, so the density values have been calculated in steps of $0.1\,°C$ in this range. The values for the density of mercury refer to an ambient pressure of $101\,325$ Pa. The tables prepared by Bigg [21] based on the results of Cook *et al* were on the ITS-48 scale. Although Ambrose [22] constructed density tables on the ITS-90 scale, the steps are only $1\,°C$, but his density table has a much wider range (-20–$300\,°C$). Moreover we have constructed a density table (see table 5.5) up to 0.001 kg m^{-3} instead of 0.01 kg m^{-3}, as was done in all other cases.

This table takes the density of mercury at $20\,°C$ to be $13\,545.848$ kg m^{-3} and the following density–temperature relation:

$$\rho = \rho_0 \times [1 + (a_0(t\,°C^{-1}) + a_1(t\,°C^{-1})^2 + a_2(t\,°C^{-1})^3 + a_3(t\,°C^{-1})^4)]^{-1}$$
$$a_0 = 1.815\,868 \times 10^{-4} \qquad a_1 = 5.458\,43 \times 10^{-9}$$
$$a_2 = 3.4980 \times 10^{-11} \qquad a_3 = 1.5558 \times 10^{-14}.$$

References

[1] Cook A H and Stone N W M 1957 Precise measurement of the density of mercury at $20\,°C$: I. Absolute displacement method *Phil. Trans. R. Soc.* A **250** 279–323

[2] Cook A H 1957 Precise measurement of the density of mercury at $20\,°C$: II. Content method *Phil. Trans. R. Soc.* A **250** 279–323

[3] Hulett G A and Minchin H D 1905 *Phys. Rev.* **21** 388

[4] Davis R S 1992 Equation for determination of density of moist air (1981–1991) *Metrologia* **29** 67–70

[5] Beattie J A, Blaisdell B E, Kaye J, Gerry H T and Johnson C A 1941 *Proc. Am. Acad. Arts Sci.* **71** 371

[6] Sommer K D and Poziemski J 1993/94 Density, thermal expansion and compressibility of mercury *Metrologia* **30** 665–8

[7] Bett K E, Weale K E and Newitt D M 1954 *Br. J. Appl. Phys.* **5** 243

[8] Bridgman P W 1911 *Proc. Am. Acad. Arts Sci.* **47** 342

[9] Hubban J C and Lomis A L 1954 *Phil. Mag.* **5** 1177

[10] Ringo G R, Fitzgerald J W and Huddle B G 1947 *Phys. Rev.* **72** 87

[11] Pena M D and McGlasan 1959 *Trans. Faraday Soc.* **55** 2018

[12] Seemann H J and Klien F H 1965 *Z. Phys.* **4** 368

[13] Davis L A and Gordan R B 1967 *J. Chem. Phys.* **46** 2650

[14] Hayward A T J 1971 *J. Phys. D: Appl. Phys.* **4** 951

[15] Grindley T and Lind J E 1971 *J. Chem. Phys.* **54** 3983

[16] Kleppa O J 1949 *J. Chem. Phys.* **17** 668

[17] Furtig M 1973 *Exp. Technol. Phys.* **21** 521

[18] Adametz H 1985 *Metrolog. Abh. ASMW* **5** 135
[19] Adametz H and Wloke M 1991 Measurement of the absolute value of density of mercury in the ASMW *Metrologia* **28** 333
[20] Patterson J B and Prowse D B 1985 *Metrologia* **21** 107
[21] Bigg P H 1964 The density of mercury *Br. J. Appl. Phys.* **15** 1111–13
[22] Ambrose D 1990 The density of mercury *Metrologia* **27** 245–7

Chapter 6

Special methods of density determination

Symbols

σ	Density of air
ρ	Density of liquids
$\delta v, \delta V$	Change in volume
$\delta V / V$	Fractional change in volume
M, m	Mass
V, v	Volume
D	Diameter of wire
Δ	Density of standard weights or wire
L	Length
γ	Coefficient of expansion of water
$\mathrm{d}f/\mathrm{d}h$	Differential coefficient of f with respect to h (height)
α	Coefficient of cubic expansion of solid
W	Apparent mass
C_p	Specific heat at constant pressure
γ_f	Compressibility of solids
γ_w	Compressibility of water
P	Pressure
A	Atmospheric pressure
n	Frequency
$\mathrm{Sp}\, t/t\,°\mathrm{C}$	Specific gravity or relative density of substance at $t\,°\mathrm{C}$ with respect of water at $t\,°\mathrm{C}$
$\mathrm{d}\rho_\mathrm{F}/\mathrm{d}t$	Density gradient of float with respect to temperature
$\mathrm{d}\rho_\mathrm{L}/\mathrm{d}t$	Density gradient of liquid with respect to temperature
$\mathrm{d}\rho_\mathrm{s}/\mathrm{d}t$	Density gradient of solid with respect to temperature
S	entropy
V_t	Volume of float at temperature $t\,°\mathrm{C}$
d_m	Density of water at the centre of the float

6.1   Use of relative density bottle/pyknometer

A relative density bottle and a pyknometer are essentially one and the same thing, both being content measures of known but constant capacity, the difference lies only in their shape. The shape of a pyknometer may change depending upon its use.

6.1.1   Relative density bottle (RD bottle)

A relative density (RD) or specific gravity bottle is a small flask-like object with a ground glass stopper. The stopper has a fine axial capillary bore. The bottle is a content measure of constant capacity. The constancy in its capacity naturally depends upon the repeatability in fitting the stopper in the same position, so the quality of an RD bottle will depend upon how snugly the stopper fits and stops always in the same position. The stopper as well as the corresponding female part of the RD bottle needs to be ground so that the stopper always fits exactly in the same position. In addition, the internal and external finish of the walls of the RD bottle are equally important. The material of the bottle should have a smaller coefficient of cubic expansion so that any error due to a change in temperature is minimal. The bottle should also be well annealed to ensure stability in its capacity with respect to time. The nominal volumes of these bottles are 10, 20, 50 and 100 cm^3 although one can have an RD bottle of any other capacity for a specific purpose. A typical RD bottle is shown in figure 6.1.

6.1.2   Density of liquids

For accurate measurement of the relative density of a liquid, an RD bottle should be used at a standard temperature, which for tropical countries is 27 °C. Sometimes for specific purposes such as petroleum liquids, 20 or even 15 °C can also be taken as the standard temperature. Measurement of the relative density of a liquid involves the following steps.

(1) Find the mass of the empty RD bottle.
(2) Fill the bottle completely with the liquid. As the relative density of a liquid depends upon its temperature, the RD bottle along with the liquid is brought to a standard temperature. The liquid should be at a slightly lower temperature when poured into the bottle, so that when the bottle is placed in a thermostatic bath and the temperature of its contents rises, the liquid completely fills the capillary part of the stopper. The bath is maintained at a standard temperature, say, 27 °C. Make sure that the bottle and its contents acquire the temperature of the bath.
(3) Take it out, clean and dry it properly. All this should be carried out at a lower temperature than the standard, otherwise some liquid will spill out. Find the total mass of the RD bottle and liquid filling it.

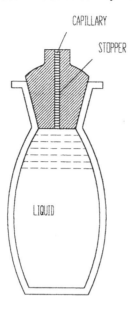

CAPILLARY

STOPPER

LIQUID

Figure 6.1. Relative density bottle (a vessel of fixed capacity).

(4) Take out the liquid, clean the bottle properly and fill it with distilled water of known isotopic composition. The water should be air free or the level of air saturation should be known or determined experimentally. The appropriate corrections for isotopic composition and level of air saturation should be applied to the density value obtained from the water density table. Find the mass of the bottle filled with the water as described earlier.

It should be remembered that the values for the density of water given in table 4.4 are for air-free Standard Mean Ocean Water (SMOW). The isotopic composition of SMOW as well as the equation for the difference in density between SMOW and sample water are given. An expression for applying a correction for dissolved air is also provided.

Calculations

Let M_0 be the mass of the empty RD bottle, i.e. filled with air, M_1, the mass of the bottle plus the mass of the liquid required to fill it completely and M_2, the mass of the bottle plus mass of water required to fill it completely. V_g and ρ_g are the volume and density of the glass of the RD bottle. Furthermore, let V be its capacity, then M_0, M_1 and M_2 can be expressed as

$$M_0 = V_g \rho_g + V\sigma \tag{6.1}$$

$$M_1 = V_g \rho_g + (V - \delta V_1)\rho_1 + \delta V_1 \sigma \tag{6.2}$$

$$M_2 = V_g \rho_g + (V - \delta V_w)\rho_w + \delta V_w \sigma \tag{6.3}$$

giving

$$M_1 - M_0 = (V - \delta V_1)(\rho_1 - \sigma) \tag{6.4}$$

$$M_2 - M_0 = (V - \delta V_w)(\rho_w - \sigma). \tag{6.5}$$

Dividing equation (6.4) by (6.5) gives

$$(M_1 - M_0)/(M_2 - M_0) = [(V - \delta V_1)/(V - \delta V_w)][(\rho_1 - \sigma))/(\rho_w - \sigma)] \tag{6.6}$$

$$(\rho_1 - \sigma)[(1 - \delta V_1/V)/(1 - \delta V_w/V)] = (\rho_w - \sigma)(M_1 - M_0)/(M_2 - M_0) \tag{6.7}$$

$$\rho_1 = (\rho_w - \sigma)(M_1 - M_0)/(M_2 - M_0) + \sigma + \text{the error term } T. \tag{6.8}$$

The term T is given by

$$T = (\rho_1 - \sigma)[\delta V_1/V - \delta V_w/V] \tag{6.9}$$

where δV_w and δV_1 are the error terms due to under-filling or acquiring different temperatures for the water and liquid respectively. The sources of errors are:

- The volumes of the liquid may not exactly correspond to the capacity of the bottle.
- The volume of the water may not exactly correspond to the capacity of the bottle.
- The liquid might not have acquired the standard temperature at which the bath is maintained.
- The water might not have acquired the standard temperature at which the bath is maintained.
- The cleaning and drying, before the bottle with the liquid was weighed, may not be perfect.
- The cleaning and drying, before the bottle with the water was weighed, may not be perfect.
- Glass is a non-conductor of electricity, so electrostatic charge may cause a problem in the weighing at each step.
- The air temperature and pressure at which the weighings are carried out may change, causing a change in air density and hence the need for a buoyancy correction.

To avoid the process of removing the RD bottle from the thermostatic bath for weighing, the weighing can be carried out in the water bath itself. This has three advantages: (1) the process of drying and incurring unnoticed errors is totally avoided; (2) the effect of electrostatic charge is greatly reduced; and (3) the apparent weight at each step will be greatly reduced, so that a small capacity balance with a finer resolution can be used.

6.1.3 Density of small solid or powder

The RD bottle may also be used to determine the density of a small size solid or solid powder (which does not react with water either physically or chemically). The steps are as follows:

(1) Find the mass M_0 of the empty bottle.
(2) Find the mass M_1 of the empty bottle plus solid.
(3) Fill the bottle with water taking all precautions and applying the corrections mentioned in section 6.1.2 and find the mass M_2 of the empty bottle plus solid plus water occupying the whole space of the bottle other than that occupied by the solid.
(4) Remove everything from the bottle and clean it. Fill it fully with water only and weigh it. Let this be M_3.

Calculations

M_0, M_1, M_2 and M_3 can be expressed as follows:

$$M_0 = V_g\rho_g + V\sigma \tag{6.10}$$

$$M_1 = V_g\rho_g + V_p\rho_p + (V - V_p)\sigma \tag{6.11}$$

$$M_2 = V_g\rho_g + V_p\rho_p + (V - V_p)\rho_w + \delta V_1\sigma - \delta V_1\rho_w \tag{6.12}$$

$$M_3 = V_g\rho_g + V\rho_w + +\delta V_2\sigma - \delta V_2\rho_w \tag{6.13}$$

$$M_1 - M_0 = V_p(\rho_p - \sigma). \tag{6.14}$$

$M_1 - M_0$ is the mass of solid or solid powder with volume V_p,

$$M_2 - M_1 = (V - V_p)(\rho_w - \sigma) - \delta V_1(\rho_w - \sigma) \tag{6.15}$$

$$M_3 - M_0 = V(\rho_w - \sigma) - \delta V_2(\rho_w - \sigma). \tag{6.16}$$

The mass of water with the same volume as that of the solid or solid powder is given by

$$(M_3 - M_0) - (M_2 - M_1) = V_p(\rho_w - \sigma) + (\delta V_1 - \delta V_2)(\rho_w - \sigma). \tag{6.17}$$

So dividing equation (6.14) by (6.17), we get

$$(M_1 - M_0)/[(M_3 - M_0) - (M_2 - M_1)]$$
$$= V_p(\rho_p - \sigma)/[V_p(\rho_w - \sigma) + (\delta V_1 - \delta V_2)(\rho_w - \sigma)] \tag{6.18}$$
$$(\rho_w - \sigma)(M_1 - M_0)/[(M_3 - M_0) - (M_2 - M_1)]$$
$$= (\rho_p - \sigma)/[1 + (\delta V_1 - \delta V_2)/V_p] \tag{6.19}$$

or

$$(\rho_p - \sigma) = (\rho_w - \sigma)(M_1 - M_0)/[(M_3 - M_0) - (M_2 - M_1)]$$
$$+ (\rho_p - \sigma)(\delta V_1 - \delta V_2)/V_p. \qquad (6.20)$$

The second term on the right-hand side is the error term due to under-filling the bottle with water, in steps 3 and 4 respectively.

   If the solid body or powder reacts with water, some other liquid, which is inert to the solid is taken and the previous steps are followed, giving the relative density of the solid with respect to that liquid. The relative density of the liquid is found as in section 6.1.2. Multiplication of the relative density of the liquid with respect to water and the relative density of solid with respect to liquid gives the relative density of the solid with respect to water.

6.2   Fully submersed floats in two liquids

We have studied hydrometers floating in a single liquid; note the unsubmerged portion of it was in air. Although air is also a fluid and exerts an upward thrust, its density is very much lower than that of most liquids, so we took the floating of a hydrometer as the floating of a body in one liquid. In this section, we are going to consider the floating of a body in two immiscible liquids.

   Let us consider a float with essentially three parts as shown in figure 6.2. The upper part A is made of a lighter material such as glass hollow tubes with a small pan. The upper part should be such that it floats in the lighter liquid and requires some weights to just float in it. The lower part is a heavier body in conical form. This itself should sink in a denser liquid. The two parts are connected with a fine wire of uniform diameter and high surface finish. The symmetry axes of the upper and lower parts must coincide with the axis of the attached wire. Two immiscible liquids are taken; the lower one has a higher density such that the system floats fully submerged and the wire intersects the interface of the two liquids. If a small solid is placed in the pan of the upper body, it will exert a downward force, slackening the wire and, as a result, the lower part B falls. This way a longer length of wire will go inside the heavier liquid, therefore exerting more upward thrust due to the heavier liquid. The downward force due to the body will be the difference between its weight and the upward thrust on it in the lighter liquid. The upward force counter-balancing it will be equal to the volume of the extra portion of wire gone down into the heavier liquid multiplied by the difference in the density of the two liquids.

   Let m and v denote the mass and volume with suffix A for the upper part and a suffix B for the lower part. Let D and Δ denote the diameter and density of the wire respectively and ρ_1 and ρ_2 be the density of the upper and lower liquids. Consider the equilibrium equations at C—at the interface between the two liquids (figure 6.2).

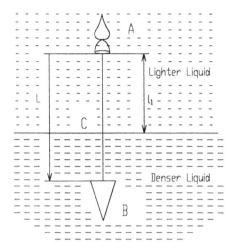

Figure 6.2. Fully submerged float in two liquids.

For the upper part A

$$m_A - V_A \rho_1 + \pi D^2 \Delta l_1/4 - \pi D^2 \rho_1 l_1/4 + T = 0. \qquad (6.21)$$

For the lower part B

$$m_B - V_B \rho_2 + \pi D^2 \Delta(L - l_1)/4 - \pi D^2 (L - l_1)\rho_2/4 - T = 0. \qquad (6.22)$$

Here T is the tension in the wire, which should always be positive. Giving the necessary and sufficient conditions for the float as

$$m_A - V_A \rho_1 + \pi D^2 \Delta l_1/4 - \pi D^2 \rho_1 l_1/4 < 0 \qquad (6.23)$$

and

$$m_B - V_B \rho_2 + \pi D^2 \Delta(L - l_1)/4 - \pi D^2 (L - l_1)\rho_2/4 > 0 \qquad (6.24)$$

and adding equations (6.21) and (6.22), we obtain

$$m_A - V_A \rho_1 + \pi D^2 \Delta l_1/4 - \pi D^2 \rho_1 l_1/4 + m_B - V_B \rho_2$$
$$+ \pi D^2 \Delta(L - l_1)/4 - \pi D^2 (L - l_1)\rho_2/4 = 0. \qquad (6.25)$$

If a small body of mass M and volume V is placed in the pan of the upper part A, the float will sink downward and the new equilibrium equation will be

$$M - V\rho_1 + m_A - V_A \rho_1 + \pi D^2 \Delta(l_1 - h)/4 - \pi D^2 \rho_1(l_1 - h)/4 + m_B - V_B \rho_2$$
$$+ \pi D^2 \Delta(L - l_1 + h)/4 - \pi D^2 (L - l_1 + h)\rho_2/4 = 0. \qquad (6.26)$$

Subtracting (6.25) from (6.26), we get

$$M - V\rho_1 = \pi D^2(\rho_2 - \rho_1)h/4. \tag{6.27}$$

By observing the distance h, through which the float has fallen, we can find the volume V of the body. The mass M of the body is determined separately and the other parameters such as D, ρ_1 and ρ_2 are known.

The sensitivity of the method depends upon $(\rho_2 - \rho_1)$—the smaller the difference is, the larger the value of h will be. However, instead of finding ρ_1 and ρ_2 and the other parameters, the system is calibrated with three samples of known mass and volume but each with a different density. If M_1, M_2 and M_3 and V_1, V_2 and V_3 are their respective masses and volumes, then from (6.27), we obtain:

$$M_1 - V_1\rho_1 = \pi D^2(\rho_2 - \rho_1)h_1/4 \tag{6.28}$$
$$M_2 - V_2\rho_1 = \pi D^2(\rho_2 - \rho_1)h_2/4 \tag{6.29}$$
$$M_3 - V_3\rho_1 = \pi D^2(\rho_2 - \rho_1)h_3/4. \tag{6.30}$$

With the help of these equations, we eliminate D, ρ_1 and ρ_2 as follows. Dividing (6.28) by (6.29) and collecting the terms containing ρ_1 we obtain:

$$M_1h_2 - M_2h_1 = (V_1h_2 - V_2h_1)\rho_1. \tag{6.31}$$

Similarly dividing (6.27) by (6.30), we obtain

$$Mh_3 - M_3h = (Vh_3 - V_3h)\rho_1. \tag{6.32}$$

Dividing (6.32) by (6.31), we obtain

$$(Mh_3 - M_3h)/(M_1h_2 - M_2h_1) = (Vh_3 - V_3h)/(V_1h_2 - V_2h_1). \tag{6.33}$$

From (6.33), we can easily calculate the value of V as all other parameters including the mass of the body are known. The only restriction is that all the time it is only the wire which intersects the interface of two liquids and this never becomes slack, i.e. T is positive. V can be expressed as

$$V = (M - hM_3/h_3)(V_1h_2 - V_2h_1)/(M_1h_2 - M_2h_1) + hV_3/h_3. \tag{6.34}$$

Schoonover and Davis of NBS [1, 2] have employed a method based upon this principle with the float shown in figure 6.3. The claimed accuracy is one part in 10^4 for very small objects weighing less than 1 g. The difficulty in determining the density of such small objects can be well visualized.

6.3   Single liquid with constant temperature gradient

Consider a long tube full of water, well insulated from the sides. The temperature, at the bottom of the tube, is maintained at T_0 while the top surface is maintained at

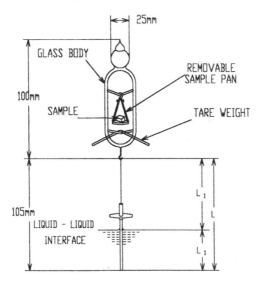

Figure 6.3. Float used at National Institute of Standards and Technology, USA (National Bureau of Standards).

temperature T_1. T_0 is lower than T_1. As the temperature at the bottom is less than that at the top, there will be no convection currents. Most of the heat exchange will be by conduction. As water is a bad conductor a linear density gradient will be established.

Let us consider a float of mass M and volume V_0. All distances are measured from the bottom of the tube and denoted by h with different subscripts. α is the effective coefficient of cubic expansion of the float. The mass of the float is adjusted in such a way that it just floats in water. The equilibrium equation is

$$M - V_t d_m = 0 \tag{6.35}$$

where d_m is the density of water at the centre of buoyancy of the float, whose distance from the bottom is h_m. Let dT/dh be the temperature gradient and γ the coefficient of expansion of water, then the density at the centre of buoyancy is given by

$$d_m = d_0/(1 + \gamma h_m \, dT/dh) = d_0(1 - \gamma h_m \, dT/dh). \tag{6.36}$$

Similarly the volume V_t of the float may be written as

$$V_t = V_0(1 + \alpha h_m \, dT/dh) \tag{6.37}$$

where α is the effective coefficient of cubic expansion of the float. Equation (6.35) can then be written as

$$M - V_0 d_0[\{1 - (\gamma - \alpha)h_m\} \, dT/dh] = 0. \tag{6.38}$$

If L is the depth of the centre of buoyancy from the top of the float and H is the height of the water column, then (6.38) becomes

$$M - V_0 d_0[1 - (\gamma - \alpha)(H - L)\,dT/dh] = 0. \tag{6.39}$$

Let us add a body of mass m_1 and volume v_1, such that the float falls to a height h_1. The new equilibrium equation will be

$$M - V_0 d_0[1 - (\gamma - \alpha)(h_1 - L)\,dT/dh] + m_1 - v_1 d_0[1 - (\gamma - \alpha_1)h_1\,dT/dh] = 0. \tag{6.40}$$

Subtracting (6.40) from (6.39), we obtain

$$V_0 d_0[(\gamma - \alpha)(H - h_1)\,dT/dh] = m_1 - v_1 d_0[1 - (\gamma - \alpha_1)h_1\,dT/dh]. \tag{6.41}$$

Normally α_1 for solids is small in comparison with γ the coefficient of cubic expansion of water, so we may take α to be constant and consider a mean value for it for different samples. Normally the metals have an expansion coefficient between 30×10^{-6} and 54×10^{-6} while that of water is 2×10^{-4}. The ratio of α to γ is 3/20, so when we take α to be constant, the variation in $\gamma - \alpha$ may vary from 17×10^{-5} to 15×10^{-5}. Furthermore, it is to be multiplied by dT/dh, which will be of the order of 10^{-2} K cm^{-1}. Similarly, there are low expansion solids for which α is small and the variation in their coefficients is negligible. Hence the error, in taking $(\gamma - \alpha_1)$ as constant, will not be significant within 1 ppm. Furthermore, the right-hand side of equation (6.41) is nothing other than the apparent mass of the body in water at temperature $T_0 + h_1\,dT/dh$. Writing this as w_1, we may rewrite equation (6.41) as

$$w_1 = V_0 d_0(\gamma - \alpha)(H - h_1)\,dT/dh. \tag{6.42}$$

Similarly when two more samples of mass m_2 and m_3 are placed one by one on the float and the float comes to rest at heights h_2 and h_3, then their respective effective weights in water w_2 and w_3 are:

$$w_2 = V_0 d_0(\gamma - \alpha)(H - h_2)\,dT/dh \tag{6.43}$$
$$w_3 = V_0 d_0(\gamma - \alpha)(H - h_3)\,dT/dh. \tag{6.44}$$

Dividing (6.43) by (6.42) gives the ratio w_2/w_1 as

$$w_2/w_1 = (H - h_2)/(H - h_1) \tag{6.45}$$

giving

$$(w_2 - w_1)/w_1 = (h_1 - h_2)/(H - h_1). \tag{6.46}$$

Similarly from (6.44) and (6.42), we get

$$(w_3 - w_1)/w_1 = (h_1 - h_3)/(H - h_1). \tag{6.47}$$

Dividing (6.47) by (6.46), we get

$$(w_3 - w_1)/(w_2 - w_1) = (h_3 - h_1)/(h_2 - h_1). \qquad (6.48)$$

If there are two samples of known mass and density, we can find the values for their apparent weights in water, hence we can calculate the value of w_3. Once we know m_3 and w_3 we can write $Sp\,t_3/t_3$, the relative density of the third sample at the temperature $t_3 = T_0 + h_3\,dT/dh$, as

$$Sp\,t_3/t_3 = m_3/(m_3 - w_3).$$

Following the analogy from the right-hand side of (6.41), w_3 may be written as

$$w_3 = m_3 - v_3 d_0[1 - (\gamma - \alpha_1)h_3\,dT/dh];$$

using this expression for w_3 in the above equation, we obtain

$$Sp\,t_3/t_3 = m_3/(m_3 - m_3 + v_3 d_0[1 - (\gamma - \alpha_1)h_3\,dT/dh])$$
$$Sp\,t_3/t_3 = m_3/v_3 d_0[1 - (\gamma - \alpha_1)h_3\,dT/dh]$$

or

$$Sp\,t_3/t_3 = [m_3/v_3(1 + \alpha_1 h_3\,dT/dh)]/[d_0(1 - \gamma h_3\,dT/dh)] \qquad (6.49)$$
$$Sp\,t_3/t_3 = [\text{density of solid at } t_3]/[\text{density of water at } t_3]$$
$$\text{density of solid at } t_3\,°C = \text{density of water at } t_3\,°C \times Sp\,t_3/t_3. \qquad (6.50)$$

The density of water is already known within 1×10^{-6}, hence the density of the solid at a temperature very near to T_0 is known.

This method is best suited for different samples of the same material with very small difference in their density values, for example silicon samples from different sources.

From equation (6.42), one can determine the maximum apparent mass of the body which can be placed on the float by replacing h_1 by the length l_f of the float, which is the minimum height from the bottom of the tube. W_m is given by

$$W_m = V_0 d_0(\gamma - \alpha)(H - l_f)\,dT/dh. \qquad (6.51)$$

Taking $\gamma = 0.0002\,K^{-1}$, $H\,dT/dh = 1\,°C$, only one degree difference between the bottom and top, the length of the float $l_f = 10$ cm, $V_0 = 10$ cm^3 and $d_0 = 1$ g cm^{-3}, the value of W_m will be about 2 mg, which means this method will be applicable only for very small samples. I know no other method for such small samples.

6.3.1 Samples of bigger mass

However, for samples of bigger size, a lighter float can be made so that it just floats in water with the sample on it. In that case equation (6.40) becomes

$$M - V_0 d_0[1 - (\gamma - \alpha)(h_1 - L)\,dT/dh] + m_1 - v_1 d_0[1 - (\gamma - \alpha_1)h_1)\,dT/dh] = 0. \qquad (6.52)$$

It should be noted that in this case the sum of the first two terms will be negative.

$$M - V_0 d_0[1 - (\gamma - \alpha)(h_1 - L) \, dT/dh] + w_1 = 0. \qquad (6.53)$$

Similarly for other samples, we may write

$$M - V_0 d_0[1 - (\gamma - \alpha)(h_2 - L) \, dT/dh] + w_2 = 0 \qquad (6.54)$$

$$M - V_0 d_0[1 - (\gamma - \alpha)(h_3 - L) \, dT/dh] + w_3 = 0. \qquad (6.55)$$

Subtraction of (6.54) from (6.53) and (6.55) from (6.53) gives

$$w_1 - w_2 = V_0 d_0(\gamma - \alpha)(h_2 - h_1) \, dT/dh \qquad (6.56)$$

$$w_1 - w_3 = V_0 d_0(\gamma - \alpha)(h_3 - h_1) \, dT/dh. \qquad (6.57)$$

Dividing (6.57) by (6.56), we get

$$(w_1 - w_3)/(w_1 - w_2) = (h_3 - h_1)/(h_2 - h_1). \qquad (6.58)$$

Having samples 1 and 2 of known mass and density, the values of w_1 and w_2 can be calculated. Equation (6.58) gives the value of w_3 and the density of sample 3 is calculated in the same manner as described before. In this case, the float is constructed according to the mass of the samples which are going to be used. The mass of the sample may be up to several grams. However it should be noted that the maximum difference in apparent weights should be no more than 2 mg.

6.3.2 Direct flotation method

Some organic liquids and aqueous solutions are available, which have densities as high as 4.9 g cm^{-3}. In such liquids, low-density solids can float by themselves. If a vertical temperature gradient is maintained in such a liquid, solids with densities that differ by very small amounts will float at different levels. A solid of mass m and volume v will float at a height h when its density is equal to the density of the liquid at that level. If γ and α are, respectively, the coefficients of cubic expansion of the liquid and solid, then applying the procedure described in section 6.3, the equilibrium equation may be written as

$$m = v_0 d_0[1 - (\gamma - \alpha)h \, dT/dh]. \qquad (6.59)$$

Putting the density of solid $d_s = m/v_0(1 + \alpha h \, dT/dh)$, d_s is given as

$$d_s = d_0[1 - \gamma h \, dT/dh] \qquad (6.60)$$

where d_s is the density of the solid at $T_0 + h \, dT/dh \, °C$. In this case, two samples of known density are chosen and the density of any other sample is determined in terms of their density. Let suffixes 1 and 2 be used for samples 1, 2 of known density.

$$d_{s1} = d_0[1 - \gamma h_1 \, dT/dh] \qquad (6.61)$$

$$d_{s2} = d_0[1 - \gamma h_2 \, dT/dh] \qquad (6.62)$$

giving

$$d_{s2} - d_{s1} = d_0\gamma(h_1 - h_2)\,\mathrm{d}T/\mathrm{d}h. \tag{6.63}$$

Similarly for a sample of unknown density d_s

$$(d_s - d_{s1}) = d_0\gamma(h_1 - h)\,\mathrm{d}T/\mathrm{d}h. \tag{6.64}$$

Dividing (6.64) by (6.63), we get

$$(d_s - d_{s1})/(d_{s2} - d_{s1}) = (h_1 - h)/(h_1 - h_2). \tag{6.65}$$

So any variation in the density of different samples of essentially the same material can be determined with high accuracy, provided $\mathrm{d}T/\mathrm{d}h$ remains constant. The sensitivity of the method can be assessed from equation (6.63); by taking $(h_1 - h_2) = 1$ mm and other parameters as in section 3, we get

$$(d_{s1} - d_{s2}) = 1.0 \times 0.0002 \times 0.001 = 2 \times 10^{-7} \text{ g cm}^3. \tag{6.66}$$

6.4 Motion of a solid body in a slightly lighter liquid

6.4.1 Motion of a non-compressible spherical body

Let us consider a sphere of radius r, mass m and density d_s. It is placed in a liquid of density ρ, which is slightly less than that of the sphere. If the sphere is placed gently with no downward motion the effective downward force acting on the body is:

$$4\pi r^3 d_s g/3 - 4\pi r^3 \rho g/3.$$

It starts moving with an acceleration $\mathrm{d}v/\mathrm{d}t$, but at the same time, a resistive force due to the viscosity η of the liquid equal to $6\pi r\eta v$ opposes the motion. The equation of motion may be written as

$$(4\pi r^3 d_s/3)\,\mathrm{d}v/\mathrm{d}t = 4\pi r^3 d_s g/3 - 4\pi r^3 \rho g/3 - 6\pi r\eta v$$
$$= 4\pi r^3 \Delta\rho g/3 - 6\pi r\eta v \tag{6.67}$$

giving

$$\mathrm{d}v/\mathrm{d}t = g\Delta\rho/d_s - kv. \tag{6.68}$$

Here

$$\Delta\rho = d_s - \rho \quad \text{and} \quad k = 9\eta/(2r^2 d_s). \tag{6.69}$$

The solution of differential equation (6.68) is

$$v = v_{max}\{1 - \exp(-kt)\} \tag{6.70}$$

where v_{max} is the maximum velocity for which $\mathrm{d}v/\mathrm{d}t = 0$ giving

$$v_{max} = g\Delta\rho/d_s k = 2gr^2\Delta\rho/9\eta. \tag{6.71}$$

From a similar expression for another body of density d_{s1}, which is again slightly denser or lighter than the liquid, we get another value for $v_{max\,1} = 2gr_1^2\Delta\rho_1/9\eta$, giving

$$v_{max\,1}/v_{max} = r_1^2\Delta\rho_1/r^2\Delta\rho$$
$$= (d_{s1} - \rho)r_1^2/(d_s - \rho)r^2. \tag{6.72}$$

In the case of the lighter body, it will move upwards, i.e. the velocity will be negative. So by measuring the terminal velocities of two solid bodies in the form of spheres, their densities can be compared. The density values differ only slightly from each other and also from that of the liquid.

The sensitivity of the method depends upon the smallest velocity one can measure. Taking the following values for the various parameters, the value of $\Delta\rho$ can be calculated from (6.71):

$$\eta = 2.3 \text{ mPa s} \qquad r = 2 \text{ cm} \qquad v_{max} = 1 \ \mu\text{m s}^{-1}$$

we get

$$\Delta\rho = 9\eta v_{max}/2gr^2 = 9 \times 2.3 \times 10^{-3} \times 10^{-6}/2 \times 9.80 \times 4 \times 10^{-4}$$
$$20.7 \times 10^{-9}/78.4 \times 10^{-4} = 2.6 \times 10^{-6} \text{ kg m}^{-3} = 2.6 \times 10^{-9} \text{ g cm}^{-3}.$$

However, there is one problem with this method—how do we know whether v_{max}, the terminal velocity, has been reached. Theoretically, v_{max} will be reached after infinite time. To appreciate the problem, let us express time t in terms of the ratio of velocities. This can be obtained by taking the natural logarithm (to the base e) of both sides of (6.70):

$$t = -\ln([1 - v/v_{max})/k. \tag{6.73}$$

For a spherical sample of $r = 2$ cm and $\Delta\rho$ of the order of 10^{-7} times d_s, then the time taken to reach

- 50% of v_{max} is 1 min,
- 90% of v_{max} is 3.4 min,
- 95% of v_{max} is 4.4 min and
- 99% of v_{max} is 6.8 min.

This problem can be solved by making ρ, the density of the liquid, exactly equal to that of the solid, so that the solid is in a stationary state. The density of the liquid can be made equal to that of the solid in three ways, by changing (1) the concentration, (2) the pressure or (3) the temperature. All these techniques will be discussed in the forthcoming sections.

A similar method has been used in determining the density differences of two bodies by Bowman and Schoonover [3]. They attained the equilibrium of upward and downward forces by constructing a compressible diver, which was more compressible than water.

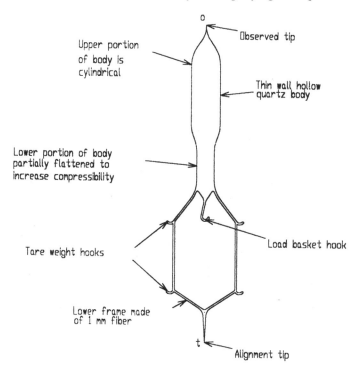

O

Observed tip

Upper portion
of body is
cylindrical

Thin wall hollow
quartz body

Lower portion of body
partially flattened to
increase compressibility

Tare weight hooks

Load basket hook

Lower frame made
of 1 mm fiber

t

Alignment tip

Figure 6.4. Compressible diver from the National Institute of Standards and Technology, USA (National Bureau of Standards).

6.4.2 Compressible diver

The special diver constructed by Bowman and Schoonover is shown in figure 6.4. It is a hollow, compressible body with a closed surface to which are attached various hooks and a load basket. The upper portion is a thin-walled cylindrical body of quartz with an observing tip O. The lower portion is partially flattened to increase compressibility. The lowermost portion of the diver is a frame made from a thin-walled tube. The tip T for alignment is at the bottom of the frame. The volume of the diver is about 10 cm^3 and its mass is 5 g. So the diver will experience a water buoyancy force greater than its weight by about 5 g. Hence samples of apparent weight in water of about 5 g can be placed on the load basket hook.

Let γ_f and γ_w be the compressibility coefficient for the diver and water respectively. Then V, the volume of the diver, and d_w, the density of water at pressure P, will be

$$V = V_0[1 - \gamma_f(P - A)]$$
$$d_w = d_0[1 + \gamma_w(P - A)].$$

Hence, M, the mass of water displaced by the diver, is

$$M = V_0[1 - \gamma_f(P - A)]d_0[1 + \gamma_w(P - A)]$$

or

$$M = V_0 d_0[1 - (\gamma_f - \gamma_w)(P - A) - \gamma_f\gamma_w(P - A)^2] \qquad (6.74)$$

where V_0 and d_0 are the volume of the float and density of water at atmospheric pressure A.

The buoyancy force is proportional to the mass of displaced water. In addition, in (6.74) the third term in the square brackets is always smaller than the second term, so if $(\gamma_f - \gamma_w)$ is positive, i.e. the float is more compressible than water, an increase in pressure will decrease the buoyancy force and a decrease in pressure will increase the buoyancy force. However, in a liquid, there is a pressure gradient due to gravity; i.e. the pressure increases with an increase in the depth of fluid. Let P_e be the pressure of the liquid at which the float is in equilibrium. Now if it is displaced slightly upward, then the hydrostatic pressure will decrease but the buoyancy force will increase, resulting in a net force acting upward and the float will accelerate upwards. Similarly displacing the float downward the buoyancy force will decrease, making the float accelerate downward. So in this case, a very slight change in pressure or displacement will move the float either downward or upward, so the equilibrium in this situation will be unstable or, in other words, the float will be most sensitive to any pressure change. Hence to find the equilibrium pressure P_e, we have to achieve a stage where the float is in unstable equilibrium. Therefore the mean value of two close values of pressures at which the float moves in opposite directions is the equilibrium pressure.

However, if $(\gamma_f - \gamma_w)$ is negative, the float will be in a stable equilibrium; a slight change in pressure will not move the float either way so we will be unable to find the equilibrium pressure exactly. In this case, there will be a finite range of pressures during which the float will remain practically stationary. Therefore, to obtain better resolution, the unstable equilibrium condition is used.

The diver motion is observed through a rigidly supported long-range microscope by observing the tip O. The vertical range of the eye-piece of the microscope is 0.13 mm. To find P_e, we find two close pressure values for which the float moves in opposite directions. The pressure change is such that the observation tip O takes 30 s to travel the range on either side of the central zero of the eye-piece reticule. At a pressure very close to P_e, we can assume that the average velocity y of the diver, during the travel of the range, changes linearly with pressure. Let y_R be the average rising velocity and y_F the fall velocity; these are almost equal and the corresponding pressures are P_R and P_F. Then if from the linear velocity–pressure graph (figure 6.5), we take only absolute values of velocities, we obtain

$$(P_e - P_R)/(P_F - P_R) = y_R/(y_F + y_R).$$

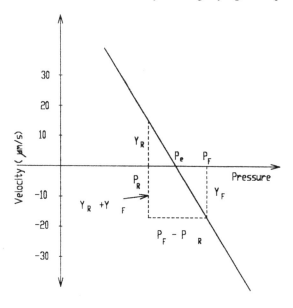

Figure 6.5. Velocity–pressure graph.

If M is the mass of water displaced by the loaded diver with a body of mass m and volume v, then

$$M = V d_w.$$

It can be seen from (6.74) that M is not a linear function of pressure and can best be represented as a quadratic function of P:

$$M = aP^2 + bP + c. \tag{6.75}$$

The constants a, b and c are functions of the atmospheric pressure, γ_f and γ_w. Bowman *et al* [3] found the value of a to be 8×10^{-8} g (cm Hg)$^{-2}$. If w represents the apparent mass of the body then

$$w = m - v d_w. \tag{6.76}$$

Taking $M1$ as the mass of the diver, the equilibrium equation may be written as

$$M1 - M + m - v d_w = 0.$$

Substituting the value of M from (6.75) and that of w from (6.76), we get

$$M1 - aP^2 - bP - c + w = 0. \tag{6.77}$$

Procedure for density determination with diver

To eliminate the constants $M1$, b and c, the values of the equilibrium pressure are determined with two standards S1 and S2 of known mass and volume, i.e. w for these samples is known. Taking the corresponding values of P and w with their respective subscripts, we may obtain the following equations:

$$M1 - aP_{S1}^2 - bP_{S1} - c + w_{S1} = 0 \tag{6.78}$$

$$M1 - aP_{S2}^2 - bP_{S2} - c + w_{S2} = 0 \tag{6.79}$$

$$M1 - aP_x^2 - bP_x - c + w_x = 0 \tag{6.80}$$

where P_x and w_x are the equilibrium pressure and apparent weight of the unknown body. Subtracting equation (6.79) from (6.78) and (6.80) from (6.78), we get

$$-a(P_{S1}^2 - P_{S2}^2) - b(P_{S1} - P_{S2}) + (w_{S1} - w_{S2}) = 0 \tag{6.81}$$

$$-a(P_{S1}^2 - P_x^2) - b(P_{S1} - P_x) + (w_{S1} - w_x) = 0. \tag{6.82}$$

In equation (6.81), everything is known except the value of b so the value of b can be calculated from (6.81). Substituting this in equation (6.82), we get the value of w_x and hence the density of the sample as its mass m_x has been determined separately. To eliminate drift in such experiments, the order of observations is P_{S1}–P_{S2}–P_{S1}–P_x–P_{S1}. The value of $(P_{S1} - P_{S2})$ required in the calculations is obtained by subtracting the value of P_{S2} from the mean of two values of P_{S1} taken in the first and third steps. The three observed values of P_{S1} give the effective value of P_{S1} during the experiment and are used for monitoring any drift.

NBS used this method for calculating the density differences of silicon crystals with a standard deviation of 4.4×10^{-7} g cm^{-3}, i.e. with a relative uncertainty of 0.22 ppm.

6.5   Flotation method for determining the density of solids

6.5.1   Principle

In the next two sections, we will be considering the methods in which the solid, whose density is to be determined, floats directly in a liquid. A mixture of two suitable liquids is taken and the density of their mixture is made approximately equal to that of the solid. The temperature of the liquid is changed to make its density exactly equal to that of the solid. At this temperature, the solid will remain stationary, i.e. it will neither move upwards nor downwards. This temperature is called the equilibrium temperature. At any other temperature, the solid will be moving either upwards or downwards. The method of flotation by pressure adjustment is similar. In this method, pressure is applied to the mixture so that its density becomes equal to that of the solid. At this pressure, the solid will remain stationary, i.e. it will neither move upwards nor downwards. This pressure

Table 6.1. Suitable liquids for the flotation method. The compressibilities β for liquids marked with an asterisk (*) have been obtained from the value of density ρ and the ultrasonic velocity c in the liquid [14] using the formula $\beta = 1/c^2\rho$, and has been expressed per bar for all liquids. The density coefficients α and β refer to the following relation: $d_t = d_0[(1 + \alpha((t - t_0)\,°C^{-1}) + \beta((t - t_0)\,°C^{-1})^2]$.

Sample no	Name	Formula	Density (g cm^{-3})	Coefficients of density ($\alpha \times 10^3$)	($\beta \times 10^6$)	Compress-ibility
1	Toluene	$C_6H_5CH_3$	0.866 9	−0.9159	0.368	90
2	Xylene	$C_6H_4(CH_3)_2$	0.870	—	—	66
3	Ethylene dichloride	$C_2H_2Cl_2$	1.282 48	−1.4217	−0.933	53.9*
4	Carbonyl chloride	CCl_2O	1.435	—	—	—
5	Ethyl bromide	C_2H_5Br	1.501 38	−2.0644	0.2673	129.4
6	Bromo-benzene	C_6H_5Br	1.522 31	−1.345	−0.24	66.8
7	Chloroform	$CHCl_3$	1.526 43	—	—	93
8	Tetrachloro-ethane	$C_2H_2Cl_4$	1.586 9	−153	—	49.1*
9	Carbon tetrachloride	CCl_4	1.632	−1.911	−0.69	105
10	Tetra chloro-ethylene	C_2Cl_4	1.647 5	−1.62	—	75.6
11	Isopropyl iodide	C_3H_7I	1.743 9	−1.948	—	66.4*
12	N-propyl iodide	C_3H_7I	1.784 4	−1.845	−1.25	64.9*
13	Iodo-benzene	C_6H_5I	1.860 59	−1.4814	−0.425	43.3*
14	Ethyl iodide	C_2H_5I	1.980 49	−2.217	−1.55	102
15	Ethylene di-bromide	$C_2H_2Br_2$	2.222 3	−2.090	−0.20	—
16	Methyl iodide	CH_3I	2.334 3	−2.67	−1.77	69
17	Tri-bromo-propane	$C_3H_5Br_3$	2.41	—	—	45.2
18	Tetra-bromo-ethane	$C_2H_2Br_4$	3.008 7	—	—	30.7*
19	Thallium formate	$HCOOTl$	3.5	−0.393	—	—
20	Thallium malonate	$CH_2(COOTl)_2$	4.9	−0.393	—	—

is called the equilibrium pressure. At any other pressure the solid will be moving either upwards or downwards. A float of known volume may be employed to determine the density of the liquid *in situ* or two solids of known density may be used to eliminate the coefficients of expansions of the liquid and solid.

6.5.2 Suitable liquids for the flotation method

Solids normally have a much higher density than liquids and thus the methods described here cannot be taken as universal. However, some organic liquids and aqueous solutions of organic salts have densities which are comparable to those of solids. A list of some such liquids, together with their density, density coefficients and isothermal compression, is given in table 6.1 [4, 5].

6.5.3   Methods for detecting the equilibrium temperature/pressure

Static method

One method is to attain the equilibrium temperature/pressure by observing the position of the solid whose density is being determined. A motion of 1 μm in 10 s may be assumed to be negligible, as it is equivalent to detecting a density difference of 1 in 10^{10}. A cathetometer, with an eye-piece with 1 μm graduations, is used and focused on a suitable edge or mark on the body. The temperature/pressure at which the body does not move even by one graduation in 10 s is taken as the equilibrium temperature/pressure. The uncertainty achieved in the relative density difference by this method is around 1 in 10^7.

Dynamic method

Instead of obtaining the exact equilibrium temperature/pressure, the velocity of the solid is measured at the temperatures/pressures around the equilibrium temperature/ pressure. The temperature or pressure values are chosen so that at one particular temperature or pressure, the velocity is positive and at the other it becomes negative; i.e. the two values surround the equilibrium temperature/ pressure.

To illustrate this method, an example of temperature adjustment is given here. From equation (6.71), we see that the terminal velocity is proportional to $\Delta\rho$, where $\Delta\rho = d_s - \rho$. But $\Delta\rho$ is a linear function of temperature, so the plot of velocity against temperature will be a straight line. If the density, ρ, of the solution is less than the density of the body, d_s, then the body will start moving downward, while it will rise if the density of the solution is increased by lowering its temperature. Hence, by measuring the velocities at least at three different temperatures, plotting the straight-line graph and finding its intersection with the line of zero velocity, we get the flotation temperature at which density of the solid is exactly equal to the density of the liquid.

Typical plots of the velocity against temperature are shown in figure 6.6 for three bodies A, B and C. The densities of all three bodies are close to one another. The value of the equilibrium temperature for bodies A, B and C are, respectively, $t_1 = 25.508$, $t_2 = 25.230$ and $t = 25.318\,°C$. If d_{s1}, d_{s2} and d_s are the respective densities of A, B and C, then substituting these values into equation (6.89), we get

$$(d_s - d_{s2})/(d_{s1} - d_{s2}) = (t_2 - t)/(t_2 - t_1)$$
$$= (25.508 - 25.318)/(25.508 - 25.230)$$
$$= 190/278.$$

Knowing the difference between the two densities and the value for d_{s2}, d_s may be calculated.

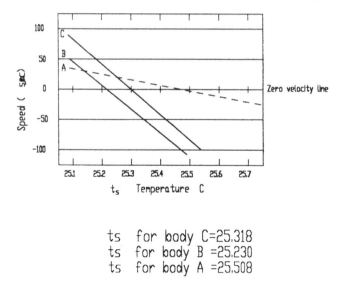

ts for body C=25.318
ts for body B =25.230
ts for body A =25.508

Figure 6.6. Velocity–temperature graph for equilibrium temperature t_f, t_s.

6.6 Determination of the density of a floating body by temperature adjustment

6.6.1 Basic theory

Let a solid, of mass m and volume v_0 at reference temperature t_0, float in a liquid at temperature t °C. Then the density of the solid at t °C is

$$m/v_0\{1 + \gamma_s(t - t_0)\} = \text{density of liquid at } t \text{ °C} = d_{L0}\{1 - \gamma_l(t - t_0)\}. \quad (6.83)$$

Hence, the density of the solid d_s at reference temperature t_0 is given by

$$d_s = m/v_0 = d_{L0}\{1 - (\gamma_l - \gamma_s)(t - t_0)\} \quad (6.84)$$

where γ_l and γ_s are the coefficients of volume expansion of the liquid and the solid, respectively, and d_{L0} is the density of liquid at the reference temperature t_0 °C. By finding the value of the density of the liquid at the reference temperature and the thermal density gradients of the solid and liquid, one can find the density of the solid. The density of diamond pieces has been determined following a similar approach [6].

However, another method uses two standards of known density, which eliminates the need to measure either the density gradients or the density of the liquid. Rewriting equation (6.84) for solid sample 1:

$$d_{s1} = m/v_0 = d_{L0}\{1 - (\gamma_l - \gamma_s)(t_1 - t_0)\}. \quad (6.85)$$

Similarly solid sample 2 of the same material but differing in density will attain equilibrium at temperature t_2, then its density d_{s2} at reference temperature t_0 °C is given by

$$d_{s2} = d_{L0}\{1 - (\gamma_l - \gamma_s)(t_2 - t_0)\}. \tag{6.86}$$

Subtracting equation (6.86) from (6.85), we get

$$d_{s1} - d_{s2} = d_{L0}(\gamma_l - \gamma_s)(t_2 - t_1). \tag{6.87}$$

If these two samples are such that density d_{s2} and their density difference are known, then any other sample of density d_s acquiring equilibrium at temperature t °C will have the following relationship:

$$d_s - d_{s2} = d_{L0}(\gamma_l - \gamma_s)(t_2 - t). \tag{6.88}$$

Dividing (6.88) by (6.87) we get

$$(d_s - d_{s2})/(d_{s1} - d_{s2}) = (t_2 - t)/(t_2 - t_1). \tag{6.89}$$

Hence, small differences in density between samples of the same material can be easily determined. It is assumed that the coefficients of thermal expansion of the solid standards and the sample used are either equal or negligibly small. This method is quite general, but is limited due to the availability of high-density liquids. The methods for detecting the equilibrium, i.e. when the solid body just floats in the liquid, may differ from user to user. The sensitivity of the method is of the order of 1 ppm, if the temperature is measured in mK.

6.6.2   Determining the density of diamonds

An 87% saturated aqueous solution of equal parts by weight of thallium formate (HCOOTl) and thallium malonate [$CH_2(COOTl)_2$] has a density of 3.515 g cm^{-3}. This is almost equal to that of natural diamond, so diamonds may be made to float in this solution by fine adjustment of the concentration, temperature or pressure of the mixture. The solution's expansion coefficient equals 3.93×10^{-4} K^{-1}, so by varying its temperature in steps of 0.1 K its density can be obtained in steps of 0.000 04 g cm^{-3}. To achieve an uncertainty of 1 ppm or better the temperature of the liquid should be maintained within 0.001 K. Mykolajewycz *et al* [6] used this mixture to measure the density of diamonds,. They used the temperature adjustment method to equalize the density of liquid and that of the diamond pieces. The equilibrium temperature was estimated dynamically. A float of known volume was employed to measure the density of the liquid *in situ*.

6.6.3   Use of the float for density determinations

To use the float method, let t_f, t_s be the respective temperatures of flotation for the float and solid. Furthermore, let ρ_L, ρ_f and ρ_s be, respectively, the density of the liquid, the float and the solid sample. Denoting the density gradients with respect to the temperature of the liquid by $d\rho_L/dt$, of the float by $d\rho_f/dt$ and of the solid by $d\rho_s/dt$, and taking $t_0 = 25\,°C$ as the reference temperature, we may write the equations for density as follows.

$$\rho_L(t) = \rho_L(25) - (d\rho_L/dt)(t - 25) \tag{6.90}$$

$$\rho_f(t) = \rho_f(25) - (d\rho_f/dt)(t - 25) \tag{6.91}$$

$$\rho_s(t) = \rho_s(25) - (d\rho_s/dt)(t - 25). \tag{6.92}$$

At the flotation temperature for the float, t_f, equating the density of the float with that of the liquid, we get

$$\rho_L(t_f) = \rho_L(25) - (d\rho_L/dt)(t_f - 25) = \rho_F(t_f) = \rho_f(25) - d\rho_f/dt\,(t_f - 25)$$

giving

$$0 = \rho_f(25) - \rho_L(25) - (d\rho_f/dt - d\rho_L/dt)(t_f - 25). \tag{6.93}$$

Similarly at the flotation temperature for the solid, t_s, at which the density of the liquid is equal to that of solid, we get

$$0 = \rho_s(25) - \rho_L(25) - (d\rho_s/dt - d\rho_L/dt)(t_s - 25). \tag{6.94}$$

Subtracting (6.94) from (6.93), we get

$$\rho_s(25) = \rho_f(25) - (d\rho_f/dt - d\rho_L/dt)(t_f - 25) + (d\rho_s/dt - d\rho_L/dt)(t_s - 25). \tag{6.95}$$

The values of $d\rho_F/dt$, $d\rho_L/dt$ and $d\rho_s/dt$ have been measured [6] to find the density of diamond pieces varying in mass from 1.3 to 429 mg.

The special float, shown in figure 6.7(*a*), is a quartz cylinder with a bulge in the upper portion containing a cylindrical soft iron piece floating in mercury with a pointer. The soft iron cylinder enables the float to be moved with a magnet from outside the thermostatic bath and its pointed tip is used as the reference to observe its movement. Mercury is used to adjust the density of the float to that of the solution. A Pyrex cylindrical vessel, shown in figure 6.7(*b*), of diameter 2.4 cm and length 28 cm is used as the flotation vessel. The upper end is closed with a ground joint cap. The side tube serves to remove gas from the solution by evacuation. A platinum screen is placed near the bottom to prevent the diamond pieces from coming into contact with the bottom where deposits might be present. A similar platinum screen is placed in the upper part to prevent the diamond pieces rising to the surface of the solution.

To prevent convection currents, the flotation vessel is placed inside another wide glass tube. The solution is stirred by bubbling it with nitrogen pre-saturated with water vapours. A thermostatic bath capable of maintaining a temperature uniformity of 0.001 K and attaining any temperature in steps of 0.1 K within a range of 1 K around the reference temperature is used in the experiment. The density values have been reported up to the fifth decimal place when expressed in g cm^{-3}.

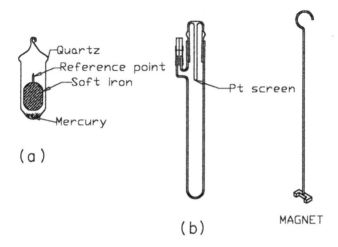

Figure 6.7. Float and flotation vessel.

6.6.4 Density differences between single silicon crystals

In 1990, Kozdon, Wagenbreth and Hoburg also equalized two densities by changing temperature [7].   For the silicon samples, a mixture of 1,2,3-tribromopropane and decalin with a density very close to that of silicon was used.   Decalin has a density of 0.88 g cm^{-3}.   Finally the density of the mixture was adjusted to that of the silicon sample by altering its temperature in terms of 1 mK. The temperature of 3 litres of this mixture was controlled with a precision of 0.1 mK. The non-linearity of the density dependence on temperature was also taken into account—$d\rho_L/dt$ for the mixture was measured as -2.22×10^{-3} g cm^{-3} °C. For silicon, $d\rho_s/dt$ was taken from the literature as 0.02×10^{-3} g cm^{-3} °C. The slow drift in the density over time was also taken into account and linear as well as parabolic models for the density drift were considered. The static method for detecting the equilibrium temperature, as described in section 6.5.3, was used. The uncertainty achieved in the relative density difference was around 1 in 10^7 and the ability to detect a density difference was 1 in 10^{10}.

6.6.5 Change in density due to selective adsorption of liquids used

With such a fine ability to detect density differences, it will be appropriate to consider the effect of adsorption of the immersion liquid. It is known that a solid silicon surface is covered with a SiO_2 layer. It is possible that this SiO_2 layer might have a better tendency of adsorbing one of the two liquids. For example in a tribrompropane and decalin mixture, it is probable that the hydrogen atoms of the decalin could be attracted more by the oxygen atom of SiO_2 than the bromine atoms in the tribromopropane. The density of decalin is only 0.88 g cm^{-3}, so the apparent density of the silicon crystal will change. Adsorption is a surface phenomenon, so the change in density will be more prominent in smaller samples as, in these cases, q, the surface-to-volume ratio, will be rather large. If m and v are the mass and volume of the crystal, then the apparent density, ρ, can be expressed as

$$\rho = (m + m_a)/(v + v_a)$$

where m_a and v_a are the mass and volume of the adsorbed liquid. If a is the thickness of the adsorbed layer and F its surface area, then

$$v_a = Fa \qquad \text{and} \qquad m_a = v_a \rho_a.$$

If ρ_{si} is the real density of the silicon then

$$\rho_{si} = m/v$$

giving

$$\rho = (\rho_{si} + aq\rho_a)/(1 + aq)$$
$$\rho_{si} - \rho = aq(\rho_{si} - \rho_a)/(1 + aq).$$

However, a, the thickness of the SiO$_2$ layer, is very small, so aq is negligibly small in comparison to 1, so

$$\rho_{si} - \rho = aq(\rho_{si} - \rho_a).$$

Let us define

$$\Delta\tau = a(\rho_{si} - \rho_a)$$

where $\Delta\tau$ is the addition of mass per unit area,

$$\rho = \rho_{si} - q\Delta\tau.$$

As q is different for different shapes, the apparent density will also differ for different shapes of the same material.

q, the ratio of surface area S and volume v

- For a parallelepiped with edges a, b, and c

$$q = S/v = 2(ab + bc + ca)/abc = 2(1/a + 1/b + 1/c).$$

- For a cylinder/disc of diameter D and length L

$$q = S/v = [\pi DL + \pi D^2/2]/(\pi D^2 L/4) = 4/D + 2/L.$$

- For a sphere of diameter D

$$q = S/v = \pi D^2/(\pi D^3/6) = 6/D.$$

6.6.6 Change in density due to the polishing effect

Polishing effects can also be interpreted in terms of adsorption. Mechanical polishing damages the surface layer of the single silicon crystal as it can remove several silicon atoms. All such vacancies cannot be penetrated and filled by the relatively large molecules of the immersion liquid so some vacancies will be left unfilled resulting in a loss of mass, say by an amount of Δm. Furthermore, the value of Δm will be proportional to the surface of the crystal exposed to the polishing. Δm may then be written as

$$\Delta m = -FC$$

where C is the constant of proportionality and F the surface of the crystal. The corresponding decrease in density $\Delta \rho$ is given by

$$\Delta \rho = -\Delta m / V = -CF/V = -Cq.$$

Here also q is the ratio of the surface area to the volume of the crystal. No such problem will arise if the surface is finally etched.

Results

Three samples of WASO K material from the same rod [7] were found to have a maximum density difference of $(0.33 \pm 0.09) \times 10^{-6}$ g cm^{-3} whereas silicon crystals from different sources, formed in different shapes and polished differently, were found to have a maximum difference in density of $(10.78 \pm 0.15) \times 10^{-6}$ g cm^{-3}. Measurements of density difference confirmed the predicted trend of adsorption by silicon crystals of different shapes with mechanically polished surfaces, but the results do not agree quantitatively. To avoid the necessity for two different types of liquid so far as an excess of hydrogen atoms is concerned, 1,2-dibromoethane has been used instead of transdecalin [8].

6.7 Determination of density by pressure adjustment

Liquids are more compressible than solids, so by applying pressure, the change in the density of a liquid is greater than that of a solid. Kozdon and Spieweck have used this method to equalize the density of the liquid to that of the float [8]. 1,2,3-tribromopropane and 1,2-dibromoethane are mixed in the ratio 2:1 to obtain a density of 2.33 g cm^{-3}. The density of the mixture was finally made equal to that of the floating silicon crystal by changing the pressure. If γ_l is the compressibility coefficient for the liquid and γ_s that for the density of the solid silicon samples, and if the first sample of density d_{s1} is brought to a stationary position by applying pressure p_1, then the density of the liquid will be equal to that of the solid, giving us

$$d_{s1}(1 + \gamma_s p_1) = d_0(1 + \gamma_l p_1)$$

giving

$$d_{s1} = (1 - \gamma_s p_1)d_0(1 + \gamma p_1)$$

or

$$d_{s1} = d_0\{1 + (\gamma - \gamma_s)p_1\}. \tag{6.96}$$

Similarly another sample of density d_{s2} is brought to a stationary position by applying a pressure p_2, giving us

$$d_{s2} = d_0\{1 + (\gamma - \gamma_s)p_2\}. \tag{6.97}$$

Subtracting equation (6.97) from (6.96), we get

$$(d_{s2} - d_{s1}) = d_0(\gamma - \gamma_s)(p_2 - p_1). \tag{6.98}$$

Putting $(d_{s2} - d_{s1}) = \Delta d_s$ and dividing (6.98) by (6.96), we obtain

$$\Delta d_s/d_{s1} = d_0(\gamma - \gamma_s)(p_2 - p_1)/d_0\{1 + (\gamma - \gamma_s)p_1\} \tag{6.99}$$

and neglecting the second term in the denominator

$$\Delta d_s/d_{s1} = (\gamma - \gamma_s)(p_2 - p_1). \tag{6.100}$$

Hence, by finding γ and γ_s, it is possible to find $\Delta d_s/d_{s1}$. However, when a liquid is compressed its temperature also changes, so to find an effective value for γ, we should do a little more calculation.

At constant entropy with symbol 's'

$$(\delta T/\delta p)_s = T(\delta V/\delta T)_p/C_p$$
$$= TV\gamma_v/C_p. \tag{6.101}$$

In the case of organic liquids such as tribromopropane, which has a relatively large molar volume and thermal coefficient of expansion γ_v, but also a smaller molar heat (C_p), $(\delta T/\delta p)_s$ is not negligible. At 20 °C, i.e. for $T = 293$ K, $V = 116.5$ cm^3 mol^{-1}, $C_p = 175$ J mol^{-1} K^{-1}, and $\gamma_v = -9.5 \times 10^{-4}$ K^{-1} [5], $(\delta T/\delta p)_s = 18.5 \times 10^{-8}$ K Pa^{-1} so the effective value of γ is $\gamma_T - \gamma_v(\delta T/\delta p)_s$, where γ_T is the coefficient of volume contraction due to change in pressure of the liquid at constant temperature. Substituting 45.2×10^{-11} Pa^{-1} for the value of γ_T, we get the following value for γ:

$$(45.2 - 9.5 \times 1.85) \times 10^{-11} \text{ Pa}^{-1} = 27.6 \times 10^{-11} \text{ Pa}^{-1}.$$

However, Peuto [7] measured the thermal effects of 1,2,3-tribromopropane directly, giving $(\delta T/\delta p)_s = (12.7 \pm 0.3) \times 10^{-8}$ KPa^{-1} so the effective value of γ should be taken as 33.1×10^{-11} Pa^{-1}.

The density of 1,2,3-tribromopropane is 2.41 g cm^{-3} and that of 1,2-dibromoethane 2.18 g cm^{-3}, so mixing them in the ratio 2:1 will give a mixture

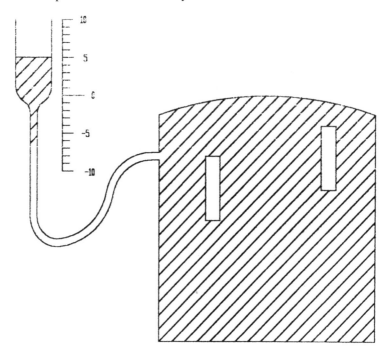

Figure 6.8. Schematic diagram for density measurement by pressure adjustment.

of density nearly 2.328 g cm^{-3}, which is roughly the same as that of silicon. By adding a little of either of the two liquids, the density of the liquid mixture can be adjusted, as far as possible, to be nearly equal to that of the standard silicon sample. Finally, by adjusting the pressure of the mixture in a thermocontrolled sealed cylinder, the density of the liquid mixture can be made equal to that of the sample by observing its movement. The pressure at which the liquid and silicon crystal have equal densities was estimated using the static method. The limit of detection is 1 part in 10^{10}. An outline of the system is shown in figure 6.8. To produce a change in the hydrostatic pressure in the liquid mixture, a small cylindrical vessel B containing the same liquid and connected with the main vessel through a rubber tube is raised or lowered as necessary. Taking into consideration the density of the liquid producing the hydrostatic pressure and other constants, the final relation for the relative density difference in equation (6.100) may be written as follows.

$$\Delta d_s / d_{s1} = C(h_2 - h_l) \qquad (6.102)$$

where h_1 and h_2 are the columns of the liquid producing the desired pressure. Although the value of C is determined experimentally, by calculation its value will be about 0.611×10^{-7} cm^{-1}, so even a measurement error of 1 cm in the height difference will produce an uncertainty of only 0.611×10^{-7}. A height difference of 100 cm will be equivalent to 1.4×10^{-5} g cm^{-3}, which may roughly be taken as the measuring range.

Results

The density difference between two samples of single crystals determined by this method has been found to be $(1.394 \pm 0.007) \times 10^{-6}$ g cm^{-3}, so a high precision of 7×10^{-9} g cm^{-3} can be obtained this way.

6.8 Method for obtaining two samples of known density difference

Let us assume that we have a standard of density d_{s1} and we wish to prepare another standard of density d_{s2}, which is very close to d_{s1}. The difference in density may be only a few ppm. For this purpose a very thin wire (20 μm diameter) of platinum is taken and wrapped round a cylinder of the material of density d_{s1}. The platinum should be of high purity (spectrally pure) so that its density may be taken from the literature. If m_1 and m_{Pt} are the mass of the body and platinum wire respectively, then

$$d_{s2} = (m_{Pt} + m_1)/(m_{Pt}/d_{Pt} + m_1/d_{s1}) \qquad (6.103)$$
$$d_{s2} - d_{s1} = m_{Pt}(1 - d_{s1}/d_{Pt})/(m_{Pt}/d_{Pt} + m_1/d_{s1})$$
$$(d_{s2} - d_{s1})/d_{s1} = (1 - d_{s1}/d_{Pt})/(m_1/m_{Pt} + d_{s1}/d_{Pt}).$$

If the relative difference in density (left-hand side) is only a few ppm, say $\Delta \times 10^{-6}$, then

$$(m_1/m_{Pt} + d_{s1}/d_{Pt}) = (1 - d_{s1}/d_{Pt})/\Delta \times 10^{-6}$$
$$m_1/m_{Pt} = [1 - (d_{s1}/d_{Pt})(1 - \Delta \times 10^{-6})]/\Delta \times 10^{-6}.$$

Neglecting $\Delta \times 10^{-6}$ as it is small in comparison with one, we get

$$m_{Pt}/m_1 = \Delta \times 10^{-6}/(1 - d_{s1}/d_{Pt}).$$

For example, if the first silicon standard has a density $2.329\,079\,37$ g cm^{-3} and mass $m_1 = 800$ g, then the mass of the required platinum wire will be

$$m_{Pt} = 800\Delta \times 10^{-6}/(1 - 2.329/21.4)$$
$$= 1.122 \times 800 \times \Delta \times 10^{-6}$$
$$= 897.698 \times \Delta \times 10^{-6} \text{ g}$$
$$= 1.795 \text{ mg} \qquad \text{for } \Delta = 2.$$

The density of the second standard with the platinum wire, d_{s2}, as calculated from equation (6.103), will be

$$d_{s2} = 2.329\,0839 \text{ g cm}^{-3}.$$

6.9   Vibrating element densitometers

If a body containing or surrounded by a fluid is made to vibrate, its resonance frequency is a function of the shape and size of the body and the elasticity of its material, induced stresses and mass distribution (stiffness). The resonant frequency will naturally also depend upon the density of the surrounding fluid. It can be basically equated to a mechanical spring damper in which the spring constant is a function of the dimensions of the body and the induced stress. Therefore the square of the resonance frequency n will be:

- directly proportional to its stiffness; and
- inversely proportional to its combined mass of the body and the fluid.

This will give

$$n^2 = K/(M + V\rho)$$

where K is the stiffness of the vibrating system, M the mass of the transducer, V the system constant, which is essentially its volume, and ρ the density of the medium. If T is the time period then

$$T^2 = M/K + V\rho/K = A + B\rho \qquad (6.104)$$

where $A = M/K$ and $B = V/K$.

The two constants are determined with the help of at least two samples of known density. Therefore each such instrument needs to be calibrated against standard materials of known density. The materials used as standards are mostly (1) water of known isotopic composition and air saturation or (2) air of known composition with specified temperature, pressure and relative humidity. The densities of both these materials are known with an adequate degree of accuracy.

There are a number of methods for utilizing vibrating elements in densitometers. Each has its own advantage in its own specific field. However, there are three types of vibrating element [10] which are used in commercial instruments: (1) tubes containing the fluid, (2) cylinders and (3) tuning forks that are immersed in the fluid whose density is to be measured.

6.9.1   Densitometers with vibrating tube as element

The tube length is normally 20 times its diameter. The tube is constrained, both mechanically and by the arrangement of the driver mechanism, to vibrate only in one plane. The mode of vibration is transverse to the axis of the tube. There are various variations in the design of the vibrating element.

Single-tube densitometers

A single-tube element is shown in figure 6.9(a) and has the advantage of being straight throughout, so it does not offer any obstruction to the flow of the fluid;

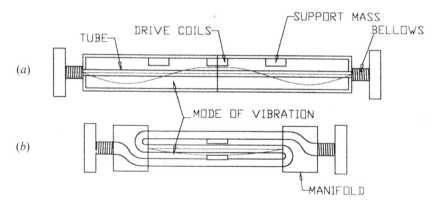

Figure 6.9. (*a*) Single straight tube as vibrating element. (*b*) Single folded tube as vibrating element.

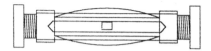

Figure 6.10. Twin-tube vibrating element.

i.e. the pressure drop across it is minimal. However, such densitometers are quite long.

To reduce the length of the densitometer, the vibrating tube is folded, as shown in figure 6.9(*b*). This is also a single-tube densitometer, but it uses heavy section side tubes which form a stiff structure to support the node of the vibrating central tube. A disadvantage of such an element is the circuitous path of the fluid resulting in a higher pressure drop.

Twin-tube densitometers

A twin-tube vibrating element is shown in figure 6.10. There is very little blockage of the fluid path as it is straight and it can also be easily cleaned. It has the distinct advantage of compactness. The twin-tube element achieves a good dynamic balance as the two tubes vibrate in opposite phases. Their nodes are fixed at the end of the tubes with minimum sensitivity to installation clamping and mass loads. The main design problems for such elements are to minimize the effects of (1) end-loading, (2) pressure and (3) temperature. To isolate these elements from external vibrations and the effects of end-loading caused by differential expansion with respect to the casing with change in temperature, they are provided with bellows.

As the fluid only runs inside the tube, the outside and inside pressures differ. This results in greater stress in the tube thus changing the stiffness of

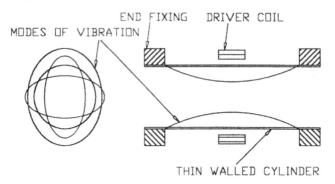

Figure 6.11. Cylinder as vibrating element.

the tube. If the stiffness changes then so do the constants connecting the time period or frequency. Furthermore, the Young's modulus of the tube sensors will change with temperature and pressure of the liquid maintained inside the tube. All manufacturers initially provide the pressure and temperature coefficients for correction purposes. However, these meters require periodic calibration against several liquids of different density so as to cover the range of density, temperature and pressure for which the meter has been designed. Densitometers with tubes as vibrating elements are used to measure the density of liquids.

6.9.2   Densitometers with cylinder as vibrating element

A thin-walled cylinder with a relatively smaller length to diameter ratio with stiff ends, as shown in figure 6.11, is used as the vibrating element. The wall thickness varies from 25 to 300 μm. The wall thickness depends upon the density range of the fluid to be used. The cylinder is excited to vibrate in hoop mode and the vibrations are maintained by a magnetic drive, which may be fixed either inside or outside the cylinder. The cylinder is made from corrosion-resistant and magnetic steel, e.g. FV 520. This material is good as regards magnetic and corrosion resistance but has a fairly large thermoelastic coefficient, so temperature correction is necessary. To minimize the error due to temperature, where possible an alloy of nickel and iron called Ni-span-C is used. With appropriate heat treatment this material can be made to have an almost zero thermoelastic coefficient. As the cylinder is completely submerged in the liquid, unlike densitometers with a tube as the vibrating element, there is no pressure difference. The change in frequency is entirely due to local loading of the fluid in contact with the cylinder.

As is evident from equation (6.104), the time period versus density relation is a parabolic curve. Hence it is only within a limited range of density that the time period may be taken as a linear function of density. The resonant frequency varies around 2–5 kHz. Such cylinders are quite expensive as a high degree of precision is required in manufacturing them.

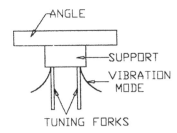

Figure 6.12. Tuning fork as vibrating element.

6.9.3   Densitometers with tuning fork as vibrating element

A low-mass tuning fork, shown in figure 6.12, driven electromagnetically can also be used as a vibrating element. It can be either placed inside a small vessel containing the fluid, whose density is to be measured, or it can be inserted directly in the line through which the fluid is flowing for online density measurements. Like any other meter with a vibrating element, it also requires periodic calibration against fluids of known density. Densitometers with a tuning fork or thin cylinder as the vibrating elements are used to measure the density of gases.

6.9.4   Availability of densitometers with vibrating elements

Messrs Solartron Mobery [11] and Messrs Onix Measurement (previous name Peek Measurement) [12] make densitometers which use vibrating elements as sensors.   Sarasota is the brand name of the densitometers manufactured by Onix Measurement. The essential particulars of their products are given in the following table.

Messrs Solartron Mobrey	Solartron density transducers				
Model no	7835	7845	7846	7847	7826
Density range (g cm^{-3})	0–3	0–3	0–3	0–3	0–3
Calibrated range (g cm^{-3})	0.3–1.1	0.6–1.6	0.6–1.6	0.6–1.6	0.6–1.25
Accuracy (g cm^{-3})	0.000 15	0.000 35	0.000 35	0.000 35	0.001
Repeatability (g cm^{-3})	0.000 02	0.0001	0.0001	0.0001	0.0001
Temperature range (°C)	−50 to +110	−50 to +110	−50 to +110	−50 to +110	−50 to +160
Temperature coefficient/K (ppm)	5	5	50	50	100
Pressure up to (bar)	150	100	50	20	207
Pressure coefficient/K (ppm)	6	6	6	6	negligible
Material of sensor	Ni-span-C	SS316L	Hastelloy	SS316L	SS316L

Messrs Onix Measurement Limited	Sarsota densitometers		
Model no	FD 910	FD 930	FD950	FD960
Density range (g cm^{-3})	0–2.1	0–2.1	0–2.1	0–2.1
Calibrated range (g cm^{-3})	0.65–1.6	0.65–1.6	0.65–1.6	0.65–1.6
Accuracy (g cm^{-3})	0.0001	0.0001	0.0001	0.0001

Repeatability (g cm^{-3})	0.00002	0.00002	0.00002	0.00002
Temperature range (°C)	−50 to +180	−20 to +180	−50 to +180	−50 to +110
Temperature coefficient/K (ppm)	5	5	5	5
Pressure up to (bar)	as per flange	10	150	as per flange
Pressure coefficient/K (ppm)	6	6	6	6
Material of sensor	SS316L	SS316L	Hastelloy	SS316L

6.9.5 Densitometers with U-tube as vibrating element

Messrs Anton Paar [13] manufacture densitometers with a U-shaped sample tube as the vibrating element. The tube is rigidly supported at its open ends and electromagnetically excited to vibrate at its natural frequency. The U-tube is made from a special glass with the brand name DURAN 50. From the change in frequency caused by a specific sample the density of the liquid can be determined with great precision. These densitometers have fairly high repeatability. Since the mass of the sample participating in the vibration is established in this way, the density can be derived from the assumption that the capacity of the tube is constant, hence the temperature stability needs to be very stable to achieve a repeatability of 1 in 10^6. Volatility or the surface tension of the sample does not affect the result. Either a thermocontrolled quartz oscillator or the period of a second measuring cell converted by an unconditionally stable phase-locked loop (PLL) is used as the time base for the time period measurement.

The sample tube is filled either by injection with a hypodermic syringe or by continuous flow. The oscillatory part of the tube has to be filled completely homogeneously without entrapping any air or solid. It takes about 3 to 4 minutes for the sample to attain the same temperature as the sample tube, which is maintained at a predetermined temperature. Then the oscillation period, T, may be read from the numerical display. The constants in equation (6.103) are determined by measuring the periods of two samples of known density.

To attain high repeatability the temperature of the liquid sample is maintained within ±0.01 °C. The maximum error will be around 1.5×10^{-6} g cm^{-3} for measuring differences in density, which are less than or equal to 0.5 g cm^{-3}. The error in the density of the sample due to its thermal expansion within 0.02 °C will be 2×10^{-6} g cm^{-3}. This leads to a usable precision of only 3.5×10^{-6} g cm^{-3}. When measuring the absolute density in the range 0–3 g cm^{-3}, the reproducible deviation from the true density is no more than 1×10^{-4} g cm^{-3}. The temperature range is normally −10 to +70 °C, but in special cases it can cover temperatures as low as −180 °C and as high as 150 °C. Similarly the normal pressure range is 0–10 bar but can be extended from 0 to 50 bar. Using an automatic sample changer, density measurements can be automatic. The samples are contained in 5 cm^3 vials, which are placed in a 10-vial magazine and inserted into the automatic sample changer. The magazines are coupled to each other by a U-shaped pin connector. The automatic measuring cycle may be optimized by varying the filling and rinsing times.

6.9.6 Drive mechanism for the vibrating elements

Magnetic drive

A magnetic drive and pickup for the vibrating elements usually consists of a single coil to produce a signal proportional to the movement of the element. This signal is amplified and fed back as the current to a drive coil producing a disturbing force on the sensor element. If the element is non-magnetic then a small armature is attached to drive it. The main advantage of this magnetic drive and pickup arrangement is that the system has no contact with it. Copper winding is generally used to enable them to work between -200 and $200\,°C$.

Piezoelectric drive

A wide range of piezoelectric materials is available and several meet the requirements for driving a vibrating element sensor. Their main advantage is their low cost and high impedance. However, the piezoelectric device is fixed to the vibrating element with an adhesive, so great attention must be paid to the design of the mount in order to reduce the strain experienced by the piezo element due to thermal and pressure stresses.

6.9.7 Advantages of densitometers with vibrating elements

- These meters can be used both for gases and liquids for measuring the density.
- The results are very accurate.
- These can be used for online, real-time, measurements of density.
- They are easily digitized as it is the frequency which is used to measure the density.
- Computers can be used to apply corrections for temperature and pressure. If necessary, the output may be shown in other density-related parameters, such as the relative density, degree Brix, degree Twaddle or API etc.
- They are relatively robust and easy to install.

The only disadvantage is that they require more frequent calibration.

References

[1] Schoonover R M 1982 The density of small solid objects by a simple float method I *J. Res. Natl Bur. Stand.* **87** 197–205
[2] Davis R S 1982 The density of small solid objects by a simple float method II *J. Res. Natl Bur. Stand.* **87** 207–9
[3] Bowman H A and Schoonover R M 1965 Cartesian diver as a density comparator *J. Res. Natl Bur. Stand.* C **69** 217–23
[4] Maston *et al* (ed) 1959 *Methods of Experimental Physics* vol 6, part A (New York: Academic) pp 287–91

[5] *International Critical Tables* vol 3 (London: McGraw-Hill) pp 28–41

[6] Mykolajewycz R *et al* 1964 High-precision density determination of natural diamonds *J. Appl. Phys.* **35** 1773–8

[7] Kozdon A, Wagenbreth H and Hoburg D 1990 Density difference measurement on silicon single crystals by temperature of flotation method, PTB-Bericht, W-43

[8] Kozdon A and Spieweck F 1992 Determination of difference in density of silicon single crystals by observing flotation at different pressure *IEEE Trans. Instrum. Meas.* **41** 420–6

[9] Peuto A M 1989 Tests on the compressibility of liquids used in density measurements by flotation method *IMGC Report* 170 (Torino: IMGC)

[10] Mason P 1992/93 Fluid density *Meas. Control* **25** 297–302

[11] Solartron density transducers, Solartron Mobery, 158 Edinburgh Avenue, Slough, Berkshire, SL1 4UE, UK

[12] Sarasota industrial density meters, Onix Measurement Limited, King's Worthy, Winchester, Hampshire, SQ23 7QA, UK

[13] High precision digital density measuring system (DMA 60 + DMA 602), Messrs Anton Paar KG, A-8054, Graz, Postfach 58, Karntnerstrasse 322, Austria

[14] Pandey J D and Ashok K 1994 Ultrasonic velocity in pure liquid *J. Pure Appl. Ultrason.* **16** 63–8

Chapter 7

Hydrometry

Symbols

Δ	Density of standard weights used
M, m	Mass
T	Surface tension of the liquid
α	Angle of contact
ρ	Density of liquids
σ	Density of air
d_s	Density of solid
d_r	Density of ring
ρ_w	Density of water
D	Diameter of stem of hydrometer
I	Density interval of a hydrometer
R	Range of a hydrometer
L	Scale length of hydrometer
V	Volume of bulb
v	Volume of stem
St_1/t_2	Specific gravity of a liquid equal to density of liquid at t_1 and density of water at t_2 °C
°F	Degrees Fahrenheit
γ	Coefficient of cubic expansion
d_w	Density of water
d_g	Density of glass
TC	Temperature correction
T_s	Reference temperature for the standard hydrometer
T_u	Reference temperature for hydrometer under test
CC	Certificate correction
STC	Surface tension correction
STF	Surface tension factor of a hydrometer
MC	Meniscus correction
A_{ij}, a_{ij}	Constants or coefficients in various relations
b_{ij}	Constants or coefficients of various powers of variable.

7.1   Introduction

The hydrometer is one of the most simple but highly effective instruments for the measurement of the density of liquids. With due care and by estimating one-fifth of the graduated interval, we can measure density to within 0.04 kg m^{-3} (0.000 04 g cm^{-3}). Hydrometers are used in the petroleum and chemical industries for determining the density or specific gravity of liquids, in breweries for assessing the strength of alcohol, in the sugar industry for measuring the percentage of sugar present in sugar cane solutions and in the dairy industry for measuring the fat content from the density of milk. Their names also change from industry to industry. They are alcoholometers in breweries, Brix hydrometers in the sugar industry and lactometers in the dairy industry. However by whatever name they may be called, hydrometers measure only the density of the liquid. The names correspond to how and in what units their scales are graduated.

7.2   Classification of hydrometers

Hydrometers are essentially classified into two classes: constant-volume hydrometers and constant-mass hydrometers.   A constant-volume hydrometer is made to float up to a specific mark in liquids of different density by placing weights on it. Thus the immersed volume of the hydrometer is kept constant and its mass is varied to keep it floating in liquids of different density. But a constant-mass hydrometer floats up to different positions in liquids of varying density. In other words, it is its immersed volume which varies, so that it floats in liquids of different density while its mass remains unchanged.

7.2.1   Constant-volume hydrometer

Construction

The most well known one in this class is the Nicholson hydrometer. It has a pan P at the top connected to a thin circular rod of uniform diameter called the stem of the hydrometer (figure 7.1). The other end of the stem is connected to a hollow body B. The body B consists of a hollow cylinder surmounted by two cones. The lower cone has another pan, which is the base of a heavily loaded cone. It may be loaded with mercury or small lead shots to keep it floating vertically in the liquids. Normally the hydrometer is made of a thin metallic non-corrodible sheet. All the parts are made symmetrical about the single axis XX. A very sharp mark G is graduated around the stem. The hydrometer is always made to float exactly up to this mark.

Every hydrometer follows the laws of flotation based on Archimedes' principle; i.e. while in equilibrium the upward thrust due to the mass of displaced liquid is equal, opposite and along the same line of action as the gravitational force due to the weight of the hydrometer.

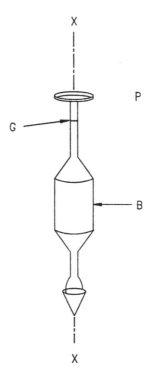

Figure 7.1. Constant-volume hydrometer.

Basic theory

Weigh the hydrometer in air of density σ against weights of density Δ and mass m with V being the volume of the hydrometer. If M is the mass of the additional weights to be placed on its pan P, so that it floats up to the fixed mark G marked on the stem in a liquid of density ρ_1, then the forces acting on it are:

Downward forces

(1) $(M + m)(1 - \sigma/\Delta)g$ is the vertically downward force due to gravity.
(2) Due to surface tension T, some liquid will rise up along the stem. As the liquid ascends around the stem, it tries to pull the hydrometer down, so effectively increasing its mass. The force due to surface tension is equal to the product of the perimeter of the stem and the vertical component of the surface tension. If α is the angle of contact, the surface tension will act along the line making an angle α with the vertical. The downward force due to surface tension is $\pi DT \cos(\alpha)$.

Upward forces

(1) The upthrust of liquid of volume V and density ρ_1 is equal to $V\rho_1 g$.

(2) Air buoyancy on the portion of the stem above the liquid is equal to $v\sigma g$, where v is the volume of hydrometer above the fixed mark G.

(3) Air buoyancy on the volume of the liquid raised due to surface tension—the weight of the liquid raised will be equal to the upward force due to surface tension. Therefore the volume of liquid raised due to surface tension is the mass of liquid raised divided by ρ_1, the density of the liquid. However, the mass of liquid is equal to the surface tension force divided by g, i.e. $\pi DT \cos(\alpha)/g$, giving the volume of liquid as $\pi DT \cos(\alpha)/g\rho_1$ so the upthrust due to air will be $\pi DT\sigma \cos(\alpha)/\rho_1$.

Equating the upward and downward forces and cancelling g from both sides,

$$(M + m)(1 - \sigma/\Delta) + \pi DT \cos(\alpha)/g = V\rho_1 + \{v + \pi DT \cos(\alpha)/g\rho_1\}\sigma. \quad (7.1)$$

To obtain reasonably repeatable readings, the liquid must wet the stem properly, for which it is necessary for α to be zero. So equation (7.1) becomes

$$(M + m)(1 - \sigma/\Delta) + \pi DT/g = V\rho_1 + \{v + \pi DT/g\rho_1\}\sigma. \quad (7.2)$$

σ is about one-thousandth of the density of common liquids met in day-to-day life and v and $\pi DT/g\rho_1$ are also very small in comparison with V, so the terms due to air buoyancy are small.

Let us consider the term due to surface tension on the left-hand side of (7.2). The surface tension of normal liquids is of the order of 50 mN m^{-1} and the diameter of the stem is 5 mm giving

$$\pi DT/g = 3.1416(0.005)50 \times 10^{-3}/9.808\,55 \text{ kg} = 0.08 \text{ g}.$$

However, normally the mass m of a hydrometer is around 80 g, so this term may be neglected if accuracy of 0.1% is required; otherwise this term should be considered. For the time being we are neglecting this. So this equation, to the first approximation, may be written as

$$(M + m)(1 - \sigma/\Delta) = V\rho_1. \quad (7.3)$$

Measurement of the density of a liquid

Let M_1 be the mass of weights placed on the pan, when the hydrometer is made to float up to the specified mark in pure distilled water of density ρ_w. This gives us

$$(M_1 + m)(1 - \sigma/\Delta) = V\rho_w. \quad (7.4)$$

If M_2 is the mass of weights placed on its pan to make it float to the same fixed mark in a liquid of density ρ, this gives us

$$(M_2 + m)(1 - \sigma/\Delta) = V\rho. \quad (7.5)$$

It is assumed that the density of air does not change in the two situations or the variation is too small to be considered vis-à-vis the necessary accuracy. Dividing the two equations, we get

$$[(M_2 + m)(1 - \sigma/\Delta)]/[(M_1 + m)(1 - \sigma/\Delta)] = V\rho/V\rho_w \qquad (7.6)$$

giving

$$\rho/\rho_w = (M_2 + m)/(M_1 + m). \qquad (7.7)$$

Hence, we obtain the specific gravity of liquid with respect to water at the temperature of measurement.

Measurement of the density of a solid

This instrument may also be used for measuring the density of small solids. Let the mass of the body be W when weighed in air against the weights of density Δ. M_1 is the mass of added weights when the solid is in the upper pan and M_2 the mass of weights when the solid is transferred to the lower pan of the hydrometer to float up to the mark G in a liquid of known density, say water. The equilibrium equations are:

$$(M_1 + W + m)(1 - \sigma/\Delta) = V\rho_w \qquad (7.8)$$
$$(M_2 + m + W)(1 - \sigma/\Delta) - V_1\rho_w = V\rho_w \qquad (7.9)$$

giving us

$$(M_1 + W + m)(1 - \sigma/\Delta) = (M_2 + W + m)(1 - \sigma/\Delta) - V_1\rho_w$$
$$(M_2 - M_1)(1 - \sigma/\Delta) = V_1\rho_w. \qquad (7.10)$$

If d_s is the density of a solid with volume V_1, then

$$W(1 - \sigma/\Delta) = V_1 d_s.$$

Dividing the two equations, we get

$$W/(M_2 - M_1) = d_s/\rho_w. \qquad (7.11)$$

However, d_s/ρ_w is the specific gravity of the solid. In other words the density of the solid is known in terms of the density of water at the temperature of measurement.

Assumptions

Here it should be noted that we have made the following assumptions.

(1) The temperature remains the same in all the measurements. If the temperature changes the volume V of the hydrometer and the density of air and water also changes, so we would need to apply the necessary corrections.

(2) The effects of surface tension and air buoyancy on the stem are negligible.

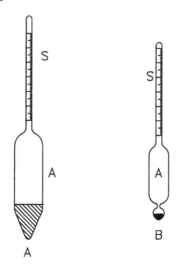

Figure 7.2. Constant-mass hydrometers.

Requirements for constant-volume hydrometer

For the instrument to function well,

- the stem diameter must be small;
- the graduation mark G must be fine;
- the surface of the body and the stem of the hydrometer must be smooth, so that wetting of the hydrometer is proper and no air bubbles remain stuck to it; and
- the hydrometer must always float vertically for which weights should be placed in the pan centrally.

The use of constant-volume hydrometers is quite limited and so these are not frequently found in industry.

7.2.2   Constant-mass hydrometer

Construction

Constant-mass hydrometers are generally used in industry, scientific research and education. They have basically two main parts: (1) the body A sometimes with bulb B; and (2) the stem C (figure 7.2).

(1) The main body is cylindrical in shape and terminates in somewhat conical portions on both ends. All transitions are perfectly smooth and continuous. The main loading occurs either in the lower conical portion or in the bulb B attached to its end, either with mercury or with lead shots with quite

small diameters. Mercury or loose lead shots keep the hydrometer floating vertically and make its centre of gravity low. Shellac or similar material is used to bind the lead shots when the hydrometer is freely floating in the vertical position. The binding material for lead shots must be such that it does not soften below 80 °C.

(2) The stem S is in the form of a thin hollow tube with a perfectly uniform diameter attached to the upper conical portion of the body. A paper with a graduated scale is fixed inside the tube of the stem. A fine reference mark is cut or etched on the stem so that any future displacement of the paper scale is noticeable. The stem and the body are hermetically sealed together. The axis of the scale inside the stem is kept vertical. All graduations should be legible and normal to the axis of the scale. The paper of the scale must be such that it does not warp or discolour with time or due to change in temperature.

7.2.3   Equilibrium equation for the floating hydrometer

Let D be the diameter of the stem and V the volume of the hydrometer up to the graduation mark to which the hydrometer is floating in a liquid of density ρ, while v is the volume of the stem exposed to air. If T is the surface tension of the liquids, then the forces acting on the freely floating hydrometer are as follows.

The downward forces are:

(1) mg, the vertically downward gravitational force due to its mass and
(2) those due to surface tension. Some liquid will rise along the stem and this will have the effect of a downward pull equal to its weight. The weight of the liquid will be equal to the upward force due to surface tension. If the liquid completely wets the stem, i.e. the angle of contact (α) is zero, the force due to surface tension is equal to πDT, the product of the perimeter of the stem and surface tension. Hence another downward force will be πDT.

The upward forces are:

(1) the upthrust of liquid of volume V and density ρ, which is equal to $V\rho g$,
(2) air buoyancy on the portion of the stem above the liquid, which is equal to $v\sigma g$ and
(3) the upthrust on the volume of the liquid raised by surface tension, which is $\pi D\sigma T/\rho$.

Equating the upward and downward forces, we obtain

$$mg + \pi DT = V\rho g + \sigma(v + \pi DT/\rho g)g$$

giving

$$m + \pi DT/g = V\rho + \sigma(v + \pi DT/\rho g). \tag{7.12}$$

Reference temperature and surface tension

Here we see that V, v, ρ, σ and T are all temperature-dependent quantities. Hence, the hydrometer reading ρ will only be correct at a particular temperature. Hence, like volumetric measures, all hydrometers have a reference or standard temperature at which its indications will be correct. This temperature may be different for different types of hydrometer. Furthermore, the hydrometer reading also depends on the surface tension; however, the term containing the surface tension is small in relation to the mass of the hydrometer, so hydrometers are qualified for a range of surface tensions. The hydrometers may belong to any of three categories of surface tension: high, medium and low. Hence, hydrometers are referred to as high, medium and low surface tension hydrometers. Therefore, every hydrometer should bear its reference temperature and its surface tension category.

7.3   Density hydrometers

The scales of these hydrometers indicate the density in terms of g cm^{-3} or kg m^{-3}. The indication is supposed to be correct at a particular temperature, so each hydrometer is marked with its reference temperature, for example 15 °C, 20 °C or 27 °C, and with the word high, medium or low to indicate the surface tension range within which it is supposed to be used. Hydrometers are used to measure the density of liquids ranging from 600 to 2000 kg m^{-3}. The entire range is not normally covered by a single hydrometer. The highest mark of the scale represent the lowest density and the lowest mark represent the highest density, which the hydrometer can read. The difference between these two densities is called the range of the hydrometer.

7.3.1   Classification of density hydrometers

The normal range of standard hydrometers is 20 or 50 kg m^{-3}. However, in commercial density hydrometers ranges of 100 kg m^{-3}, 200 kg m^{-3} or 1000 kg m^{-3} are not uncommon. Density hydrometers are classified according to (a) the width of the scale, i.e. the value of density between two graduation marks, (b) the range and (c) the maximum permissible error [1, 2].

There are five main series namely L20, L50, M50, M100 and S50 and three sub-series—L50SP, M50SP and S50SP. The letters L, M and S stand for the length of the scale of hydrometer, i.e. long, medium or short, and the numbers 20, 50 and 100 indicate the range, i.e. 20, 50 or 100 kg m^{-3} respectively. The letters SP stand for special tolerance, i.e. smaller maximum permissible errors.

7.3.2 Range and value of the density interval between consecutive graduation marks

Depending upon the scale spacing between two consecutive graduation marks, i.e. the total scale length and range, hydrometers are designated as follows:

- L20, longer scale length, range 20 kg m^{-3}, density interval 0.2 kg m^{-3} with 100 graduations,
- L50, longer scale length, range 50 kg m^{-3}, density interval 0.5 kg m^{-3} with 100 graduations,
- M50, medium scale length, range 50 kg m^{-3}, density interval 1.0 kg m^{-3} with 50 graduations,
- M100, medium scale length, range 100 kg m^{-3}, density interval 2.0 kg m^{-3} with 50 graduations,
- S50, short scale length, range 50 kg m^{-3}, density interval 2.0 kg m^{-3} with 25 graduations and
- S50SP, short scale length, range 50 kg m^{-3}, density interval 1.0 kg m^{-3} with 50 graduations.

7.3.3 Relation between the consecutive volumes immersed with equal density intervals

Let V_{n-1}, V_n and V_{n+1} be the volumes immersed with three consecutive graduation marks representing densities of $n-1$, n and $n+1$ times I, I being the density interval. Leaving aside the small buoyancy effect on the exposed portion of the stem [3], the equilibrium equations are:

$$m + \pi DT/g = V_{n-1}\{(n-1)I + \rho_1\} \tag{7.13}$$
$$m + \pi DT/g = V_n\{(nI) + \rho_1\} \tag{7.14}$$
$$m + \pi DT/g = V_{n+1}\{(n+1)I + \rho_1\} \tag{7.15}$$

giving

$$1/V_{n-1} = \{(n-1)I + \rho_1\}/(m + \pi DT/g) \tag{7.16}$$
$$1/V_n = \{nI + \rho_1\}/(m + \pi DT/g) \tag{7.17}$$
$$1/V_{n+1} = \{(n+1)I + \rho_1\}/(m + \pi DT/g). \tag{7.18}$$

Subtracting one from the next in pairs we get

$$1/V_n - 1/V_{n-1} = 1/V_{n+1} - 1/V_n \tag{7.19}$$

showing thereby that V_{n-1}, V_n and V_{n+1} are in harmonic progression. Equal density intervals do not correspond to equal volume intervals. From this also

$$V_{n-1} - V_n = V_{n-1}(V_n - V_{n+1})/V_{n+1}. \tag{7.20}$$

As V_{n-1} is bigger than V_{n+1}, we see that

$$V_{n-1} - V_n > V_n - V_{n+1}. \tag{7.21}$$

This shows that the graduation marks corresponding to equal intervals of density at the lower end of the scale are closer to one another.

7.3.4   Volume of the stem in terms of bulb volume

Let ρ_1, ρ_2 be the densities corresponding to the upper- and lowermost graduation marks respectively, v the volume of the stem between these two marks and V the volume of the bulb below the lowest graduation mark. Then

$$M' = \rho_1(V + v) \tag{7.22}$$

where $M' = m + \pi DT/g$

$$M' = \rho_2 V. \tag{7.23}$$

Subtracting we get

$$v = V(\rho_2 - \rho_1)/\rho_1 \tag{7.24}$$
$$v = VR/\rho_1 \tag{7.25}$$

where R is the range of the hydrometer.

For hydrometers with equal ranges, i.e. equal R, the stem volume is inversely proportional to the density corresponding to the highest graduation mark. That is, hydrometers with the same range but starting from a higher density will have smaller stem volumes. The diameter of the stem for higher density hydrometers, which have the same bulb volume and scale length, will be smaller. For example, for a hydrometer measuring between 600 and 650 kg m^{-3}, v is given by

$$v = (50/600)V = (1/12)V \tag{7.26}$$

while for a similar hydrometer which starts from 1950 kg m^{-3}, v is given by

$$v = (50/1950)V = (1/35)V. \tag{7.27}$$

Hence, the volume of the stem of the highest density hydrometer is almost one-third that with the lowest density. In other words, if we wish to have the same stem volume, the hydrometer with the higher density will have a bulb volume almost three times that of a hydrometer with a smaller density. Furthermore, we can see that the volume of the stem is smaller for a smaller range R. The ratio of the volume of the body to that of the stem is large if the range is made small. However, if V/ρ_1 is kept constant then v, the volume of the stem, will also remain unchanged for different hydrometers. In order to keep the diameter of the stem constant, V/ρ_1 should be kept constant.

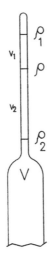

Figure 7.3. Scale graduation marks.

7.3.5 Scale graduation marks

The exact spacings of the graduation marks on a density hydrometer are calculated as follows: Referring to figure 7.3, let ρ be the density corresponding to any intermediate mark, while ρ_1 and ρ_2, respectively, correspond to the uppermost and lowermost graduation marks of the hydrometer.

If v_1 is the volume of the stem between ρ and ρ_1, and v_2 is the volume of the stem between ρ and ρ_2, then

$$M' = V\rho_2 \tag{7.28}$$
$$M' = (V + v_2)\rho \tag{7.29}$$
$$M' = (V + v_2 + v_1)\rho_1. \tag{7.30}$$

From (7.28) and (7.30)
$$(V + v_2 + v_1)\rho_1 = V\rho_2 \tag{7.31}$$

giving
$$V(\rho_2 - \rho_1) = (v_1 + v_2)\rho_1. \tag{7.32}$$

From (7.28) and (7.29) $V\rho_2 = (V + v_2)\rho$, giving
$$V(\rho_2 - \rho) = v_2\rho. \tag{7.33}$$

Dividing (7.33) by (7.32), we get
$$v_2/(v_2 + v_1) = \rho_1(\rho_2 - \rho)/(\rho_2 - \rho_1)\rho. \tag{7.34}$$

Similarly,
$$v_1/(v_2 + v_1) = \rho_2(\rho - \rho_1)/(\rho_2 - \rho_1)\rho.$$

If v is the volume of the stem then

$$v_1/v = [\rho_2/(\rho_2 - \rho_1)][1 - \rho_1/\rho] \tag{7.35}$$
$$v_2/v = [\rho_1/(\rho_2 - \rho_1)][\rho_2/\rho - 1]. \tag{7.36}$$

If the stem has a uniform diameter throughout its scale, v_1 may be replaced by l_ρ and v by L. Here l_ρ is the distance between the graduation marks of density ρ and ρ_1 and L is the total length of the scale:

$$l_\rho = L\{\rho_2/(\rho_2 - \rho_1)\}[(\rho - \rho_1)/\rho] \tag{7.37}$$

or, in words, L (the highest density on the scale divided by the range) multiplied by (the difference in densities from the lowest density divided by the density corresponding to the mark).

7.3.6   Distance of graduations from the top graduation

Let

$$\rho_2/(\rho_2 - \rho_1) = n \tag{7.38}$$

then

$$\rho_1/(\rho_2 - \rho_1) = -[(\rho_2 - \rho_1 - \rho_2)/(\rho_2 - \rho_1)] = -[1 - \rho_2/(\rho_2 - \rho_1)]$$

giving

$$\rho_1/(\rho_2 - \rho_1) = n - 1. \tag{7.39}$$

In addition, if there are p graduations, and I is the value of the density interval then

$$pI = (\rho_2 - \rho_1) = \rho_1/(n - 1)$$

giving

$$\rho_1 = pI(n - 1). \tag{7.40}$$

Furthermore, if ρ is the density represented by the rth graduation, then

$$\rho - \rho_1 = rI$$

giving

$$\rho = \rho_1 + rI \tag{7.41}$$
$$= pI(n - 1) + rI$$
$$= \{p(n - 1) + r\}I. \tag{7.42}$$

Substituting the values

$$\rho_2/(\rho_2 - \rho_1) = n$$
$$(\rho - \rho_1) = rI$$

and

$$\rho = \{p(n-1)+r\}I$$

in (7.37), we obtain

$$l_r = nLr/[p(n-1)+r]. \tag{7.43}$$

Here l_r is the distance of the rth graduation from the top graduation and r takes values $1, 2, \ldots, p$.

Case I. L20 hydrometers

For L20 hydrometers with range 20 kg m^{-3} and total scale length 100 mm, the formula for the distances of the p graduation marks becomes

$$l_r = 100nr/[5(n-1)+r] \qquad \text{for } p = 5$$

where r can take values 1–5 and n will take a value 31 for a hydrometer with a range from 600 to 620 kg m^{-3} and will increase by one for each subsequent hydrometer. n will be 100 for the hydrometer from 1980 to 2000 kg m^{-3}. The total number of graduation marks of equal density intervals will be five.

$$l_r = 100nr/[10(n-1)+r] \qquad \text{for } p = 10$$

where r takes values 1–10 and n will take a value 31 for a hydrometer from 600 to 620 kg m^{-3} and will increase by one for each subsequent hydrometer. n will be 100 for the hydrometer from 1980 to 2000 kg m^{-3}. The total number of graduation marks of equal density intervals will be 10.

$$l_r = 100nr/[20(n-1)+r] \qquad \text{for } p = 20$$

where r takes values 1–20 and n will take a value 31 for a hydrometer from 600 to 620 kg m^{-3} and will increase by one for each subsequent hydrometer. n will be 100 for the hydrometer from 1980 to 2000 kg m^{-3}. The total number of graduation marks of equal density intervals will be 20.

Similarly we can take p equal to 50 or 100 for calculating the relative positions of the 50 or 100 graduation marks.

Case II. L50, M50 or S50 hydrometers

For L50, M50 hydrometers, each hydrometer with range 50 kg m^{-3} and a total scale length of 100 mm, the formula for graduation distances from the top of the graduation becomes

$$l_r = n100r/[5(n-1)+r] \qquad \text{for } p = 5$$

where r takes values 1–5 and n will take a value 13 for a hydrometer from 600 to 650 kg m^{-3} and will increase by one for each subsequent hydrometer. n will be

40 for the hydrometer from 1950 to 2000 kg m^{-3}. The total number of graduation marks of equal density intervals will be five.

$$l_r = nr/[10(n-1)+r]100 \qquad \text{for } p = 10$$

where r takes values 1–10 and n will take a value 13 for a hydrometer from 600 to 650 kg m^{-3} and will increase by one for each subsequent hydrometer. n will be 40 for the hydrometer from 1950 to 2000 kg m^{-3}. The total number of graduation marks of equal density intervals will be 10.

$$l_r = nr/[50(n-1)+r]100 \qquad \text{for } p = 50$$

where r takes values 1–50 and n will take a value 13 for a hydrometer from 600 to 650 kg m^{-3} and will increase by one for each subsequent hydrometer. n will be 40 for the hydrometer from 1950 to 2000 kg m^{-3}. The total number of graduation marks of equal density intervals will be 50.

Similarly we can take p equal to 100 to give the relative positions of the 100 graduation marks.

Case III. M100 hydrometers

For M100 hydrometers with range 100 kg m^{-3} and a total scale length of 100 mm, the formula for the distances of p graduation marks becomes

$$l_r = 100nr/[100(n-1)+r] \qquad \text{for } p = 100$$

where r takes values 1–100 and n will take a value 7 for a hydrometer from 600 to 700 kg m^{-3} and will increase by one for each subsequent hydrometer. n will be 20 for the hydrometer from 1900 to 2000 kg m^{-3}. The total number of graduation marks of equal density intervals will be 100.

These formulae have been discussed in detail for the benefit of manufacturers of hydrometers and to emphasize that distances for the same scale range are not the same for all ranges of hydrometers; for example, for a 100 mm scale length ($L = 100$), the distances for two L50 hydrometers with ranges 600–650 and 1950–2000 kg m^{-3} are given in table 7.1.

7.3.7 Surface tension category of density hydrometers

From the point of view of surface tension, all density hydrometers are divided into three categories, namely low, medium and high. Density hydrometers with low surface tension are calibrated for 15 mN m^{-1} at 600 kg m^{-3}. The surface tension increases by 1 mN m^{-1} for every change of 20 kg m^{-3} until the density of 980 kg m^{-3} is reached. That is, the surface tension for a low surface tension category hydrometer at 980 kg m^{-3} is 34 mN m^{-1}; for hydrometers with a density above 980 kg m^{-3} to 1300 kg m^{-3}, the surface tension will be 35 mN m^{-1}.

Table 7.1. Distances of equal density intervals from the top for two extreme hydrometers in the L50 series.

Mark	Range 600–650 kg m^{-3} (mm)	Mark	Range 1950–2000 kg m^{-3} (mm)
610	21.3	1960	20.4
620	41.9	1970	40.6
630	61.9	1980	60.6
640	81.2	1990	80.4
650	100	2000	100

For density hydrometers in the medium surface tension category, up to a density of 940 kg m^{-3}, the value of the reference surface tension is the same as that for the low surface tension category of hydrometers. The values of the reference surface tension for these hydrometers are, respectively, taken as 35, 40, 45, 50 mN m^{-1} for 960, 970, 980 and 990 kg m^{-3} and it is uniformly 55 mN m^{-1} for hydrometers with a density between 1000 and 2000 kg m^{-3}.

High surface tension hydrometers range from 1000 to 2000 kg m^{-3}. The surface tension is uniformly 75 mN m^{-1}. In fact the categorization has been made according to the surface tension of the liquids for which the hydrometers are normally going to be used.

7.4   Specific gravity

The specific gravity is the ratio of the masses of equal volumes of the substance and water. In other words it is the ratio of the density of a substance to that of water. As the volume or density of water as well as that of any substance depends upon temperature, the definition of specific gravity contains two temperatures, i.e. the temperature of the substance and that of the water. These two temperatures may also either be the same or different. So S, the symbol for specific gravity, is accompanied by two reference temperatures and is written as St_1/t_2 or the mass of the substance with a certain volume at t_1 °C divided by the mass of water with a volume equal to that of the substance at t_2 °C.

If $t_1 = t_2 = t$, then St/t is the ratio of the mass of the substance of a certain volume at temperature t °C to the mass of water with the same volume at t °C. Here t_1 and t_2 or t are called the standard or reference temperatures.

According to an earlier concept, the density of water was taken to be 1 g cm^{-3} at 4 °C. So when t_2, the temperature of water taken, is 4 °C, the specific gravity of a substance would be numerically equal to its density at t °C. However, at the moment, the density of water is at a maximum not at 4 °C but at 3.983 035 °C and its value is not 1 g cm^{-3} but 0.999 975 g cm^{-3}. In order to convert $St/4\,^\circ$C to density, it should be multiplied by 0.999 975 to give the density in g cm^{-3}.

The use of the term 'specific gravity' is on the decline. The term must have originated from the use of Archimedes' Principle for measuring density, because in this case the loss in weight of a body divided by its volume gives the specific gravity unless we divide it by the actual density of water at the temperature of measurement. This is also the case when a specific gravity bottle or pyknometer is used to measure the density of a liquid. In this case also the ratio of the mass of the liquid to that of water required to fill the bottle or pyknometer gives the specific gravity and not the density. Another reason might have been the use of a large number of different units for the mass and volume in those days, all of which were not well defined and which gave different numerical values for the density of the same substance. However, the specific gravity is independent of the system of units, giving a unique value for a given substance, hence it was better suited for measurements at that time.

7.5   Specific gravity hydrometers

The specific gravity hydrometer is very similar to a density hydrometer. The only difference is that the indications for specific gravity are pure numbers unlike those in density hydrometers for which the indications are in g cm^{-3} or kg m^{-3}. A specific gravity hydrometer will bear two reference temperatures in the form t_1/t_2: t_1 is the reference temperature of liquid and t_2 the reference temperature for water. Therefore, in addition to surface tension, two reference temperatures should also be defined for a specific gravity hydrometer. As in the guidelines from the International Union of Pure and Applied Chemistry (IUPAC—Compendium of Analytical Nomenclature—Definitions and Rules 1997), the term 'relative density' is preferred in place of specific gravity.

7.6   Different types of scale [4]

7.6.1   Baume scale

Baume developed two arbitrary scales. The first scale is for liquids lighter than water, the point at which the hydrometer floats in water is assigned 10 degrees Baume and the point of 15% salt solution corresponds to 0 degrees Baume. The distance between these two points is equally divided and the scale continues beyond 10 degrees by similarly spaced graduation marks. The second scale is for liquids heavier than water. In this scale '0 degree' is assigned to pure water and 10 degrees to 15% salt solution and the scale is similarly extrapolated.

The main drawback is that one Baume hydrometer can indicate 0 degrees in water while another will indicate 10 degrees. The relationship between density and Baume degree is not unique—I have found five different relations in one reference. However, Baume scale hydrometers were extensively used in India in the latter part of the 19th century. The most commonly used scale is:

For liquids denser than water

$$\text{Degree Baume} = 145 - 145/S60/60\,^{\circ}\text{F}. \tag{7.44}$$

(Some people have used 144 or 144.3 in place of 145.)

For liquids lighter than water

$$\text{Degree Baume} = 140/S60/60\,^{\circ}\text{F} - 130. \tag{7.45}$$

However, these figures are not unique as stated earlier; there is a good amount of multiplicity in these figures; for example, the American Petroleum Institute (API) scale for liquids lighter than water differs only slightly from the Baume scale.

7.6.2 American Petroleum Institute (API) degree

The American Petroleum Institute introduced the degree API, which is similar to the Baume degree and is defined as

$$\text{Degree API} = (141.5/S60/60\,^{\circ}\text{F}) - 131.5. \tag{7.46}$$

7.6.3 Twaddle scale

The Twaddle scale is also essentially based on the specific gravity at $60/60\,^{\circ}\text{F}$ in which a zero degree Twaddle corresponds to a specific gravity of 1 and 1 degree Twaddle is equivalent to 0.005 sp.gr. Therefore Twaddle degree and specific gravity $S60/60\,^{\circ}\text{F}$ are related as follows:

$$T \text{ degree Twaddle} = 200(S60/60\,^{\circ}\text{F} - 1.000)$$

or

$$1.000 + 0.005T = S60/60\,^{\circ}\text{F}. \tag{7.47}$$

7.6.4 Sikes scale

Sikes hydrometers are commonly used to measure the strength of spirits. The scale was first introduced to levy excise duty on the production of alcohol. Originally Sikes hydrometers were made of gilded brass and were replicas of the original hydrometer, made by Sikes at the beginning of the 19th century. It has an arbitrary scale numbered from 0 to 10, sub-divided into five parts. Associated with this scale are nine weights known specifically as 'poises'. These are marked with the numbers 10, 20, 30, ..., 90. In order to load the hydrometer, one poise is hung at a time on a stem below the bulb. The zero of the Sikes scale corresponds to 70% over proof of an alcohol–water mixture, while point 100 of the Sikes scale corresponds to pure water. In pure water the Sikes hydrometer loaded with the ninth poise (number 90) below the bulb corresponds to a scale reading of 10. Later on, to read higher percentages of alcohol in water mixtures, two more scales were introduced. These two scales have slightly different nomenclatures in Britain and India. In British practice these are known as A10 and B10: A10 equals 0 Sikes and B10 equals A0. In India, we have simply extended the scale by 20 points and called these two scales A10 and A20. Thus, B0 on the British scale is equivalent to A0 on the Indian scale and B10, equalling A0 on the British scale, is equivalent to A10 on the Indian scale. A10 on the British scale is equal to 0 Sikes, which is equivalent to A20 on the Indian scale.

The basis of the Sikes scale was purely arbitrary. The scale depended solely upon the original gilded hydrometer. Unfortunately because of the higher density and softness of the gold, it was difficult to maintain the original gilded hydrometer. Inevitably the original instrument changed slowly with a consequent uncertainty in the true basis. The scale was later defined in terms of specific gravity (60 °F) of the solutions, in which the metal hydrometer had specific indications under certain conditions of surface tension and wetness of stem. These conditions were originally derived empirically but were later adopted as conventions. The scales of glass Sikes hydrometers were originally derived by comparison with the standard metal instrument at 51 °F. However, their use is declining because of their ill defined scale.

7.6.5 Alcoholometers

In the International Recommendation [5], there are three types of alcoholometer:

(1) glass hydrometers indicating the percentage of alcoholic strength by mass, normally referred to as mass alcoholometers,
(2) glass hydrometers indicating the percentage of alcoholic strength by volume, normally referred to as volume alcoholometers and
(3) glass hydrometers indicating the density in kilogram per cubic metre of alcohol–water mixtures called alcohol hydrometers.

It should be noted that the numbering on the scale in alcoholometers whether by volume or mass will be the reverse of those of alcohol hydrometers. The scale numbering in alcoholometers will increase from bottom to top, while in alcohol hydrometers it will increase from top to bottom, as is the case in density hydrometers. In the OIML Recommendation [5] the definition of the words 'alcoholic strength' is rather vague. However, the Indian Standard specification [6] defines alcohol by its chemical name of ethanol. Moreover, there is only one category of scale, which defines alcoholic strength by volume. However, the scale is based on the density of ethanol–water mixtures as given in OIML Recommendation—IR 22—and IS: 3506 1989—Tables for Alcoholometry (By the Pyknometer Method) [7]. The reference temperature for these instruments is 20 °C.

7.6.6   Brix degree

The Brix degree is the percentage by mass of sucrose in a pure sucrose solution at 20 °C. Zero degree Brix corresponds to pure water [8] at a reference temperature of 20 °C. The specific gravity ($S20/4$ °C) of a sucrose solution has been determined by various authors but the values of specific gravity accepted by the Fourth International Congress of Chemistry, Paris, are generally used [9]. I have used these values and by employing a least-squares method derived a fourth-degree polynomial in the percentage of sucrose, i.e. degrees Brix. The polynomial is:

$$S20/4\,^{\circ}C = A_0 + A_1 B + A_2 B^2 + A_3 B^3 + A_4 B^4$$

where A_0, A_1, A_2, A_3 and A_4 have the following values:

$$A_0 = 0.998\,234$$
$$A_1 = 3.859\,969\,911 \times 10^{-3}$$
$$A_2 = 1.252\,761\,461 \times 10^{-5}$$
$$A_3 = 6.998\,000\,385 \times 10^{-8}$$
$$A_4 = -\,2.766\,519\,146 \times 10^{-10}.$$

The density of the sugar solution also decreases with a rise in temperature but the decrease is not constant and it depends upon the percentage of sugar present in the solution. Jose and Remy [10] have experimentally determined the cubic expansion coefficients of sucrose solutions which have sucrose ranging from 6%wt to 67%wt. Schonrock [11] has given a more complete formula for the cubic expansion coefficient of sucrose solutions involving temperature and degree Brix B, the percentage of sugar:

$$\gamma = 0.000\,291 + 0.000\,0037(B - 23.7) + 0.000\,0066(t\,^{\circ}C^{-1} - 20)$$
$$-\,0.000\,000\,19(B - 23.7)(t\,^{\circ}C - 20).$$

Therefore the specific gravity $S20/4$ °C of a sucrose solution at any temperature t is given by

$$St/4\,^{\circ}C = S20/4\,^{\circ}C[1 - \gamma(t\,^{\circ}C - 20)]$$

where S is the specific gravity at $20/4$ °C, B is the percentage by weight of sucrose and t is the measurement temperature in °C. To change the temperature correction given in g cm^{-3} to a percentage of sucrose, we have to divide it by the density difference for consecutive percentages by weight of sucrose.

7.7   Calibration of a hydrometer by the sinker method

One of the oldest methods of calibrating hydrometers is by simultaneous observation of the reading of the hydrometer and a determination of the density of the liquid in which the hydrometer is floating. The liquid should have the appropriate surface tension and density. A sinker of known mass and volume is hung from the pan of the balance. Its apparent weight in the liquid gives the density of the liquid. All measurements are carried out at the reference temperature of the hydrometer [3].

7.7.1 Experimental set-up

The hydrometer under calibration is floated in a liquid contained in a rectangular jar A. The front and back walls of the jar are parallel to each other. Made from plane optical glass, they are polished so that the hydrometer scale does not appear to be distorted. The jar is placed under a table on which a balance is placed as shown in figure 7.4. A rod C, passing through the hole in the base of the balance and the tabletop, is attached to the pan of the balance and has a hook on its lower end. A sinker D of known volume and mass is suspended through a fine platinum wire, which is attached to the hook. A thermometer T is also suspended in the liquid through a fixed hook at the bottom of the tabletop. The thermometer is suspended so that its bulb is equally distant from the sinker and hydrometer. The jar should be big enough to allow the liquid to be stirred vigorously.

There are two ways of maintaining the reference temperature. The first is to maintain the contents of the jar exactly at the reference temperature during all measurements. An alternative method is for the temperature of the jar to be raised or lowered in such a way that the reference temperature lies between the temperatures of the room and jar. In this way, while the temperature of the liquid gradually approaches room temperature, it passes through the reference temperature. Then one observer takes the reading of the hydrometer and thermometer and watches that nothing touches the hydrometer or sinker while another observer takes the reading of the balance for the apparent mass of the sinker. The liquid is well stirred and allowed to settle down before each set of observations. Several such readings are taken at temperatures above and below the standard temperature. Graphs are drawn from the readings of the hydrometer and that of the apparent mass of the sinker against temperature. From these graphs, the readings for the hydrometer and the apparent mass of the sinker at reference temperature are determined. From the apparent mass of the sinker and its known volume, the density of the liquid can be determined. The density of liquid minus the corrected reading of the hydrometer is the correction at the point on the scale of the hydrometer. The level of liquid in the jar is kept fixed so that the length of the platinum wire under the liquid remains the same for all graduation settings.

7.7.2 Pyknometer method

Instead of using a sinker to determine the density of the liquid, a pyknometer may be used. In this method, the hydrometer to be calibrated is floated in a liquid of appropriate density so that the liquid surface level intersects the graduation mark under test. The jar containing the liquid, hydrometer, thermometer, stirrer

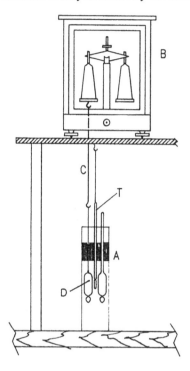

Figure 7.4. Experimental set-up for calibrating hydrometers by the sinker method.

etc is maintained at the reference temperature. Filling an appropriate pyknometer of known volume and finding the mass of the liquid contained at the standard temperature gives the density of the liquid. From a knowledge of the mass and volume of the liquid contained in the pyknometer, the density of the liquid can be calculated. The calculated value of the density minus the corrected indication of the hydrometer under test gives the correction to be applied at the point under examination. Sources of error are:

- the position of the stopper does not always reach exactly the same position, i.e. the volumes of the water and liquid are not same;
- the differences in the heights of the liquid and water in the capillary of the stopper due to surface tension means the volumes are not always equal;
- unequal temperatures inside and outside the specific gravity bottle or pyknometer; and
- the non-uniformity of the temperature inside the specific gravity bottle or pyknometer.

Most of these problems have been tackled beautifully by Masui *et al* [12] who was able to measure the difference in density of two water samples with a standard uncertainty of 0.2 ppm.

7.7.3   Density with the help of two pyknometers

If a pyknometer is weighed in water instead of air, the errors due to the first two causes are considerably reduced. If the pyknometer is underfilled by an amount v cm^3, the error will be equal to $v(\rho - \sigma)$ if weighed in air, but if the pyknometer is weighed in water, then the error due to same underfilling will be $v(\rho - d_w)$, which is much less. Because $(\rho - d_w)$ is much smaller than $(\rho - \sigma)$, the errors will be further reduced if two pyknometers of almost the same volume and material are used. This method has been successfully used for detecting very small differences in the density of water samples. The difference in density of two water samples was due to dissolved air. The procedure for this is given in the next section.

7.7.4   Procedure for comparing two samples of nearly equal density

(1) Two pyknometers are filled with the same sample and are weighed in water of density d_w and their mass difference Δm_0 is determined.
(2) The two pyknometers are filled with two different samples of density ρ_1 and ρ_2 and the mass difference Δm_1 is determined.

Let the volumes of two pyknometers be V_1 and V_2 and their respective masses m_1 and m_2, the density of their material d_g and their temperatures T_1 and T_2 at the time of the two measurements, then

$$M_{10} = m_1(1 - d_w/d_g) + V_1(\rho - d_w) \tag{7.48}$$
$$M_{20} = m_2(1 - d_w/d_g) + V_2(\rho - d_w) \tag{7.49}$$

where M_{10} and M_{20} are the effective masses of the two pyknometers when weighted in water. Subtracting, we get

$$\Delta m_0 = (m_1 - m_2)(1 - d_w/d_g) + (V_1 - V_2)(\rho - d_w). \tag{7.50}$$

Similarly, when filled with different samples at temperature T_2, we get

$$M'_{10} = m_1(1 - d_{w1}/d_g) + V_1(\rho_1 - d_{w1}) \tag{7.51}$$
$$M'_{20} = m_2(1 - d_{w1}/d_g) + V_2(\rho_2 - d_{w1}). \tag{7.52}$$

Subtracting (7.52) from (7.51) and re-arranging, we get

$$\Delta m_1 = (m_1 - m_2)(1 - d_{w1}/d_g) + V_1(\rho_1 - \rho_2) + (V_1 - V_2)(\rho_2 - d_{w1}). \tag{7.53}$$

Subtracting (7.50) from (7.53), we get

$$\Delta m_1 - \Delta m_0 = (m_1 - m_2)(d_w - d_{w1})/d_g + V_1(\rho_1 - \rho_2) \\ + (V_1 - V_2)(d_w - d_{w1} + \rho_2 - \rho).$$

But $d_w - d_{w1} = \alpha(T_1 - T_2)d_w$, giving

$$(\rho_1 - \rho_2) = (\Delta m_1 - \Delta m_0)/V_1 - \alpha(T_1 - T_2)d_w(m_1 - m_2)/V_1 d_g \\ + (V_1 - V_2)\{d_w\alpha(T_1 - T_2) + \rho_2 - \rho\}/V_1. \tag{7.54}$$

The second term is small as $\alpha(T_1 - T_2)$ is small. While in practice the two chosen pyknometers are such that $V_1 - V_2$ is small the third term also becomes very small, giving us

$$(\rho_1 - \rho_2) = (\Delta m_1 - \Delta m_0)/V_1. \tag{7.55}$$

7.8 Calibrating hydrometers by Cuckow's method

The method given in section 7.7.1, though very accurate and reliable, has the drawback of requiring a large number of liquids, as each point of the scale to be calibrated needs a separate liquid with the appropriate density. If all the L20 hydrometers starting from 600 to 2000 kg m^{-3} are calibrated, the number of liquids required will be several hundred.

A method in which only one liquid covers the whole calibration range from 600 to 2000 kg m^{-3} was developed initially at the NPL, UK [13] and suitably adapted by us in India. My colleagues, at the NPL, India, developed an open-scale hydrometer with a very small range but with a scale interval of 0.01 kg m^{-3} and also designed and fabricated a thermostatic bath. The bath was examined for constancy and uniformity of temperature, throughout the working space. The hydrometer is calibrated by submerging it to the required level and finding its apparent mass in the liquid. The density of the liquid is simultaneously determined with the help of the open-scale hydrometer.

Using this method, a hydrometer of any range can be calibrated by finding the apparent mass of the hydrometer when it is immersed to a desired scale mark (graduation) in a single liquid of known density. The density of the liquid is measured either by hydrostatic weighing of a body of known volume or by an open-scale hydrometer.

7.8.1 Theory

Let m be the mass of the hydrometer, V its volume submerged in the liquid when made to float in the liquid of density d, v the volume of stem above the liquid and exposed to air of density σ. The subscripts will indicate the values at different times.

When weighed in air, against standards of mass W_1 and of density Δ, the equilibrium equation is

$$W_1(1 - \sigma/\Delta) = m - (V + v)\sigma. \tag{7.56}$$

When freely floating in a liquid of density ρ,

$$V\rho + v\sigma = m + \pi D T_1/g \tag{7.57}$$

where T_1 is the surface tension of the liquid. If the hydrometer is again made to float up to the desired graduation mark, so that V remains unchanged in a liquid of density d and surface tension T_2, its apparent mass is W_2, then

$$Vd + v\sigma + W_2(1 - \sigma/\Delta) = m + \pi DT_2/g. \tag{7.58}$$

Here we have assumed that the corrections due to a change in the air density and volume of hydrometer are negligible, so that the air density, V and v have been taken to be the same in all equations. Subtracting (7.57) from (7.56)

$$W_1(1 - \sigma/\Delta) + \pi DT_1/g = V(\rho - \sigma). \tag{7.59}$$

Subtracting (7.58) from (7.56)

$$(W_1 - W_2)(1 - \sigma/\Delta) + \pi DT_2/g = V(d - \sigma). \tag{7.60}$$

Dividing (7.59) by (7.60),

$$\rho = (d - \sigma)[\{W_1(1 - \sigma/\Delta) + \pi DT_1/g\}/\{(W_1 - W_2)(1 - \sigma/\Delta) + \pi DT_2/g\}] + \sigma. \tag{7.61}$$

7.8.2   Open-scale hydrometer

All weighings are carried out at around the standard temperature, so any change in temperature is small and hence any change in the density of the hydrostatic liquid is very small. Therefore a hydrometer with an open scale in which one graduation corresponds to a density of 0.01 kg m^{-3} is used. The bulb of the hydrometer has a diameter of 45 mm and length 350 mm; its stem diameter is 3 mm with a scale length of 30 mm. There are only 30 graduations on the scale. The scale runs from 860.90 to 861.20 kg m^{-3}. The hydrometer floats in xylene at the 15 mm graduation at 20 °C. The hydrometer is calibrated against a solid body of known volume by hydrostatic weighing. The temperature range is very small. The hydrometer with its scale graduations is shown in figure 7.5.

Figure 7.5. Open-scale hydrometer (1 graduation $= 0.01$ kg m^{-3}).

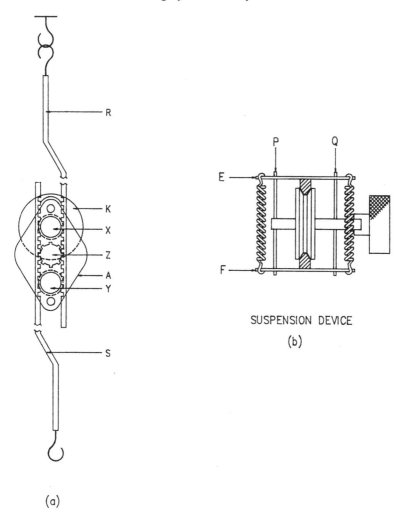

Figure 7.6. Suspension and adjustment device for hydrometer.

7.8.3   Suspension device

The suspension device consists of two brass rods R and S each with rakes cut at opposing faces which are engaged by a common pinion. The system is held together with the help of two plates, together with some spacers and springs. The arrangement is shown in figure 7.6. The pinion is moved through a knob, which moves the rods R and S in opposite directions. Each rod has a small hook, so that the hydrometer may be hung from the lower hook through a special clamp and upper hook may be used to suspend the system from the pan of the balance.

Figure 7.7. Special clamp for hydrometer.

7.8.4   Special clamp for hydrometer

The special clamp consists of a V-shaped strip attached to another light plate carrying a small eye with the help of a set of spring-loaded screws. The clamp is shown in figure 7.7. The portions of the strip and the plate, which come in contact with the stem of the hydrometer are properly padded.

7.8.5   Procedure for calibration

Apparent mass of hydrometer in air

The apparent mass W_1 of the hydrometer may be determined separately by weighing it in a balance with the appropriate capacity. However, the length of the hydrometer is too large to be accommodated in any balance so the doors of the balance are kept open. Therefore it is better for the hydrometer to be weighed while suspended in air with the clamp and adjustment device in figure 7.8. In order to use the substitution method, sufficient weights are kept on the pan so that when the hydrometer is suspended from the pan in air, weights equivalent to the mass of the hydrometer are removed from the same pan. The double substitution method is used and the mean of three such weighings taken. Let W_1 be the mean of the mass of the standard weights of density Δ, which balances the hydrometer in air of density σ.

The complete set-up with bath, clamping device and balance as used at the NPL, New Delhi, is shown in photographs P7.1 and P7.2. All components of the system shown in photograph P7.1 were indigenously fabricated at the NPL. The density of the xylene is determined with the help of a specially designed open-scale hydrometer. This special hydrometer is calibrated against a Pyrex glass pre-calibrated sinker for its volume. The system shown in photograph P7.2 was imported from Germany and here the density of the liquid used is measured separately by finding the upthrust of the liquid on a solid (sphere) of known volume. To calculate the density of air, the temperature, pressure and humidity

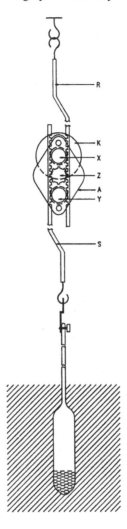

Figure 7.8. Clamp and adjustment device.

are measured simultaneously. A sample set of observations is given in table 7.2. The mean value W_1 of the mass of weights of density Δ balancing the hydrometer in air of density σ is 106.641 25 g.

Apparent mass of hydrometer in reference liquid when immersed to different levels

The suspension device along with the hydrometer clamp is suspended through the hook from the balance pan. A fixed weight is placed in the other pan of the balance. The value of the weight is such that even if the upthrust on the

Photograph P7.1. Calibration of hydrometers by Cuckow's method (courtesy NPL, India).

hydrometer is less than the mass of the hydrometer in air even then some extra weights are required to balance it. In the case of a single-pan balance, the same effect is obtained by lifting the maximum capacity weight. The standard weights are placed on the pan from which the hydrometer is suspended so that equilibrium is reached in the middle of the scale and the hydrometer is substituted by the standard weight of known mass and density Δ.

The hydrometer under calibration is clamped to the suspension device and immersed in a thermostat containing xylene at the standard temperature of the hydrometer. The open-scale hydrometer also floats simultaneously. Sufficient weights are placed so that the hydrometer dips further when the balance is released. The vertical position of the hydrometer is adjusted by turning the knob

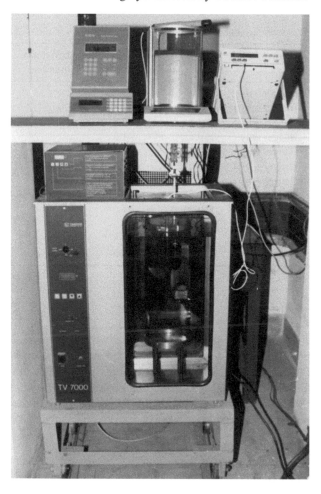

Photograph P7.2. Calibration of hydrometer by hydrostatic weighing (courtesy NPL, India).

of the suspension device so that the graduation mark of the hydrometer under calibration intersects the liquid surface level of the xylene. This adjustment can be easily done when the balance is in the arrested position. For a single-pan balance the vertical motion is quite small so it can be ignored. Alternatively, the vertical motion of the pan can be assessed through a microscope in terms of length of the scale graduation and appropriate allowance is made at the time of initial adjustment so in the case of a two-pan balance it has to be pre-estimated. The weights are so adjusted that the rest points are practically the same when the weighings are carried out with and without the hydrometer. The apparent weight W_2 of the hydrometer immersed up to the graduation mark is the difference in the

Table 7.2. Observations and calculations of the apparent mass of the hydrometer in air.

Dial setting (g)	Below the pan	Std weights on the pan (g)	Indication (mg)	Difference in mass	Corrected mass
180	Hanger only	150.25	41.6		
	Hanger + hyd.	43.6	32.8	106.6412	106.641 172
	Hanger + hyd.	43.6	32.8	106.6412	
	Hanger only	150.25	41.6		
180	Hanger only	150.25	41.5		
	Hanger + hyd.	43.6	32.8	106.6413	106.641 272
	Hanger + hyd.	43.6	32.8	106.6413	
	Hanger only	150.25	41.5		
180	Hanger only	150.25	41.6		
	Hanger + hyd.	43.6	32.8	106.6412	106.641 172
	Hanger + hyd.	43.6	32.8	106.6412	
	Hanger only	150.25	41.6		

mass of weights placed on the pan without and with the hydrometer suspended from the pan. After every determination, the xylene pump is operated for 30 s; a period of 1 min is given for the liquid to settle down. The observations are repeated after disturbing and readjusting the position of the hydrometer so that the graduation mark again intersects the liquid surface level of the xylene. The mean of three such observations is taken at a given graduation mark.

The hydrometer is calibrated at other graduation points by adjusting its position, with the help of the suspension device, to that graduation point and finding its apparent mass. D, the diameter of the stem, is measured at every graduation mark under test. The surface tension, T_2, of xylene is either measured *in situ* or taken from the literature. Therefore, the numerical value of ρ can be calculated by observing the density of xylene from the open-scale hydrometer. A typical set of observations for a L50 series hydrometer range 1.000–1.050 g cm^{-3} is given in table 7.3. The calculations are based upon the following values: $T_1 = 35$ mN m^{-1}, $T_2 = 28$ mN m^{-1}, $g = 9.806\,55$ m s^{-2}, $D = 0.737$ cm and $\sigma = 0.001\,1685$ g cm^{-3} and are given in table 7.4.

Hydrometers which do not sink in xylene, i.e. with a maximum density value less than that of xylene, are calibrated with the help of a sinker sufficient in mass to overcome the upthrust of xylene. The difference between the apparent weights when only the sinker is suspended and when the sinker with the hydrometer is suspended gives the apparent weight W_2 of the hydrometer, which will be negative in this case. Liquids like nonane, which have a similar density range but slightly more friendly properties, may also be used instead of xylene.

Table 7.3. Observations and calculations of apparent mass of the hydrometer in the reference liquid. Observations marked with an asterisk (*) should be repeated three times. For each observation: start the xylene pump for 30 s, wait for a minute for the liquid to settle down. Lower the hydrometer slightly and readjust it so that the graduation mark intersects the liquid surface level.

Immersion level	Below the pan	Std weights on the pan	Indication (mg)	Difference (g)	Corrected mass (g)
	Hanger only	150.25	32.8		
*1000	Hanger + hyd.	135.40	20.5	14.8377	14.837 675
Temperature of the liquid 20.010 °C					
Reading of the open-scale hydrometer 0.861 05 g cm^{-3}					
*1010	Hanger + hyd.	134.51	36.8	15.7440	15.743 955
Temperature of the liquid 20.000 °C					
Reading of the open-scale hydrometer 0.861 00 g cm^{-3}					
*1020	Hanger + hyd.	133.61	29.6	16.6368	16.636 775
Temperature of the liquid 20.005 °C					
Reading of the open-scale hydrometer 0.861 02 g cm^{-3}					
*1030	Hanger + hyd.	132.73	31.2	17.5184	17.518 405
Temperature of the liquid 19.995 °C					
Reading of the open-scale hydrometer 0.861 02 g cm^{-3}					
*1040	Hanger + hyd.	131.88	37.7	18.3748	18.374 765
Temperature of the liquid 20.005 °C					
Reading of the open-scale hydrometer 0.861 05 g cm^{-3}					
*1050	Hanger + hyd.	131.05	52.5	19.2196	19.219 625
Temperature of the liquid 20.005 °C					
Reading of the open-scale hydrometer 0.861 03 g cm^{-3}					
	Hanger only	150.25	32.9		

Table 7.4. Calculation of the density and corrections applicable to the hydrometer. $\rho - R$ is the correction to be applied to the hydrometer at the scale reading R.

Immersion level	Numerator	Denominator	$\rho - \sigma$	ρ	Correction $\rho - R$
1.000	91.757 129	91.856 905	0.998 914	1.000 082	+0.000 08
1.010	91.751 179	90.950 750	1.008 801	1.009 969	−0.000 03
1.020	91.753 928	90.058 055	1.018 825	1.019 993	−0.000 07
1.030	91.753 928	89.176 547	1.028 902	1.030 070	+0.000 07
1.040	91.757 129	88.320 307	1.038 913	1.040 082	+0.000 08
1.050	91.954 995	87.475 564	1.048 921	1.050 090	+0.000 09

7.9   High-precision calibration method by measuring dimensions of the stem

The method [14] described in this section is as good as Cuckow's with regard to precision. It is suitable for reference hydrometers and requires only one liquid with a density which lies within the range of the hydrometer, i.e. each hydrometer will require a different liquid. It involves the following steps:

(1) Make four hat-shaped weights with the appropriate mass values, which can be easily but snugly placed on the top of the hydrometer such that the hydrometer continues to float in a vertical position. The apparent mass values of the hat-type weights depend upon the mass, volume and range of the hydrometer.

(2) The apparent mass, W_h of the hydrometer is determined on a laboratory balance.

(3) The distances $L_1, L_2, L_3, L_4, \ldots, L_N$, of all cardinal points on the scale of the hydrometer are measured on a linear comparator or through a travelling microscope from some arbitrary zero point. The distance L_T of the top of the hydrometer is also measured from the same reference point.

(4) Diameters $D_1, D_2, D_3, \ldots, D_N$ of the stem at all the corresponding cardinal points are also measured with the help of an indicating micrometer.

(5) The density of the liquid ρ in which the hydrometer floats freely is measured using the hydrostatic method with the help of a solid body of known volume. The levels L_R, L_A, L_B, L_C and L_D, are observed when the hydrometer is floating freely and small hat-type weights are placed on the top of the stem. The apparent mass of hats as determined in air is W_A, W_B, W_C and W_D, respectively.

(6) Let m_T be the additive mass effect due to surface tension T, then $m_T = \pi DT/g$.

(7) The accuracy and precision of the method primarily depends upon the accuracy with which level L_R of the hydrometer at which it floats freely in the liquid of density ρ is observed. Instead of repeating the observations at L_R, which has limited advantage, levels L_A, L_B, L_C and L_D with the respective hat-shaped weights of apparent mass W_A, W_B, W_C and W_D placed on the hydrometer, are observed and the position of L_R is calculated. The equilibrium equations and further calculations are given in the next section.

7.9.1   Determination of the position for the freely floating hydrometer

$$m + m_R = \rho V_R + \sigma(V - V_R). \tag{7.62}$$

The ms with suffixes A, B, C, D or R indicate the additive mass effect due to surface tension T for the diameter at which the observation is taken, i.e. they are the masses of the meniscus liquid. M and V are the mass and volume of the hydrometer, σ is the density of air and V_R is the immersed volume. A similar equation is obtained when the hydrometer is loaded with a hat of apparent mass W_A and the flotation level is L_A.

$$m + W_A + m_A = \rho(V_R + V_A) + \sigma(V - V_R - V_A).\tag{7.63}$$

Subtraction of (7.62) from (7.63) gives

$$V_A = (W_A + m_A - m_R)/(\rho - \sigma).\tag{7.64}$$

If D_A is the average diameter in between L_R and L_A, then volume V_A can also be written as

$$V_A = \pi D_A^2 (L_R - L_A)/4.\tag{7.65}$$

Equating (7.64) and (7.65), we get

$$L_R = L_A + 4(W_A + m_A - m_R)/[\pi D_A^2(\rho - \sigma)].\tag{7.66}$$

Similarly observing L_B, L_C and L_D with hats W_B, W_C and W_D on the top of the hydrometer respectively, we get three more values for L_R. Altogether we get four values of L_R but from different flotation points, so any bias from repeating the observation at the same flotation level is removed. Let the average value of L_R be L.

We next tabulate the average values of the diameters from each reference level L to every cardinal point N. Let us denote the average value of the diameter between levels L and L_N as D_N. Therefore the volume between the reference level and the Nth cardinal point can be written as

$$V_N = \pi D_N^2 (L - L_N)/4.\tag{7.67}$$

Then equation (7.62) may be rewritten as

$$m_R + (m - \sigma V) = (\rho - \sigma)V_R.\tag{7.68}$$

However, as $m - \sigma V$ is the apparent mass W_h of the hydrometer, V_R may be written as

$$V_R = (W_h + m_R)/(\rho - \sigma).\tag{7.69}$$

Similarly the equilibrium at flotation level L_N can be written as

$$m + m_N = \rho_N(V_R + V_N) + \sigma(V - V_R - V_N)\tag{7.70}$$

which can be rewritten as

$$(m - \sigma V) + m_N = (\rho_N - \sigma)(V_R + V_N) \quad \text{or} \quad W_h + m_N = (\rho_N - \sigma)(V_R + V_N)\tag{7.71}$$

giving

$$(\rho_N - \sigma) = (W_h + m_N)/[\{(W_h + m_R)/(\rho - \sigma)\} + \pi D_N^2(L - L_N)/4].\tag{7.72}$$

From this equation, we can find the density values for each cardinal point. However, these density values are subject to the condition that the hydrometer is floating in a liquid of surface tension T and $t\,°C$ is the temperature of measurement. If the hydrometer is to be calibrated for surface tension T_2 and for reference temperature t_r, the necessary corrections, given in the next section, have to be applied.

7.9.2   Correction for temperature and surface tension

Writing $V_i = V_N + V_R$ for the total immersed volume and $V_e = V - V_R - V_N$ for the exposed volume, equation (7.70) may be written as

$$m + m_N - \rho_N V_i - \sigma V_e = 0. \tag{7.73}$$

Any changes in the temperature or surface tension will cause changes in V_i, V_e, m_N and ρ_N. If ΔV_e, ΔV_i, Δm_N and $\Delta \rho_N$ are the corresponding changes then (7.73) may be rewritten as

$$m + (m_N + \Delta m_N) - (\rho_N + \Delta \rho_N)(V_i + \Delta V_i) - \sigma(V_e + \Delta V_e) = 0. \tag{7.74}$$

All the terms in this equation are in terms of mass. Therefore all gravitational forces are vertical—downward for positive terms and upward (buoyancy forces) for negative terms. As we want the hydrometer to float at the same Nth cardinal point, the sum of all terms containing Δ should be zero. So subtracting (7.73) from (7.74), we get

$$\Delta m_N - \rho_N \Delta V_i - \Delta \rho_N \Delta V_i - \Delta \rho_N V_i - \sigma \Delta V_e = 0 \tag{7.75}$$

$$\Delta \rho_N = -(\rho_N \Delta V_i + \sigma \Delta V_e - \Delta m_N)/(V_i + \Delta V_i). \tag{7.76}$$

Considering the volumes as linear functions of temperature, we may write

$$\Delta V_i = \gamma (t_r - t) V_i \tag{7.77}$$

$$\Delta V_e = \gamma (t_r - t) V_e \tag{7.78}$$

while $\Delta m_N = \pi D_N (T_2 - T)/g$. Substituting these values into equation (7.76), we get

$$\Delta \rho_N = \{\gamma (t_r - t)[\rho_N V_i + \sigma V_e] - \pi D_N (T_2 - T)/g\}/\{V_i(1 + \gamma (t_r - t))\}. \tag{7.79}$$

From equation (7.73), the sum of all the terms in square brackets in equation (7.79) is equal to $m + m_N$ so

$$\Delta \rho_N = \{\gamma (t_r - t)(m + m_N) - \pi D_N (T_2 - T)/g\}/\{V_i(1 + \gamma (t_r - t))\}. \tag{7.80}$$

Once we know the values of T_2 and t_r, all terms on the right-hand side are known. We can calculate the value of $\Delta \rho_N$. Hence the new density value corresponding to the Nth cardinal point R_N which is the sum of ρ_N and $\Delta\rho_N$ may be also be calculated.

If S_N is the numerical value of the density engraved on the Nth cardinal point then the correction for this point may be written as

$$C_N = R_N - S_N. \tag{7.81}$$

We therefore see that by inserting different values for T_2 and t_r we may obtain the corrections applicable to a single hydrometer, with different reference temperatures and for different categories of surface tension. Therefore, a single hydrometer may work as a standard with different reference temperatures and for different categories of surface tension. Hence one hydrometer can be used for different purposes.

To summarize, this method consists of finding the position on the scale at which it freely floats in a liquid of known density and surface tension, and measuring the diameters and length from that position. If we know the apparent mass of the hydrometer, then the density corresponding to any other point of the scale can be calculated for any other value of surface tension and reference temperature.

7.10 Calibration of hydrometer by comparison method

The hydrometers are calibrated for any one of the three categories of surface tensions, i.e. low (up to 35 mN m^{-1}), medium (up to 55 mN m^{-1}) and high (75 mN m^{-1}). During the calibration process, the corrections to be applied to the indications of the hydrometer at different points of its scale are determined. Calibrating a hydrometer by the comparison method is easier and so it is the most commonly employed for routine calibration. In this method the hydrometer is calibrated by comparing its readings against those of a standard hydrometer, whose scale errors are already known.

7.10.1 Apparatus

Liquid jar

The liquid is contained in a rectangular jar, in which the front and back walls are parallel. The front wall is made from optical glass, which is plane and polished from outside. This prevents the image of the scale of the hydrometer from becoming distorted. Apart from abnormal hydrometers, a jar about 112 mm wide, 62 mm deep and 360 mm tall will be able to accommodate two hydrometers with a stirrer and a thermometer.

Stirrer

The stirrer is made of a rectangular stainless steel sheet, whose dimensions are 2 mm less than those of the cross section of the jar and which has a large number

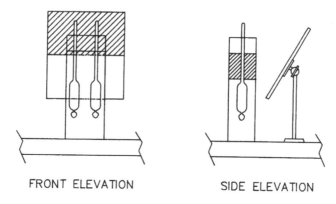

FRONT ELEVATION   SIDE ELEVATION

Figure 7.9. Arrangement of jar with hydrometers and screen.

of holes. A stout rod is screwed into the centre of the stainless steel sheet and normal to it. The rod works as a handle.

Screen

This comprises a wooden rectangular board slightly wider than the width of the jar. The top half is painted black and the bottom half white. The line separating the black and white portions is horizontal. The screen is moveable about its horizontal axis and it is placed behind the jar at an angle of 45° in such way that the line separating the black and white portions is just below the liquid surface. The screen helps in keeping the correct line of sight.

Standard hydrometer

The standard hydrometer means one for which corrections to be applied at different points of its scale are known. The reference temperature and surface tension category of the standard hydrometer should match, as far as possible, those of the hydrometer under test. An arrangement with hydrometers and screen is shown in figure 7.9.

7.10.2   Handling of hydrometer

The hydrometers are cleaned with a fibre-less cloth or tissue paper and alcohol and finally with the liquid in which they are to be used. After cleaning, the hydrometers should not be touched either at the bulb or on the stem portion containing the scale. The hydrometer is always lifted vertically from the stem. It is held with the help of the thumb and finger. It is inserted in the liquid, being held at the top of the stem and is released when it is only a few mm above the graduation mark to which it is supposed to float. This way it will make several

Figure 7.10. Holding a hydrometer.

up and down excursions and then come to rest. Take this approximate reading and for proper wetting, press it a further few mm below the mark around which it was floating with the help of thumb and forefinger as shown in figure 7.10. At this stage it does not need to be held but it should be allowed to rest within the vee formed by the thumb. On withdrawing the hand without disturbing the hydrometer, it rises and falls slowly. Observe the meniscus carefully. The meniscus should not be deformed during this up-and-down motion. If the meniscus is deformed at any time, it shows that the hydrometer stem is not clean. Take it out and clean the stem very carefully and then use it. Normally the hydrometer settles down after a few oscillations.

7.10.3   Adjustment of the density of the liquid so that the hydrometer floats up to the desired mark

The hydrometer to be calibrated is floated in a liquid of appropriate density as described earlier. The density of the liquid is finely adjusted by adding the appropriate component of the mixture and stirring the liquid vigorously. While stirring, care should be taken that air is not dissolved. The density of the liquid should be adjusted so that the liquid surface level exactly intersects the scale graduation mark to be tested. If not, the density of the liquid should be adjusted again. A standard hydrometer similarly cleaned and lifted is also floated along its side. The standard temperature and surface tension category of the standard hydrometer should match as far as possible with those of the hydrometer under test. The indication on its scale is observed as follows.

7.10.4   Observation method

An accurate hydrometer reading will be obtained only if the line of sight is normal to the stem of the hydrometer and is at the same height as that of the liquid surface in the jar. This condition is obtained with the help of the screen (figure 7.9). If we look from much below the level of the liquid surface, a white rectangle with the meniscus around the stem of the hydrometer, which will appear as an ellipse, will appear. Conversely if the line of sight is higher than the level of the liquid surface, we will see a black rectangle. The level of the eye is slowly raised from below the liquid surface; the width of the rectangle and ellipse appears to be shrinking. Continuing this way a position is reached where both the rectangle shrinks to a white line and the ellipse of the meniscus simultaneously shrinks to a line coinciding with the white line. This is the correct position for the eye for taking the reading. Observe the reading of the standard hydrometer in the same way and make sure that the graduation mark of the hydrometer under test still crosses the liquid surface. Efforts should be made to estimate one-fifth of the distance between consecutive graduations.

Normally the hydrometers are calibrated at four points, including the top and bottom graduation marks. The liquid for each graduation mark is changed and its density is adjusted so that the liquid surface level always intersects exactly at the numbered graduation mark under test. The readings are taken on the standard hydrometer. It should be noted that the actual density of the liquid need not be known—the liquid acts only as a convenient medium. However, you need to know the surface tension of the liquid to apply corrections.

7.10.5   Correction to be applied to the hydrometer under test

All the necessary corrections are applied to the reading of the hydrometer under test. Only certificate and temperature corrections are applied to the standard hydrometer. The corrected indication of the hydrometer under test is subtracted from that of the standard hydrometer. The result gives the correction to be applied to the hydrometer under test at the graduation mark.

7.10.6   Precautions for use of hydrometers in liquids of high surface tension

In order to obtain reliable readings with liquids of high surface tension, such as is the case for most aqueous solutions, it is necessary to make sure that the liquid surface is truly clean—slight contamination can change the surface tension abruptly and unpredictably. The easiest way of ensuring that the surface tension does not change is to always take observations with a fresh surface. A fresh surface of the liquid is obtained by the use of an overflow jar. One such jar is shown in figure 7.11. The jar is filled to almost the overflow level and the density of the liquid is pre-adjusted to the desired value. The hydrometer is inserted into the liquid and, if necessary, further liquid is added through the side tube so that the liquid is approximately equal in volume to the 4 cm length of the jar that has overflowed. The basic dimensions of an overflow jar, according to IS: 3104-1965, are given in table 7.5 while others are shown in figure 7.11.

7.10.7   Calibration of a hydrometer with two standard hydrometers (cyclic method)

To obtain more reliable results and the uncertainty estimate, the hydrometer is calibrated with the help of two standard hydrometers S1 and S2 using the cyclic

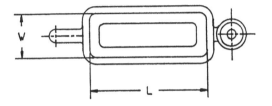

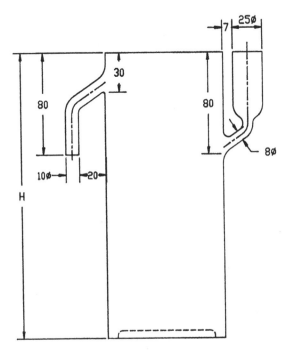

Figure 7.11. Overflow jar for automatic renewal of liquid surface.

Table 7.5. Dimensions of an overflow jar. All dimensions are in mm.

L	W	H	Suitable for hydrometers of categories
135 ± 5	55 ± 3	360 ± 5	L20 and L50
100 ± 5	45 ± 3	295 ± 5	M100 and M50
85 ± 5	35 ± 3	220 ± 5	S50

method. The cyclic method is one in which every hydrometer occupies every space position exactly the same number of times and then occupies the same positions but in reverse order.

After ensuring the density of the liquid is such that X, the hydrometer under test, floats to a numbered graduation and, after proper stirring, the hydrometer under test and the standard hydrometer S1 are floated side by side. Then take observations following the method described earlier. The standard S1 is removed and the hydrometer under test is put in its place. The second standard S2 is inserted in place of X. The observations are taken again. Now X is taken out and S2 is shifted to its place and the other standard S1 is again inserted into the place occupied by X, the hydrometer under test, and a third set of observations taken. The two standards are then interchanged and the whole process of inter-comparison repeated but in reverse order. No stirring takes place during the whole process. This process cancels out errors due to lack of symmetry in the reading conditions and also gives an uncertainty estimate in the taking of observations. The comparison scheme may be written as follows.

Position 1	Position 2
S1	X
X	S2
S2	S1
S1	S2
S2	X
X	S1

Here each hydrometer has occupied each of the two positions twice but in reverse order, so any time-dependent variation in density or temperature tends to cancel out. If the readings of the standard hydrometers are S1 and S2 and that of the hydrometer under test X and if the true corrections are s_1, s_2 and x respectively, then

the mean of	S1 − X	should be	$-s_1 + x$
the mean of	X − S2	should be	$-x + s_2$
the mean of	S2 − S1	should be	$-s_2 + s_1$.

The sum of these values should be zero. Any residual value is the uncertainty estimate.

7.10.8 Liquids used in the comparison method

The liquids used for calibrating hydrometers by the comparison method should preferably be in the low surface tension category so that the errors due to erratic surface tension variations due to slight contamination are minimal and the hydrometer stem can be wetted properly. For better visibility, the liquids should be clear and colourless. If the hydrometer under test belongs to another

Table 7.6. Density ranges for petroleum liquid mixtures.

Liquid solutions	Density range $(kg\ m^{-3})$
Mixture of petroleum ethers with boiling ranges up to 40, (40–60), (60–80), (80–100) °C	600–690
Mixtures of petroleum ether with a boiling range 80–100 °C and xylene	690–860
Mixtures of xylene and tetrachloro-ethylene	850–600
Mixtures of tetrachloro-ethylene and ethylene di-bromide	1600–2000

category of surface tension, then the appropriate surface tension correction should be applied separately.

The organic liquids most commonly used at NPL, India are:

(1) petroleum ether,
(2) xylene,
(3) tetrachloro ethylene and
(4) ethlylene di-bromide.

The use of these liquids virtually eliminates any error due to the meniscus. The density ranges over which they can be used are given in table 7.6.

Surface tension, against density of the mistures of the liquids indicated in table 7.6, is given in table 7.7.

Because of the toxic nature of these liquids the atmosphere should be kept reasonably vapour free by using an efficient exhaust system, which draws the vapours downwards. This is easier as the vapours of these liquids are denser than air.

Alternative solutions

(1) For a hydrometer to be used in high surface tension liquids, aqueous sulphuric acid solutions provide good alternatives. Although sulphuric acid is highly corrosive, for which necessary precautions and safeguards must be taken, it is a good alternative, as it is cheap, gives good wetting and covers a wide range of density—1000–1840 $kg\ m^{-3}$. For densities greater than 1840 $kg\ m^{-3}$, mercuric nitrate solutions are used at NPL, UK. The NIST, USA uses Thoulet (K_2HgI_4) solutions instead of aqueous solutions of sulphuric acid and mercuric nitrate.
(2) For a hydrometer to be used in low surface tension liquids, an aqueous solution of ethyl alcohol is a good choice, as it covers a range of 800 to 970 $kg\ m^{-3}$.
(3) For hydrometers to be used in milk, sodium carbonate dissolved in 10% of ethyl alcohol and water solution works fine.

Table 7.7. Surface tension of petroleum liquid mixtures.

Density ($kg\ m^{-3}$)	Surface tension ($mN\ m^{-1}$)	Density ($kg\ m^{-3}$)	Surface tension ($mN\ m^{-1}$)
600	15	800	25
620	16	820	26
640	17	840	27
660	18	860	28
680	19	880	29
700	20	900	30
720	21	920	31
740	22	940	32
760	23	960	33
780	24	980	34

Density from 1.000 to 1.300 g cm^{-3} inclusive
Surface tension 35 mN m^{-1}

7.11 Corrections due to the influence of factors in the calibration of hydrometers by comparison

7.11.1 The effect of the temperature of the liquid

As long as the two hydrometers under comparison have the same reference temperature, the temperature of the liquid during comparison is immaterial but uniformity in temperature is important. The calibration, therefore, should be carried out in a temperature-controlled room. The test liquids, hydrometers under calibration and the standard hydrometers should be stored in the same room. This ensures stable and uniform temperature for the liquids and hydrometers.

If the two hydrometers being compared do not have the same reference temperature, then their indications will differ by an amount proportional to the change in their volumes caused by thermal expansion of one with respect to the other. Therefore a temperature correction, TC, is necessary. The TC, given below, is to be applied to the indication of the standard hydrometer.

$$TC = \rho\gamma(t_s - t_u) \tag{7.82}$$

where

(1) the TC is the correction to be applied to the indication of the standard hydrometer,
(2) γ is the coefficient of cubic thermal expansion of glass and is assumed to be the same for both hydrometers,
(3) t_s is the reference temperature for the standard hydrometer in °C,

(4) t_u is the reference temperature for the hydrometer under test in °C and
(5) ρ is the indication of density by the standard hydrometer.

Here it was assumed that γ is the same for both hydrometers. If γ_s and γ_u are the coefficients of expansion for the standard and test hydrometers respectively, then TC is given by

$$TC = -\rho\gamma_s(t - t_s) + \rho\gamma_u(t - t_u). \tag{7.83}$$

7.11.2   Temperature correction for specific gravity hydrometers

A specific gravity hydrometer calibrated for temperature t_s gives the ratio of the density of the liquid at t_s °C to that of water at t_s °C indicated as St_s/t_s °C. Now, if it is used at temperature t, a correction c has to be applied to the reading S, so that it gives the specific gravity St/t °C, the formula for which may be derived as follows.

The equilibrium equation at t_s is given by

$$m = V S \rho_{ts} \tag{7.84}$$

where ρ_{ts} is the density of water at the standard temperature t_s.

If the same hydrometer indicates S at temperature t °C and c is the correction to be applied to the reading S to give the specific gravity of the liquid St/t at t °C, then c the correction to be applied is given by

$$m = V_t(S + c)\rho_t. \tag{7.85}$$

Dividing (7.85) by (7.84) gives $(S + c)/S = V\rho_{ts}/V_t\rho_t = \rho_{ts}/\rho_t\{1 + \gamma(t - t_s)\}$ giving

$$c = S[\rho_{ts}/\rho_t\{1 + \gamma(t - t_s)\} - 1]. \tag{7.86}$$

However, for the density hydrometer the ρ_{ts}/ρ_t term will be unity, and S will be replaced by ρ, the density indication of the density hydrometer, so correction c will be given by

$$c = -\rho\gamma(t - t_s).$$

The TC for specific gravity hydrometers is, therefore, larger than that for density hydrometers. For example a specific gravity hydrometer standardized at 20 °C and read correctly at 1.030 will require a correction of 0.0015 if used at 27 °C. The coefficient of expansion of the material of the hydrometer has been taken to be 30×10^{-6} °C^{-1}, whereas a density hydrometer with similar parameters would require a correction of -0.0002 g cm^{-3} only.

Temperature correction for density hydrometers when using the comparison method

Let C_1 be the correction for a standard density hydrometer, which is given by

$$C_1 = -\rho\gamma(t - t_s).$$

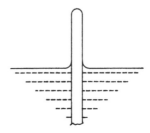

Figure 7.12. Rising of liquid around the stem of the hydrometer.

Similarly C_2, the correction applicable to the hydrometer under test, will be

$$C_2 = -\rho\gamma(t - t_u).$$

Hence the net correction applicable to the hydrometer under test will be

$$C_2 - C_1 = TC = -\rho\gamma(t - t_u) - \rho\gamma(t - t_s) = \rho\gamma(t_s - t_u). \tag{7.87}$$

This is what was given in equation (7.82).

Temperature correction for specific gravity hydrometers when using the comparison method

If the two specific gravity hydrometers have different standard temperatures, t_s and t_u, and are compared at temperature t, then the TC applicable to the hydrometer under test is given by

$$TC = S[\rho_{tu}/\rho_t\{1 + \gamma(t - t_u)\} - \rho_{ts}/\rho_t\{1 + \gamma(t - t_s)\}]. \tag{7.88}$$

7.11.3   The effect of surface tension

The surface tension of the liquid, in which the hydrometer is immersed, causes the liquid to rise around its stem. The liquid raised around the stem is called the meniscus. This is shown in figure 7.12.

The weight of the meniscus liquid results in a small additional downward force and makes the hydrometer float a little lower in the liquid. In effect, this force adds to the weight of the hydrometer and is equal to the weight of the meniscus, which attaches itself to the stem. Therefore, if a hydrometer is used in a liquid whose surface tension is different from the one for which the hydrometer was calibrated, and the same graduation is read, a correction c should be applied to the indicated density. The same may be calculated as follows.

Let ρ be the indication with the standard surface tension, then the equilibrium equation gives

$$m + \pi D T_s/g = V\rho. \tag{7.89}$$

Similarly the equilibrium equation, when the hydrometer reads ρ with correction c in a liquid of surface tension T, is given by

$$m + \pi DT/g = V(\rho + c). \tag{7.90}$$

Dividing (7.90) by (7.89) gives

$$(m + \pi DT/g)/(m + \pi DT_s/g) = (c + \rho)/\rho = 1 + c/\rho \tag{7.91}$$

$$c/\rho = (\pi D(T - T_s)/g)/(m + \pi DT_s/g). \tag{7.92}$$

The term $\pi DT_s/g$ in the denominator can be neglected in comparison to the mass m of the hydrometer, giving us

$$c = (\rho \pi D)(T - T_s)/mg. \tag{7.93}$$

For the purpose of distinguishing the correction for surface tension, c is replaced by STC. Therefore we have

$$STC = (\rho \pi D)(T - T_s)/mg.$$

If ρ is in g cm^{-3}, D in cm, T and T_s in mN m^{-1}, m in g and g in cm s^{-2}, then STC will be in g cm^{-3}, i.e. in terms of density. In SI units STC will be written as

$$STC = 10^3 (\rho \pi D)(T - T_s)/mg.$$

In this case, ρ is in kg m^{-3}, D in m, T and T_s in N m^{-1}, m in kg and g in m s^{-2}, then STC will be in kg m^{-3}, again in terms of density.

The term $\rho \pi D/mg$ is called the surface tension factor (STF). The STF depends upon the actual value of ρ, but as the STF is itself very small, the middle value of the range of the scale is taken as ρ.

If the hydrometer under test has T_u as its standard surface tension while the standard was calibrated for surface tension T_s and if the two are compared in a liquid with surface tension T, then the surface tension correction is to be applied individually to each hydrometer: $STF_s(T - T_s)$ and $STF_u(T - T_u)$, where T is the surface tension of the liquid used.

7.11.4 Surface tension correction on Brix hydrometers

When a Brix hydrometer is compared with another standard Brix hydrometer but in a liquid of surface tension T, which differs from the standard surface tension T_s for either of the Brix hydrometers, then the combined correction which should be applied to the hydrometer under test is given by

$$STC = [STF_u - STF_s](T - T_s).$$

This will be in kg m^{-3}, if the surface tension T is in N m^{-1} and STF is in s^2 m^{-3}. If STF is in s^2 cm^{-3} and the surface tension is in mN m^{-1} then the correction is in g cm^{-3}.

7.11.5   Meniscus correction

In the case of opaque liquids, the reading is taken where the top of the meniscus appears to meet the stem of hydrometer; i.e. the hydrometer will show a lower value for the density of the liquid. In order to obtain the correct indication at the level of the horizontal liquid surface, a correction MC is added to the observed indication ρ. The correction is calculated from the following equation due to Langburg.

$$\text{MC} = 1000(TR/D\rho Lg)[(1 + 2gD^2\rho/T)^{1/2} - 1] \text{ kg m}^{-3} \qquad (7.94)$$

where

- T is the surface tension of the liquid in N m^{-1},
- R is the nominal range of the hydrometer in kg m^{-3},
- D is the diameter in m,
- ρ is the indication in kg m^{-3},
- L is the scale length in m and
- g is the acceleration due to gravity (9.81 m s^{-2}).

For easy reference, the values of MC have been tabulated in table 7.8 for various series of density hydrometers conforming to BS 718: 1979. The scale lengths are given in the third row.

7.12   Examples of corrections applicable for the comparison method

7.12.1   Density hydrometers with the same reference temperature and surface tension

The hydrometer under test has the same reference temperature (RT) and surface tension (ST) as those of the standard hydrometer and the liquid employed also has a surface tension equal to that of the reference one. Then, only the correction given in the calibration certificate (CC) of the standard hydrometer will be applied to its indication.

	Indication of standard	Indication of hydrometer under test
	0.790 44	0.790
CC	−0.000 10	
Corrected indication	0.790 34	0.790

Therefore the correction applicable to the hydrometer under test is +0.000 34 g cm^{-3}.

Table 7.8. Meniscus correction MC in kg m^{-3} for various hydrometers. The surface tension is in mN m^{-1}.

Series of hydrometer		L20		L50 and L50SP		M50 and M50SP		M100		S50		S50SP	
Scale interval (kg m^{-3})		0.2		0.5		1		2		2		1	
Scale length (mm)		113	127	125	145	78	99	87	102	50	62	50	62
Indication I (kg m^{-3})	Surface tension	The values of MC (kg m^{-3})											
600	15	0.32	0.28	0.8	0.7	1.2	1.0	2.0	2.0	2.0	1.6	1.8	1.6
800	25	0.36	0.32	0.8	0.7	1.4	1.0	2.4	2.0	2.0	1.6	2.0	1.6
1000	35	0.36	0.32	0.8	0.7	1.4	1.0	2.4	2.0	2.0	1.6	2.2	1.6
	55	0.44	0.40	1.0	0.8	1.6	1.2	2.8	2.4	2.4	2.0		
	75	0.48	0.44	1.0	0.9	1.8	1.4	3.2	2.8	2.8	2.4		
1500	35	0.32	0.28	0.7	0.6	1.0	0.8	2.0	1.6	2.0	1.2		
	55	0.36	0.32	0.8	0.7	1.2	1.0	2.4	2.0	2.0	1.6		
	75	0.40	0.36	0.9	0.8	1.4	1.2	2.8	2.4	2.4	2.0		
2000	55	0.32	0.28	0.7	0.6	1.0	1.0	2.0	1.6	2.0	1.6		
	75	0.36	0.32	0.8	0.7	1.2	1.0	2.4	2.0	2.4	1.6		

7.12.2 Density hydrometers with different reference temperatures but the same surface tension

Let us take an example in which the hydrometer under test has a reference temperature of 27 °C and the standard's reference temperature is 20 °C. The correction is applied to the indication of the standard hydrometer, by using $-\rho\gamma_s(t_u - t_s)$—here it is assumed that γ is the same for both hydrometers. When the two hydrometers have different coefficients of expansion, then

$$-\rho\gamma_s(t - t_s) + \rho\gamma_u(t - t_u)$$

is used. Here t is the temperature of the liquid used.

Let us consider that γ is the same for both hydrometers and equals 0.000 030 °C^{-1}, then the correction applicable, in addition to the certificate correction CC, will be a TC given by

$$TC = -0.000\,030(27.0 - 20)1.050 = -0.000\,24 \text{ g cm}^{-3}.$$

	Indication of standard	Indication of hydrometer under test
	0.790 44	0.790
CC	−0.000 10	
TC	−0.000 24	
Corrected indication	0.790 10	0.790

Therefore the applicable correction is $+0.0001$ g cm^{-3}.

7.12.3   Density hydrometers with different surface tensions and reference temperatures

The corrections applicable in this case will be CC, TC and STC, the surface tension correction. Find the mass and diameter of the stems of both hydrometers and calculate the surface tension factor, STF, for each and apply the correction separately to the indications of each hydrometer using

$$ST\,F_s(T - T_s) \qquad \text{and} \qquad ST\,F_u(T - T_u)$$

where T is the surface tension of the liquid used.

Normally $T = T_s$. Then the correction is only applicable to the hydrometer under test. If the calculated value of $ST\,F_u$ is $0.000\,021$ s^2 cm^{-3}, $T = 35$ and T_u is 55 mN m^{-1}, then $STC = -0.000\,42$ g cm^{-3}.

	Indication of standard	Indication of hydrometer under test
	0.790 44	0.790
CC	−0.000 10	
TC	−0.000 24	
STC		−0.000 42
Corrected indication	0.790 10	0.789 58

Therefore the applicable correction is $+0.0005$ g cm^{-3}.

7.12.4   Lactometer calibrated against density hydrometers

A lactometer is essentially a specific gravity hydrometer to be used in milk, so it can be tested in petroleum liquids against a density hydrometer. The surface tension of the liquid used is 35 mN m^{-1}, which is suitable for low surface tension hydrometers. Therefore the corrections to be applied to the standard density hydrometer are CC and TC. Finally the corrected density indication of the standard hydrometer should be divided by the density of the water at the base temperature of the lactometer.

The surface tension correction, STC, and the meniscus correction, MC, should be applied to the lactometer under test. The lactometer reading is taken at the liquid surface level observed through the liquid. However, when the lactometer is used in milk it is observed at the level at which the meniscus meets the stem, so MC should be applied to the lactometer. Similarly, the surface tension, T_u, for the lactometer is 50 mN m^{-1}, which may change slightly with its reference temperature while that of the liquid used as the medium is 35 mN m^{-1}, so STC also has to be applied to the lactometer. The hydrometers to be used in milk, which are made according to Indian Standard Specifications, are specific gravity hydrometers at 27/4 °C. There are hydrometers to be used in milk with reference temperature of 28.89 °C (84 °F) or 15.5 °C (60 °F). The ranges of these hydrometers are (1) 1.000–1.040; (2) 1.020–1.040; (3) 1.020–1.035; and (4) 1.015–1.040—all are in terms of specific gravity. Let us consider a lactometer with $t_u = 27$ °C and an STF, as calculated from its mass and stem diameter, of 0.000 034 s^2 cm^{-3}. A standard hydrometer has $t_s = 15$ °C and $\gamma = 0.000\,030$ °C^{-1}.

(1) A lactometer at 27/27 °C

	Indication of standard	Indication of hydrometer under test
	1.015 5	1.020
CC	+0.000 02	STC −0.000 51
TC	−0.000 36	MC −0.000 75
Corrected indication	1.015 16	1.018 74

Dividing 1.015 16 by the density of water at 27 °C (0.996 5155) gives 1.018 71. The correction applicable to the lactometer under test at 1.020 is −0.000 03, which is almost zero.

(2) A lactometer at 28.89/28.89 °C (84/84 °F)

	Indication of standard	Indication of hydrometer under test
	1.023 6	1.030
CC	+0.000 08	STC −0.001 35
TC	−0.000 42	MC −0.002 31
Corrected indication	1.023 26	1.026 34

Dividing 1.023 26 by the density of water at 28.89 °C (0.995 972) gives 1.027 40. Note that the STF for the lactometer is 0.000 09 s^2 cm^{-3}. The correction applicable to the lactometer under test at 1.030 is +0.0011.

(3) Lactometer at 15.5/15.5 °C (60/60 °F)

	Indication of standard		Indication of hydrometer under test
	1.039 9		1.040
CC	+0.000 05	STC	−0.000 48
TC	−0.000 02	MC	−0.001 04
Corrected indication	1.039 93		1.038 48

Dividing 1.039 93 by the density of water at 15.5 °C (0.999 0258) gives 1.040 96. The correction applicable to the lactometer under test at 1.020 is +0.0025.

7.12.5   Brix hydrometer calibrated against a standard Brix hydrometer

There are two types of Brix hydrometer in circulation: one has a reference temperature of 20 °C and the other one of 27 °C. The reference temperature of the standard hydrometer available at NPL, New Delhi is 20 °C. Moreover, Brix hydrometers are tested in petroleum liquid with a surface tension of 35 mN m^{-1} but the reference surface tension for the Brix hydrometer is 75 mN m^{-1}. So the surface tension correction is to be applied to the hydrometer under test as given in section 7.11.4. In addition if the reference temperatures of the two hydrometers are different then the temperature correction (TC) should be applied. As both the aforesaid corrections are in units of density, each should be divided by the difference in density for 0.1 °Brix at the point of the scale under test to get the correction in °Brix.

(1) A Brix hydrometer against a standard Brix hydrometer with the same reference temperature

Let the STFs for the standard and test hydrometer be 0.000 018 s^2 cm^{-3} and 0.000 027 s^2 cm^{-3} respectively and let them both range from 0 to 10 °Brix. The density difference, as given in table 7.12 of section 7.17, for 0.1 °Brix at 5 °Brix is 0.000 40 g cm^{-3}. Therefore the STC is given by

$$STC = [STF_u - STF_s](T - T_s) \text{ g cm}^{-3}$$
$$(0.000\,027 - 0.000\,018)(35 - 75) = -(0.000\,009)(40) \text{ g cm}^{-3}$$

or

$$-(0.000\,009)(40)(0.1)/0.000\,40 = -0.09 \text{ Brix.}$$

This correction is applicable to the hydrometer under test.

	Indication by standard (°Brix)		Indication by hydrometer under test (°Brix)
	0.08		0.10
CC	−0.10	*STC*	−0.09
Corrected indication	−0.02		0.01

The correction applicable to the Brix hydrometer at 0.1 °Brix is −0.03 °Brix.

(2) A Brix hydrometer against a standard Brix hydrometer with different reference temperatures

Let the *STF*s for the standard and test hydrometers be 0.000021 s^2 cm^{-3} and 0.000028 s^2 cm^{-3} respectively and the range of each 10–20 °Brix. The reference temperature for the standard hydrometer is 20 °C and that of the test one 27.5 °C. The density difference, as given in table 7.12 of section 7.17, for 0.1 °Brix at 15 °Brix, is 0.00043 g cm^{-3}. *TC*, the temperature correction applicable to the standard hydrometer with a reference temperature of 20 °C, is given in table 7.13 of section 7.17. This is 0.46, 0.47, 0.48 and 0.50 °Brix respectively at 10, 14, 17 and 20 °Brix.

Observations and calculation for calibrating a Brix hydrometer at 27.5 °C.

Indication standard	*CC*	*TC*	Corrected indication standard	Indication under test	*STC*	Corrected indication under test	Correction at 27.5 °C
9.38	−0.05	+0.46	9.79	10.0	−0.10	9.9	−0.11
13.40	0.00	+0.47	13.87	14.0	−0.07	13.93	−0.06
16.40	0.00	+0.48	16.88	17.0	−0.07	16.83	+0.05
19.44	0.00	+0.50	19.94	20.0	−0.07	19.93	+0.01

It should be noted that a standard of range 0–10 °Brix has been used, when the indication by the test hydrometer is 10 °Brix and the *STF* of the 0–10 °Brix standard hydrometer is 0.000018 s^2 cm^{-3}.

7.12.6 Brix hydrometer calibrated against standard density hydrometer

At times, the laboratory may not have a corresponding standard Brix hydrometer. In this case a density hydrometer from the L50 series or, if possible, the L20 series may be used. From the Brix against specific gravity tables (*S*20/4 °C) given in section 7.17, table 7.12, one may see that 0.0004 g cm^{-3} is equivalent to 0.1 °Brix. One sub-division of the L50 hydrometer corresponds to 0.0005 g cm^{-3}, so if an L50 density hydrometer is used, the standard reading should be taken to a fifth of the sub-division. The certificate correction, *CC*, and temperature correction, *TC*, for 20 °C should be applied to the standard. The final corrected indication is divided by 0.999 975 to get the specific gravity at 20/4 °C. The STC is applied only to the Brix hydrometer as the calibration was done in a petroleum liquid of the low surface tension category and the standard used belongs to the same category. The table of Brix versus specific gravity at 20/4 °C given in table 7.12 of section 7.17 should be referred to and the correction applicable to the Brix hydrometer can finally be calculated. An example for a Brix hydrometer in the range 0–10 °Brix with a reference temperature of 20 °C follows. Standard density hydrometers from the L50 series with reference temperature 15 °C are used. The temperature of the liquid is kept at 20 °C.

Indication standard	CC $(\times 10^5)$	TC $(\times 10^5)$	Corrected indication	Sp.gr. at 20/4 °C	Equivalent °Brix	Indication under test—STC (°Brix)	Correction (°Brix)
0.9972	+05	−15	0.997 10	0.997 12	−0.28	$0.0 - 0.28 = -0.28$	0.00
1.0130	−10	−15	1.012 75	1.012 77	3.82	$4.0 - 0.28 = 3.72$	+0.10
1.0245	00	−15	1.024 35	1.024 37	6.63	$7.0 - 0.28 = 6.72$	−0.09
1.0370	+05	−15	1.036 90	1.036 92	9.70	$10.0 - 0.28 = 9.72$	−0.02

It should be noted that the STF of the hydrometer under test is 0.000 028 s^2 cm^{-3}. The surface tension of the liquid and that of the the hydrometer under test are, respectively, 35 and 75 mN m^{-1}.

7.12.7 API hydrometer against a density hydrometer

The API scale is related to the specific gravity at 60/60 °F by the following expression. The API degree is defined by the following formula:

$$\text{API degree} = (141.5/S60/60 \,°\text{F}) - 131.5.$$

This gives

API for $(0.927\,92) = 141.5/0.927\,92 - 131.5 = 152.49 - 131.5 = 20.99$

API for $(0.953\,27) = 141.5/0.953\,27 - 131.5 = 148.44 - 131.5 = 16.94$

API for $(0.979\,07) = 141.5/0.979\,07 - 131.5 = 144.52 - 131.5 = 13.02$

API for $(1.007\,09) = 141.5/1.007\,09 - 131.5 = 140.50 - 131.5 = 9.00.$

Therefore these hydrometers may also be tested against density standards in petroleum liquids suitable for low surface tension hydrometers. The corrections applicable to the standard hydrometer will be the certificate correction, CC, and temperature correction, TC. The final corrected indication of the standard is divided by the density of water at 60 °F (15.5 °C) to get specific gravity at 60/60 °F. An example of the calibration of such a hydrometer at four points follows.

Indication standard	CC $(\times 10^5)$	TC $(\times 10^5)$	Corrected indication standard	Specific gravity at 60/60 °F	API degree	Indication under test	Correction to API hydrometer
0.926 88	+15	−2	0.927 01	0.927 92	20.99	21	−0.01
0.952 26	+9	−2	0.952 33	0.953 27	16.94	17	−0.06
0.978 02	+10	−2	0.978 10	0.979 07	13.02	13	+0.02
1.006 02	+10	−2	1.006 10	1.007 09	9.00	9	0.00

7.12.8 Alcoholometers compared with density hydrometers

The ethanol content, which we used to call the alcoholic degree, is the number of parts by volume of ethanol at 20 °C contained in 100 parts by volume of the ethanol water mixture at 20 °C. The density of alcohol water mixtures is related to the ethanol content but decreases with an increase in ethanol content. The density of the ethanol water mixture depends upon temperature as well as ethanol content (alcoholic degree). The reason is that the addition of ethanol affects the contraction of the mixture due to an exothermic reaction. Moreover, the surface tension of the ethanol water mixture changes rapidly with ethanol content. An expression for the density with temperature and ethanol content in water is given in OIML R-22. The values of the density against ethanol content in steps of 1% along with the corresponding surface tension is given in IS: 3608 (part I)-1987, which is given in table 7.14, section 7.17, for ready reference.

When an alcoholometer is calibrated using a standard density hydrometer, three corrections are applied to the standard, namely the certificate correction CC, the temperature correction TC if necessary and the surface tension correction STC. One has to be cautious, however, as the surface tension will change for both the liquid and standard hydrometer. In low surface tension hydrometers, the surface tension varies by 1 mN m^{-1} for every 0.02 g cm^{-3}.

Indication standard	CC $(\times 10^6)$	TC $(\times 10^6)$	STC $(\times 10^6)$	Corrected indication standard	Ethanol content (%)	Indication under test (%)	Correction at 20 °C (%)
0.984 10	+260	0.0	+363	0.984 72	10.01	10	+0.02
0.981 40	+40	−147	+185	0.981 48	13.24	13	+0.24*
0.977 06	0	−146	+133	0.977 05	17.24	17	+0.24*
0.973 55	+15	−146	+103	0.973 53	19.97	20	−0.03

Please note that the density change for 1% change in the ethanol content is only about 0.0013 g cm^{-3}. So to calculate 0.1% of the ethanol content correctly, we should be able to read up to 0.0001 g cm^{-3}, hence only the L20 series of hydrometers should be used. Normally at NPL, India, we use two standards from the L20 series. The mean value of the corrected density is used for calculating the percentage of ethanol by volume and comparing it with the observed reading of the alcoholometer under test. The reference temperature of the hydrometer used for the first observation, in this example, is 20 °C so no temperature correction is applied.

Similar to alcoholometers in which the scale is in percentage of ethanol by volume, there are alcoholometers, which have the scale in percentage of ethanol by mass. The density and surface tension of ethanol water mixtures corresponding to alcoholic strength by mass are given in table 7.15, section 7.17. Therefore, alcoholometers indicating the percentage of ethanol content can be calibrated against a standard density hydrometer in the same way as explained before. To calibrate an alcoholometer, the working liquid should be maintained at the reference temperature of the alcoholometer.

7.13   Mandatory dimensions of hydrometers

7.13.1   Mandatory dimensions of density hydrometers according to IS: 3104-1993

	L20	L50	M50	M100	S50
Overall length (mm, max.)	335	335	270	270	190
Scale length (mm, min.)	105	125	70	85	40
Distance of lowest mark from the top of bulb (mm, min.)	5	5	5	5	5
Distance of upper mark from top of the stem (mm, min.)	15	15	15	15	15
Bulb diameter					
Maximum (mm)	40	27	24	20	20
Minimum (mm)	36	23	20	18	18
Stem diameter (mm, min.)	4	4	4	4	4
Volume of bulb below the lowest mark					
Maximum (cm^3)	132	65	45	26	26
Minimum (cm^3)	109	50	30	18	18
Nominal range of scale ($kg\ m^{-3}$)	20	50	50	100	50
Number of divisions	100	100	50	50	25
Value of one division ($kg\ m^{-3}$)	0.2	0.5	1.0	2.0	2.0
Tolerance ($kg\ m^{-3}$ normal)	±0.2	±0.5	±1.0	±2.0	±2.0
Tolerance ($kg\ m^{-3}$ special)	—	±0.3	±0.6	—	—

7.13.2   Mandatory dimensions of Brix hydrometers and thermo-hydrometer according to IS: 7324-1990

Dimensions	Hydrometer	Thermo-hydrometer
Overall length (mm, max.)	335	400
Length of scale (mm, min.)	125	125
Diameter of bulb		
Maximum (mm)	27	23
Minimum (mm)	23	19
Diameter of stem		
Maximum (mm)	5.0	5.0
Minimum (mm)	4.4	4.4
Uniform diameter of stem		
below lowest graduation line	5	5
The stem shall extend at least 15 mm above the upper most graduation line on the scale		
The volume of bulb below the lowest graduation line shall be 50 and 65 cm^3		
The error at any point on the scale shall not exceed $\pm 0.1\,°$Brix		

7.13.3   Mandatory dimensions for alcoholometers according to IS: 3608-1987

Dimensions and tolerances	Class A Hydrometers of range (°)			Class B Hydrometers of range (°)		
	0–10 to 50–60	60–70 & 70–80	80–90 & 90–100	0–10 to 50–60	60–70 & 70–80	80–90 & 90–100
Overall length (mm)	400	400	400	340	340	340
Scale length (max.)	150	150	150	105	105	105
Scale interval (degree)	0.1	0.1	0.1	0.1	0.1	0.1
Stem diameter (mm)	3.5	3.5	3.5	3	4	4
Bulb diameter						
Maximum (mm)	38	36	28	36	30	27
Minimum (mm)	36	32	26	32	28	23
Volume below scale						
Maximum (cm^3)	180	150	85	140	110	63
Minimum (cm^3)	160	130	65	120	90	50
Tolerance (degree)	± 0.05	± 0.05	± 0.05	± 0.1	± 0.1	± 0.1

The cross-section of the stem shall remain unchanged for at least 5 mm below the lowest graduation line
The stem shall extend at least 15 mm above the uppermost graduation line on the scale

7.13.4 Mandatory dimensions for a lactometer according to IS: 1183-1965

There are two types of hydrometer, namely long and short ones. Each has two specific gravity ranges: 1.015–1.025 and 1.025–1.035.

	Long	Short
1. Type		
2. Value of one sub-division (kg m^{-3})	0.2	0.5
3. Tolerance	±0.2	±0.5
4. Number of sub-divisions	50	20
5. Number of sub-divisions beyond scale	Nil	Nil
6. Distance of top graduation line from the bottom of bulb (mm)	220	140
7. Length of stem above the top graduation line (mm)	25 ± 5	25 ± 5
8. Scale length (mm)	64 ± 4	32 ± 2
9. Distance below the lowest graduation mark of uniform stem (mm)	—	—
10. External diameter of stem (mm)	3.1	2.6
11. External diameter of bulb (mm)	25 ± 1	17 ± 1
12. Length of uniform stem (mm)	100	70
13. Volume of bulb below bottom graduation line		
Maximum (cm^3)	56	20
Minimum (cm^3)	43	15

7.14 Facilities for calibration at NPL, India

NPL has the facilities to calibrate all types of glass hydrometer. The density of solids and liquids in special cases can be measured with a precision of 1 part in 10^5. A solid glass cylinder with an arrangement for suspension in a liquid is used as a standard. Its value has been established with a precision of a few parts in 10^6. In addition, there are three zerodur spheres of very high polish and uniformity in diameter, which have been calibrated by PTB, Germany against their solid-based density standards. These in conjunction with a special sample holder and electronically maintained temperature bath can be used to measure the density of solids. The system is computer controlled and no human intervention is required once the desired bath temperature has been obtained.

Hydrometers are calibrated at four points of the scale against the appropriate standards. A large number of lactometers, Brix hydrometers, density and specific gravity hydrometers are received for calibration. The master density hydrometers can be calibrated at several points by hydrostatic weighing (Cuckow's method).

7.15 Comparison of hydrometers by the ring method

Liquids or their mixtures with appropriate densities are necessary to calibrate hydrometers by the comparison method. Quite often liquids in the range 600–

650 kg m^{-3} are not available. The fraction of petroleum ether collected at a temperature of or less than 40 °C has a density of about 600 kg m^{-3}, but with storage its lighter component evaporates and its density becomes around 630 kg m^{-3}. To circumvent this difficulty, Gupta and Mohindernath [15] have developed a method for loading the hydrometer under test with a ring of known mass and volume. This way the loaded hydrometer of the range 600–650 kg m^{-3} floats at its different graduation marks in liquids of density 700 kg m^{-3} and above.

7.15.1 Theory

Let a hydrometer of mass m with a ring of mass M of density d_r float in a liquid of density d_L. The volume below the desired graduation mark is V and the volume of the exposed portion of stem is v. Let σ be the density of air. Then the equilibrium equation is

$$m + M(1 - d_L/d_r) + \pi DT_1/g = Vd_L + v\sigma. \qquad (7.95)$$

If W is the apparent mass of the hydrometer in air then

$$m - (V + v)\sigma = W(1 - \sigma/\Delta). \qquad (7.96)$$

Here Δ is the density of the standard weights used. If the density of the liquid in which the hydrometers float freely to the same desired mark is ρ then

$$m + \pi DT_2/g = V\rho + v\sigma. \qquad (7.97)$$

Subtracting (7.96) from (7.95),

$$M(1 - d_L/d_r) + \pi DT_1/g + W(1 - \sigma/\Delta) = V(d_L - \sigma). \qquad (7.98)$$

Subtracting (7.96) from (7.97),

$$\pi DT_2/g + W(1 - \sigma/\Delta) = V(\rho - \sigma). \qquad (7.99)$$

Dividing (7.99) by (7.98) and rearranging, we get

$$\rho = (d_L - \sigma)[\{\pi DT_2/g + W(1 - \sigma/\Delta)\}$$
$$\times \{\pi DT_1/g + M(1 - d_L/d_r) + W(1 - \sigma/\Delta)\}^{-1}] + \sigma. \qquad (7.100)$$

Here also the air density has been taken to be the same, as corrections due to a change in air density are too small to be taken cognisance of. The density d_L of the liquid is simultaneously determined with a standard hydrometer. As the reading on the scale was R when the hydrometer was floating with its ring on it, the correction C is given by

$$C = \rho - R. \qquad (7.101)$$

If the observations are taken at a temperature other than the reference temperature, then a small correction due to a change in volume of the hydrometers, as discussed earlier, is also applied.

7.15.2   Choice for the mass of the ring

If the mass of the ring is such that the hydrometer floats up to the topmost graduation (ρ_1 the minimum value of density) and d the density of the lightest liquid available, then (7.95) and (7.97) may be written as

$$m + M(1 - d/d_r) + \pi D T_1/g = (V + v)d$$
$$m + \pi D T_2/g = (V + v)\rho_1.$$

Subtracting, we get

$$M(1 - d/d_r) + \pi D(T_1 - T_2)/g = (V + v)(d - \rho_1). \qquad (7.102)$$

We also know that

$$(V + v)\rho_1 = V\rho_2 + v\sigma$$

giving

$$V(\rho_2 - \rho_1) = v(\rho_1 - \sigma). \qquad (7.103)$$

So the mass of the ring M may be calculated from the following equation:

$$M(1 - d/d_r) + \pi D(T_1 - T_2)/g = \{V(\rho_2 - \sigma)/(\rho_2 - \rho_1)\}(d - \rho_1). \quad (7.104)$$

Furthermore, once the mass of the ring is known, the density of the liquid at which the hydrometer with its ring on will float at the lowest graduation mark (maximum density ρ_2) may be calculated. In this case equations (7.95) and (7.97) can be written as

$$m + M(1 - d_1/d_r) + \pi D T_1/g = V d_1 + v\sigma$$
$$m + \pi D T_2/g = V\rho_2 + v\sigma$$

giving

$$V(d_1 - \rho_2) = [M(1 - d_1/d_r) + \pi D(T_1 - T_2)/g]. \qquad (7.105)$$

Therefore the mass of the ring will depend upon the density d of the available liquid, upon the volume V of the bulb of the hydrometer and the minimum and maximum densities on its scale.   All hydrometers are made to certain specifications, but the volume of the bulb is allowed to vary within a certain range. Hydrometers also have different ranges and scale lengths etc. However, it is fortunate that the essential requirements in Indian Standards, British Standards and the ISO are the same, which makes the task of choosing the mass of the ring a bit easier. The mass of the ring for each hydrometer will be different. To make a ring for each hydrometer which is to be tested is rather difficult. To find the mass of the ring which can cater for all hydrometers in the range 600–650 kg m^{-3}, the value of d, for the 600 kg m^{-3} marks, has been arbitrarily chosen as 700 kg m^{-3}. The calculated mass was then rounded off upward to the nearest gram. Rings of 8, 5, 3 and 2 g, in copper, were made with a view to finding which ring will suit the largest number of hydrometers. The density values for the liquids in which the hydrometer, with the ring, will float at its lowest and uppermost graduation marks have been calculated for the minimum and maximum values of the volume of the bulb and are given in table 7.9.

From columns 5 and 6 of table 7.9, it may be observed that a ring of mass 5 g may be used for all types of hydrometer such as L50, M50, M100 and S50, in the density ranges 600–650 and 650–700 kg m^{-3}. The only exception is the L50 hydrometer in the range 600–650 kg m^{-3}, which has the maximum permitted volume of 65 cm^3. For the 5 g copper ring, liquids with densities between 700 and 950 kg m^{-3} are required. These liquids are readily available, easy to store and also required for calibrating hydrometers from higher density ranges. So a 5 g copper ring is good enough for all categories of hydrometer. The mass of the ring should be determined just before it is used.

The most commonly used liquids between 600 and 1000 kg m^{-3} have surface tensions between 15 and 35 mN m^{-1}. The value of the surface tension, therefore, should be known for a precise calibration of the lower density hydrometers. However, as has been seen, if in the calculations the surface tensions differ by 5 mN m^{-1}, then the error would be no more than one-fifth of the smallest graduation which, for practical purposes, may be ignored.

To illustrate the use of the ring method, the calibration of an L50 hydrometer with a range of 600 to 650 kg m^{-3} is given in table 7.10. A platinum ring of mass 7.160 g was used. The density of each liquid is determined simultaneously with the help of another calibrated hydrometer from the L20 series.

7.16   Hydrometers with variable mass and volume

We first discussed constant-volume but variable-mass hydrometers, then constant-mass but variable-volume hydrometers. In the second class, a number of hydrometers are required to cover the measuring range say from 600 to 2000 kg m^{-3}. Now consider a hydrometer, which has a certain range of densities say ρ_1 to ρ_2, but measures density from ρ_3 to ρ_4 when loaded with another weight. Such a hydrometer with a set of weights may be termed a variable-mass hydrometer. However, with a particular weight loaded on the hydrometer, it works as a constant-mass hydrometer and measures the density of various liquids by floating up to different immersed volumes. With different weights, it will measure density starting from different density values. By an appropriate choice of the value of the mass of the attached weights, it is possible that the least density measured by one is the maximum density measured by its predecessor. In addition, it is possible that the difference between the maximum and minimum densities that it measures is the same for all different weights attached to it.

Let us consider a hydrometer of mass M, stem volume v and volume of bulb up to the lowest graduation mark V. If it measures densities starting from ρ_1 to

Table 7.9. Relation between the mass of rings and density of liquids.

Hydrometer designation	Range (kg m^{-3})	Volume of bulb (cm^3)	Mass of ring (g)	Density of liquids (kg m^{-3})	
				Lowest mark	Highest mark
L50	600–650	50	8	735	795
		65	8	704	762
		50	5	685	742
		65	5	666	721
	650–700	50	8	785	844
		65	8	747	804
		50	5	735	790
		65	5	709	763
M50	600–650	30	8	822	889
		45	8	755	811
		30	5	741	802
		45	5	695	752
		30	3	686	742
		45	3	657	712
	650–700	30	8	872	937
		45	8	800	860
		30	5	791	850
		45	5	745	801
		30	3	727	792
		45	3	702	762
		30	2	708	762
		45	2	689	742
M100	600–700	18	8	968	1147
		26	8	838	972
		18	5	816	947
		26	5	751	873
		18	3	732	851
		26	3	692	805
S50	600–650	18	8	963	1039
		26	8	855	924
		18	5	832	899
		26	5	762	824
		18	3	742	802
		26	3	699	756
		18	2	696	753
		26	2	667	722
	650–700	18	8	1012	1086
		26	8	905	972
		18	5	863	947
		26	5	812	873
		18	3	791	851
		26	3	749	806
		18	2	746	802
		26	2	717	771

Table 7.10. Observations and calculations for calibrating hydrometers by the ring method. The values are rounded to the nearest 0.00005 g cm^{-3}. The other parameters are $m = 53.994$ g, $D = 7.37$ mm and $d_r = 21.5$ g cm^{-3}.

Scale reading R (g cm^{-3})	Density of liquid d (g cm^{-3})	Density of air σ (g cm^{-3})	Surface tension of liquid T_1 (mN m^{-1})	Surface surface tension of hydrometer under test T_2 (mN m^{-1})	Calculated value of density indication ρ (g cm^{-3})	Correction (g cm^{-3})
0.600	0.676 88	0.001 107	19	15	0.599 91	−0.000 10
0.620	0.699 38	0.001 106	20	16	0.619 93	−0.000 05
0.640	0.721 86	0.001 104	21	17	0.639 93	−0.000 05
0.650	0.733 00	0.001 104	22	18	0.649 85	−0.000 15

ρ_2, then

$$M = V\rho_2 \quad \text{and} \quad M = (V + v)\rho_1. \tag{7.106}$$

If a weight of mass m is attached to it such that it starts measuring from ρ_2 to ρ_3, then by an appropriate choice of weights, each hydrometer may have the same range, and also the highest density of one and the lowest density of the second are equal. In other words two consecutive hydrometers will have a continuous density measuring range and, in addition, the ranges of each hydrometer will be equal. If R is the range of each hydrometer then

$$R = \rho_2 - \rho_1 = \rho_3 - \rho_2 = \rho_4 - \rho_3 = \rho_5 - \rho_4 \quad \text{and so on.} \tag{7.107}$$

If m_1, m_2, m_3 etc are the masses and v_1, v_2 and v_3 are, respectively, the volumes of the weights attached to the hydrometer, then the set of equilibrium equations is

$$
\begin{aligned}
M + m_1 &= (V + v_1)\rho_3 \quad &\text{and} \quad & M + m_1 = (V + v_1 + v)\rho_2 \\
M + m_2 &= (V + v_2)\rho_4 \quad &\text{and} \quad & M + m_2 = (V + v_2 + v)\rho_3 \quad (7.108) \\
M + m_3 &= (V + v_3)\rho_5 \quad &\text{and} \quad & M + m_3 = (V + v_3 + v)\rho_4 \\
M + m_4 &= (V + v_4)\rho_6 \quad &\text{and} \quad & M + m_4 = (V + v_4 + v)\rho_5.
\end{aligned}
$$

Considering the right-hand sides of equations in each row of (7.108), we get

$$
\begin{aligned}
-VR + v\rho_1 &= 0 \\
-(V + v_1)R + v\rho_2 &= 0 \\
-(V + v_2)R + v\rho_3 &= 0 \\
-(V + v_3)R + v\rho_4 &= 0 \\
-(V + v_4)R + v\rho_5 &= 0.
\end{aligned}
\tag{7.109}
$$

Subtracting the first from each subsequent equation in the set (7.109), we get

$$
\begin{aligned}
-v_1 R + v(\rho_2 - \rho_1) &= 0 \\
-v_2 R + v(\rho_3 - \rho_1) &= 0 \\
-v_3 R + v(\rho_4 - \rho_1) &= 0 \\
-v_4 R + v(\rho_5 - \rho_1) &= 0.
\end{aligned}
\tag{7.110}
$$

But $(\rho_5 - \rho_1)$ can be written as

$$(\rho_5 - \rho_4 + \rho_4 - \rho_3 + \rho_3 - \rho_2 + \rho_2 - \rho_1) = 4R.$$

Similarly $(\rho_4 - \rho_1)$ can be written as

$$(\rho_4 - \rho_3 + \rho_3 - \rho_2 + \rho_2 - \rho_1) = 3R \tag{7.111}$$

and $(\rho_3 - \rho_1)$ can be written as

$$(\rho_3 - \rho_2 + \rho_2 - \rho_1) = 2R.$$

This gives us

$$v_1 = v$$
$$v_2 = 2v$$
$$v_3 = 3v \tag{7.112}$$
$$v_4 = 4v.$$

Therefore the volume of each subsequent weight is increased by v. Substituting the values of v_1, v_2 etc into (7.106) and the set of equations in (7.108), we get

$$
\begin{aligned}
M &= (V + v)\rho_1 & M &= V\rho_2 \\
M + m_1 &= (V + v)\rho_3 & M + m_1 &= (V + 2v)\rho_2 \\
M + m_2 &= (V + 3v)\rho_3 & M + m_2 &= (V + 2v)\rho_4 \\
M + m_3 &= (V + 3v)\rho_5 & M + m_3 &= (V + 4v)\rho_5 \\
M + m_4 &= (V + 5v)\rho_5 & M + m_4 &= (V + 4v)\rho_6.
\end{aligned}
\tag{7.113}
$$

Subtracting in pairs, we get

$$m_1 = 2(V + v)R = T_1 \tag{7.114}$$
$$m_2 - m_1 = 2R(V + 2v) = T_2$$
$$m_2 = m_1 + 2R(V + 2v)$$

or

$$m_2 = 2R[(V + v) + (V + 2v)] = T_1 + T_2. \tag{7.115}$$

Therefore

$$m_3 - m_2 = 2(V + 3v)R = T_3$$

giving

$$m_3 = 2R[(V + v) + (V + 2v) + (V + 3v)] = T_1 + T_2 + T_3 \tag{7.116}$$

and

$$m_4 - m_3 = 2R(V + 4v) = T_4$$

giving

$$m_4 = 2R[(V+v)+(V+2v)+(V+3v)+(V+4v)] = T_1+T_2+T_3+T_4. \tag{7.117}$$

In general the nth weight m_n will be given by

$$m_n = 2R[(V + v) + (V + 2v) + (V + 3v) \ldots (V + nv)]$$
$$m_n = 2R[nV + n(n + 1)v/2]. \tag{7.118}$$

Here we observe that T_1, T_2, T_3 etc form an arithmetic progression with a common difference of $2Rv$. Therefore we can calculate the value of m_1, m_2, m_3 etc in terms of V, R and v. This way, we get a hydrometer with a set of different attachable weights. By attaching different weights, we get different hydrometers. Each hydrometer will have an equal range but with different starting values. Furthermore, the highest density of one will be the lowest density of the second hydrometer, thus providing a continuous measuring range.

The easiest way of attaching the weights is to have these weights in the form of circular rings, which can rest on the bulb of the hydrometer or having an eye at the lowest point of the bulb and a hook on the upper part of the weight.

Mass of weights attached, V, the volume below the lowest graduation mark, and the initial and final density of different hydrometers

Let us consider a hydrometer in the range $\rho_1 = 0.600$ g cm^{-3}, $\rho_2 = 0.650$ g cm^{-3}, $R = 0.050$ g cm^{-3}, and bulb $V = 60$ cm^3, then from (7.113) we get

$$M = 39 \text{ g} \quad \text{and} \quad v = 5 \text{ cm}^3.$$

R	T_r (g)	m_r (g)	V (cm^3)	v (cm^3)	$\rho_{initial}$ (g cm^{-3})	ρ_{final} (g cm^3)
0	0	0	60	5	0.600	0.650
1	6.5	6.5	65	5	0.650	0.700
2	7.0	13.5	70	5	0.700	0.750
3	7.5	21.0	75	5	0.750	0.800
4	8.0	29.0	80	5	0.800	0.850
5	8.5	37.5	85	5	0.850	0.900
6	9.0	46.5	90	5	0.900	0.950
7	9.5	56.0	95	5	0.950	1.000
8	10.0	66.0	100	5	1.000	1.050
9	10.5	76.5	105	5	1.050	1.100
10	11.0	87.5	110	5	1.100	1.150
11	11.5	99.0	115	5	1.150	1.200
12	12.0	111.0	120	5	1.200	1.250
13	12.5	123.5	125	5	1.250	1.300
14	13.0	136.5	130	5	1.300	1.350
15	13.5	150.0	135	5	1.350	1.400
16	14.0	164.0	140	5	1.400	1.450
17	14.5	178.5	145	5	1.450	1.500
18	15.0	193.5	150	5	1.500	1.550
19	15.5	209.0	155	5	1.550	1.600
20	16.0	225.0	160	5	1.600	1.650
21	16.5	241.5	165	5	1.650	1.700
22	17.0	258.5	170	5	1.700	1.750
23	17.5	276.0	175	5	1.750	1.800
24	18.0	294.0	180	5	1.800	1.850
25	18.5	312.5	185	5	1.850	1.900
26	19.0	331.5	190	5	1.900	1.950
27	19.5	351.0	195	5	1.950	2.000

Here we see that the total mass of the last hydrometer becomes 390 g and its volume 195 cm³, which is rather too high. Moreover to attach 27 weights may not be easy. However, we may select smaller measuring ranges say 0.600–1.000 g cm⁻³ and 1.000–2.000 g cm⁻³ or a further two ranges 1.000–1.500 g cm⁻³ and 1.500–2.000 g cm⁻³.

Therefore, let us consider a hydrometer with a range of 0.050 g cm⁻³ and starting from 1.000 g cm⁻³, and a bulb volume V of 60 cm³, giving $M = 63$ g and $v = 3$ cm³. Then the mass of weights to be attached to it will be as follows:

R	T_r (g)	m_r (g)	V (cm³)	v (cm³)	$\rho_{initial}$ (g cm⁻³)	ρ_{final} (g cm³)
0	0.0	0.0	60	3	1.000	1.050
1	6.3	6.3	63	3	1.050	1.100
2	6.6	12.9	66	3	1.100	1.150
3	6.9	19.8	69	3	1.150	1.200
4	7.2	27.0	72	3	1.200	1.250
5	7.5	34.5	75	3	1.250	1.300
6	7.8	42.3	78	3	1.300	1.350
7	8.1	50.4	81	3	1.350	1.400
8	8.4	58.8	84	3	1.400	1.450
9	8.7	67.5	87	3	1.450	1.500
10	9.0	76.5	90	3	1.500	1.550
11	9.3	85.8	93	3	1.550	1.600
12	9.6	95.4	96	3	1.600	1.650
13	9.9	105.3	99	3	1.650	1.700
14	10.2	115.5	102	3	1.700	1.750
15	10.5	126.0	105	3	1.750	1.800
16	10.8	136.8	108	3	1.800	1.860
17	11.1	147.9	111	3	1.860	1.900
18	11.4	159.3	114	3	1.900	1.950
19	11.7	171.0	117	3	1.950	2.000

Therefore, one to three hydrometers will be sufficient to cover the entire range of 0.600 to 2.000 g cm⁻³.

A point of caution

The hydrometer will read correctly at the highest and lowest graduation mark for all weights, but intermediary marks will be correct only for the first hydrometer. The position for the intermediate marks from the top mark is given by equation (7.37):

$$l_\rho = L\{\rho_2/(\rho_2 - \rho_1)\}[(\rho - \rho_1)/\rho].$$

In addition to range, the position of l_ρ also depends upon the value of the density represented by the lowest graduation mark. In the example after section 7.3.5, the positions from the top graduation mark corresponding to various values of the density are given. This is the case for two hydrometers in the ranges 600–650 and 1950–2000 kg m^{-3}. We see that the maximum difference between the positions corresponding to equal density difference on the two scales is 1.4 mm. The scale length is 100 mm, i.e. 100 mm corresponds to a difference of 50 kg m^{-3}. Therefore, this difference in position corresponds to a density value of 0.7 kg, which is quite large. However, for two hydrometers with smaller differences in their respective upper and lower values of density the differences will become smaller and hence negligible.

7.16.1 Value of density at numbered graduation marks for various hydrometers

An alternative method is for the scale of the hydrometer to have only cardinal numbers starting from the uppermost mark. The scale may bear only numbers like 10, 20, 30, 40, 50, ..., 100. After re-arranging, equation (7.37) can be used to find the density value for each graduation mark:

$$\rho = L\rho_1\rho_2/\{(L\rho_2 - l_n R\}. \tag{7.119}$$

We can calculate the values of the density for any graduation mark, which is l_n unit of length below the uppermost mark. ρ_1 represents the lowest density of the hydrometer with a particular weight attached to it.

Let the hydrometer have a scale of length 100 mm with the graduation marks numbered from 20, 40, 60, 80 and 100. The densities corresponding to these graduation marks in different measuring ranges are given in table 7.11 for various hydrometers.

7.16.2 Corrections at the numbered graduation mark of hydrometers in various ranges

Let us consider the corrections applicable due to imperfections in the scale graduations, stem and bulb volumes. Let the correction at the lowest graduation mark be $^1E_2$ due to an incorrect volume for the bulb in the first hydrometer. Let this be $V - \Delta V$. Then the equilibrium equations are

$$M' = (V - \Delta V)(\rho_2 + {}^1E_2) = V\rho_2 \tag{7.120}$$

giving

$$-\Delta V\rho_2 + {}^1E_2 V - \Delta V {}^1E_2 = 0.$$

Furthermore, $\Delta V {}^1E_2$ is small, hence it can be neglected, so we get

$${}^1E_2 = \rho_2\Delta V/V. \tag{7.121}$$

Table 7.11. Density values corresponding to numbered graduation marks of hydrometers in various ranges.

Graduation mark	Hydrometers in different measuring ranges							
	600–650	800–850	1000–1050	1200–1250	1400–1450	1600–1650	1800–1850	1950–2000
00	600	800	1000.00	1200.00	1400.00	1600.00	1800.00	1950.00
20	609.38	809.52	1009.62	1209.68	1409.72	1609.76	1809.78	1959.80
40	619.05	819.28	1019.42	1219.51	1419.58	1619.63	1819.67	1969.70
60	629.03	829.27	1029.41	1229.51	1429.58	1629.63	1829.67	1979.70
80	639.34	839.51	1039.60	1239.67	1439.72	1639.75	1839.781	1989.80
100	650.00	850.00	1050.00	1250.00	1450.00	1650.00	1850.00	2000.00

However, V_n, the effective volume below the lowest graduation mark of the nth hydrometer, is given as

$$V_n = V + (n - 1)v = V[1 + (n - 1)R/\rho_1] \qquad (7.122)$$

while the final density $^n\rho_2$ and initial density $^n\rho_1$ of the nth hydrometer is given by

$$^n\rho_2 = \rho_2 + (n - 1)R \qquad ^n\rho_1 = \rho_1 + (n - 1)R. \qquad (7.123)$$

$^nE_2$, the correction at the lowest graduation mark of the nth hydrometer, will be given by

$$^nE_2 = {}^n\rho_2 \Delta V/[V + (n - 1)v]$$
$$= [\rho_2 - (n - 1)R](\Delta V/V)/[1 + (n - 1)R/\rho_1]. \qquad (7.124)$$

Substituting the value of $\Delta V/V$ from (7.121), we get

$$^nE_2 = {}^1E_2(\rho_1/\rho_2)[\rho_2 + (n - 1)R]/[\rho_1 + (n - 1)R]$$
$$= {}^1E_2(\rho_1/\rho_2)[\rho_2/R + (n - 1)]/[\rho_1/R + (n - 1)]. \qquad (7.125)$$

We know that $\rho_2/R = 13$, $\rho_1/R = 12$ and $\rho_1/\rho_2 = 12/13$ and this gives us

$$^nE_2 = [(144 + 12n)/(143 + 13n)] \, {}^1E_2. \qquad (7.126)$$

The term within the square brackets on the right-hand side is 1 for $n = 1$ and 18/19 for $n = 27$. Hence $^nE_2$ will always be between $^1E_2$ and 18/19 of $^1E_2$.

Correction at the topmost graduation mark due to an inaccuracy in the stem volume

If $^1E_1$ is the correction at the topmost graduation mark of the first hydrometer due to an inaccuracy in the stem volume by Δv, then it can be obtained by considering (7.113) as follows:

$$M' = (V + v - \Delta v)(\rho_1 + {}^1E_1) = (V + v)\rho_1 \qquad (7.127)$$

giving

$$(V + v) \, {}^1E_1 - {}^1E_1 \Delta v - \rho_1 \Delta v = 0.$$

Neglecting $^1E_1 \Delta v$ again, we get

$$^1E_1 = \rho_1 \Delta v/(V + v) = \rho_1 \Delta v/(V + VR/\rho_1) = \rho_1 \Delta v/[V(1 + R/\rho_1)].$$

However, $1 + R/\rho_1 = \rho_2/\rho_1$ giving

$$^1E_1 = (\rho_1)^2(\Delta v/V)\rho_2. \qquad (7.128)$$

Similarly $^n E_1$ is given by

$$^n E_1 = {}^n \rho_1 \Delta v / [V_n + v]$$

and using (7.122) and (7.123), we get

$$
\begin{aligned}
&= [\rho_1 + (n-1)R]\Delta v / [V + nv] \\
&= [\rho_1 + (n-1)R](\Delta v / V) / [1 + nR/\rho_1] \\
&= [\rho_1 + (n-1)R]\rho_1(\Delta v / V) / [\rho_1 + nR].
\end{aligned}
$$

Substituting the value of $(\Delta v / V)$ from (7.128), we get

$$
\begin{aligned}
^n E_1 &= [\rho_1 + (n-1)R]/[\rho_1 + nR](\rho_2/\rho_1)\,^1 E_1 \\
&= (\rho_2/\rho_1)\,^1 E_1[\rho_1/R + (n-1)]/[\rho_1/R + n].
\end{aligned}
$$

As $\rho_1 = 0.600$, $R = 0.05$ and $\rho_2 = 0.650$, we get

$$
\begin{aligned}
^n E_1 &= (13/12)\,^1 E_1[12 + (n-1)]/[12 + n] \\
&= {}^1 E_1[156 + 13(n-1)]/[144 + 12n]. \tag{7.129}
\end{aligned}
$$

The term within square brackets is 1 for $n = 1$ and 19/18 for $n = 27$, so $^n E_1$ will vary between $^1 E_1$ and $19/18\,^1 E_1$. Hence for all practical purposes, the correction due to an inaccuracy in the bulb volume or in the stem volume remains the same for all hydrometers.

However, if we are able to get the correct mass and respective volume for each attached weight, the corrections for hydrometers with any other attached weight can be found from the corresponding numbered graduation marks. Therefore it is only necessary to calibrate the first hydrometer. The corrections for every hydrometer with any other attached weight can be found.

Significant value of $\Delta V / V$ or $\Delta V / (V + v)$

The volume of stem v corresponds to 0.050 g cm^{-3}. There are 100 graduations on the scale and at best we can estimate one-fifth of the graduation interval, which corresponds to 1/500th of the volume of the stem. Hence any change in the relative volume of this order can be detected. This gives us $\Delta v = 65/500$ cm^3, as $V + v$ is 65 cm^3, giving $\Delta v = 0.13$ cm^3.

7.17   Important tables

Four important tables (tables 7.12–7.15) are given here. The first (table 7.12) gives the relative density at 20/4 °C of sucrose solutions at different °Brix in steps of 0.1 °Brix. The following polynomial was used to obtain the relative density $S20/4$ °C:

$$S20/4\,°C = A_0 + A_1 B + A_2 B^2 + A_3 B^3 + A_4 B^4$$

where A_0, A_1, A_2, A_3 and A_4 have the following values:

$$A_0 = 0.998\,234$$
$$A_1 = 3.859\,969\,911 \times 10^{-3}$$
$$A_2 = 1.252\,761\,461 \times 10^{-5}$$
$$A_3 = 6.998\,000\,385 \times 10^{-8}$$
$$A_4 = -2.766\,519\,146 \times 10^{-10}$$

and B is the degree Brix.

The second table (table 7.13) gives the temperature corrections at different observed readings in °Brix (reference temperature 20 °C) (see [16]). The third table (table 7.14) gives the density and surface tension at 20 °C of ethanol mixtures against alcoholic degree defined in terms of volume/volume. The final table (table 7.15) gives the density and surface tension at 20°C of ethanol mixtures against alcoholic degree defined in terms of mass/mass.

7.18 International inter-comparisons of hydrometers between National Metrology Laboratories

Inter-comparison of solid-based density standards between various national laboratories has been in vogue for quite some time but recently [17] the Instituto di Metrologia 'G Colonnetti' (IMGC, Italy) took up two independent inter-comparisons of hydrometers, one with the Laboratoire d'Essais (LNE, France) and the other with the Mittatekniikan Keskus (MIKES, Finland). Density hydrometers in ranges 600–650, 850–900, 1300–1350 and 1950–2000 kg m^{-3} were used in inter-comparisons between IMGC and MIKES, while an alcoholometer of 20–30 degrees alcoholic strength (volume/volume) was used with the LNE. Hydrostatic weighing in a reference liquid was used in each case. The reference liquids were n-nonane by IMGC, Kerdanne oil–petroleum liquid by LNE and ethanol by MIKES. The density of the reference liquid was determined by the sinker method. IMGC used solid-based density standards made from silica, SL1 and SL2, LNE used a glass solid of volume 100 cm^3 with an uncertainty of ±0.32 mm^3 and MIKES used a stainless steel sphere (mass 174 g). Each laboratory determined the volume of its solid standard in terms of the density of double distilled water at 20 °C. The general arrangement for determining the density was similar to the one given in figure 3.5. The particulars of the equipment used by each laboratory are given in table 7.16. The system for calibrating the hydrometers by hydrostatic weighing used by MIKES is shown in photograph P7.3.

Table 7.12. Relative density at 20/4 °C of sucrose solutions at different °Brix in steps of 0.1 °Brix.

Brix	0	0.1	0.2	0.3	0.4	0.5	0.6	0.7	0.8	0.9
0	0.998 234	0.998 620	0.999 006	0.999 393	0.999 780	1.000 167	1.000 554	1.000 942	1.001 330	1.001 718
1	1.002 107	1.002 495	1.002 884	1.003 273	1.003 663	1.004 052	1.004 442	1.004 832	1.005 223	1.005 614
2	1.006 005	1.006 396	1.006 787	1.007 179	1.007 571	1.007 963	1.008 356	1.008 749	1.009 142	1.009 535
3	1.009 928	1.010 322	1.010 716	1.011 111	1.011 505	1.011 900	1.012 295	1.012 691	1.013 087	1.013 482
4	1.013 879	1.014 275	1.014 672	1.015 069	1.015 466	1.015 864	1.016 262	1.016 660	1.017 058	1.017 457
5	1.017 856	1.018 255	1.018 654	1.019 054	1.019 454	1.019 854	1.020 255	1.020 655	1.021 057	1.021 458
6	1.021 860	1.022 261	1.022 664	1.023 066	1.023 469	1.023 872	1.024 275	1.024 679	1.025 082	1.025 487
7	1.025 891	1.026 296	1.026 701	1.027 106	1.027 511	1.027 917	1.028 323	1.028 729	1.029 136	1.029 543
8	1.029 950	1.030 358	1.030 765	1.031 173	1.031 582	1.031 990	1.032 399	1.032 808	1.033 218	1.033 628
9	1.034 038	1.034 448	1.034 859	1.035 269	1.035 681	1.036 092	1.036 504	1.036 916	1.037 328	1.037 741
10	1.038 154	1.038 567	1.038 980	1.039 394	1.039 808	1.040 222	1.040 637	1.041 052	1.041 467	1.041 883
11	1.042 299	1.042 715	1.043 131	1.043 548	1.043 965	1.044 382	1.044 800	1.045 217	1.045 636	1.046 054
12	1.046 473	1.046 892	1.047 311	1.047 731	1.048 151	1.048 571	1.048 992	1.049 412	1.049 833	1.050 255
13	1.050 677	1.051 099	1.051 521	1.051 944	1.052 367	1.052 790	1.053 213	1.053 637	1.054 061	1.054 486
14	1.054 910	1.055 335	1.055 761	1.056 186	1.056 612	1.057 039	1.057 465	1.057 892	1.058 319	1.058 747
15	1.059 175	1.059 603	1.060 031	1.060 460	1.060 889	1.061 318	1.061 748	1.062 178	1.062 608	1.063 038
16	1.063 469	1.063 900	1.064 332	1.064 764	1.065 196	1.065 628	1.066 061	1.066 494	1.066 927	1.067 361
17	1.067 795	1.068 229	1.068 664	1.069 099	1.069 534	1.069 969	1.070 405	1.070 841	1.071 278	1.071 715
18	1.072 152	1.072 589	1.073 027	1.073 465	1.073 903	1.074 342	1.074 781	1.075 220	1.075 660	1.076 100
19	1.076 540	1.076 981	1.077 421	1.077 863	1.078 304	1.078 746	1.079 188	1.079 631	1.080 074	1.080 517
20	1.080 960	1.081 404	1.081 848	1.082 292	1.082 737	1.083 182	1.083 628	1.084 073	1.084 519	1.084 966
21	1.085 412	1.085 860	1.086 307	1.086 755	1.087 202	1.087 651	1.088 099	1.088 548	1.088 998	1.089 447
22	1.089 897	1.090 347	1.090 798	1.091 249	1.091 700	1.092 152	1.092 604	1.093 056	1.093 509	1.093 961
23	1.094 415	1.094 868	1.095 322	1.095 776	1.096 231	1.096 686	1.097 141	1.097 596	1.098 052	1.098 508
24	1.098 965	1.099 422	1.099 879	1.100 337	1.100 794	1.101 253	1.101 711	1.102 170	1.102 629	1.103 089

25	1.103 549	1.104 009	1.104 469	1.104 930	1.105 391	1.105 853	1.106 315	1.106 777	1.107 240	1.107 702
26	1.108 166	1.108 629	1.109 093	1.109 557	1.110 022	1.110 487	1.110 952	1.111 418	1.111 884	1.112 350
27	1.112 816	1.113 283	1.113 751	1.114 218	1.114 686	1.115 155	1.115 623	1.116 092	1.116 562	1.117 031
28	1.117 501	1.117 972	1.118 442	1.118 913	1.119 385	1.119 856	1.120 329	1.120 801	1.121 274	1.121 747
29	1.122 220	1.122 694	1.123 168	1.123 643	1.124 117	1.124 593	1.125 068	1.125 544	1.126 020	1.126 497
30	1.126 974	1.127 451	1.127 929	1.128 406	1.128 885	1.129 363	1.129 842	1.130 322	1.130 801	1.131 281
31	1.131 762	1.132 242	1.132 724	1.133 205	1.133 687	1.134 169	1.134 651	1.135 134	1.135 617	1.136 101
32	1.136 585	1.137 069	1.137 553	1.138 038	1.138 524	1.139 009	1.139 495	1.139 982	1.140 468	1.140 955
33	1.141 443	1.141 930	1.142 418	1.142 907	1.143 396	1.143 885	1.144 374	1.144 864	1.145 354	1.145 845
34	1.146 336	1.146 827	1.147 319	1.147 811	1.148 303	1.148 796	1.149 289	1.149 782	1.150 276	1.150 770
35	1.151 265	1.151 760	1.152 255	1.152 750	1.153 246	1.153 742	1.154 239	1.154 736	1.155 233	1.155 731
36	1.156 229	1.156 728	1.157 226	1.157 725	1.158 225	1.158 725	1.159 225	1.159 726	1.160 226	1.160 728
37	1.161 229	1.161 731	1.162 234	1.162 737	1.163 240	1.163 743	1.164 247	1.164 751	1.165 256	1.165 760
38	1.166 266	1.166 771	1.167 277	1.167 784	1.168 290	1.168 797	1.169 305	1.169 813	1.170 321	1.170 829
39	1.171 338	1.171 848	1.172 357	1.172 867	1.173 378	1.173 888	1.174 399	1.174 911	1.175 423	1.175 935
40	1.176 447	1.176 960	1.177 473	1.177 987	1.178 501	1.179 015	1.179 530	1.180 045	1.180 561	1.181 077
41	1.181 593	1.182 109	1.182 626	1.183 144	1.183 661	1.184 179	1.184 698	1.185 216	1.185 736	1.186 255
42	1.186 775	1.187 295	1.187 816	1.188 337	1.188 858	1.189 380	1.189 902	1.190 424	1.190 947	1.191 470
43	1.191 994	1.192 518	1.193 042	1.193 567	1.194 092	1.194 617	1.195 143	1.195 669	1.196 196	1.196 723
44	1.197 250	1.197 777	1.198 305	1.198 834	1.199 363	1.199 892	1.200 421	1.200 951	1.201 481	1.202 012
45	1.202 543	1.203 074	1.203 606	1.204 138	1.204 671	1.205 203	1.205 737	1.206 270	1.206 804	1.207 338
46	1.207 873	1.208 408	1.208 944	1.209 480	1.210 016	1.210 552	1.211 089	1.211 627	1.212 164	1.212 702
47	1.213 241	1.213 780	1.214 319	1.214 858	1.215 398	1.215 939	1.216 479	1.217 020	1.217 562	1.218 104
48	1.218 646	1.219 189	1.219 731	1.220 275	1.220 818	1.221 363	1.221 907	1.222 452	1.222 997	1.223 543
49	1.224 089	1.224 635	1.225 182	1.225 729	1.226 276	1.226 824	1.227 372	1.227 921	1.228 470	1.229 019
50	1.229 569	1.230 119	1.230 669	1.231 220	1.231 772	1.232 323	1.232 875	1.233 428	1.233 980	1.234 533
51	1.235 087	1.235 641	1.236 195	1.236 750	1.237 305	1.237 860	1.238 416	1.238 972	1.239 529	1.240 086
52	1.240 643	1.241 200	1.241 759	1.242 317	1.242 876	1.243 435	1.243 995	1.244 554	1.245 115	1.245 676
53	1.246 237	1.246 798	1.247 360	1.247 922	1.248 485	1.249 048	1.249 611	1.250 175	1.250 739	1.251 304

	0	1	2	3	4	5	6	7	8	9
54	1.251 868	1.252 434	1.252 999	1.253 565	1.254 132	1.254 699	1.255 266	1.255 833	1.256 401	1.256 970
55	1.257 538	1.258 107	1.258 677	1.259 247	1.259 817	1.260 387	1.260 958	1.261 530	1.262 102	1.262 674
56	1.263 246	1.263 819	1.264 392	1.264 966	1.265 540	1.266 114	1.266 689	1.267 264	1.267 840	1.268 416
57	1.268 992	1.269 569	1.270 146	1.270 724	1.271 301	1.271 880	1.272 458	1.273 037	1.273 617	1.274 196
58	1.274 777	1.275 357	1.275 938	1.276 519	1.277 101	1.277 683	1.278 265	1.278 848	1.279 431	1.280 015
59	1.280 599	1.281 183	1.281 768	1.282 353	1.282 939	1.283 525	1.284 111	1.284 698	1.285 285	1.285 872
60	1.286 460	1.287 048	1.287 637	1.288 226	1.288 815	1.289 405	1.289 995	1.290 585	1.291 176	1.291 767
61	1.292 359	1.292 951	1.293 543	1.294 136	1.294 729	1.295 323	1.295 917	1.296 511	1.297 106	1.297 701
62	1.298 296	1.298 892	1.299 488	1.300 085	1.300 682	1.301 279	1.301 877	1.302 475	1.303 074	1.303 673
63	1.304 272	1.304 872	1.305 472	1.306 072	1.306 673	1.307 274	1.307 876	1.308 478	1.309 080	1.309 683
64	1.310 286	1.310 890	1.311 494	1.312 098	1.312 702	1.313 308	1.313 913	1.314 519	1.315 125	1.315 732
65	1.316 338	1.316 946	1.317 554	1.318 162	1.318 770	1.319 379	1.319 988	1.320 598	1.321 208	1.321 818
66	1.322 429	1.323 040	1.323 652	1.324 264	1.324 876	1.325 489	1.326 102	1.326 715	1.327 329	1.327 944
67	1.328 558	1.329 173	1.329 789	1.330 404	1.331 021	1.331 637	1.332 254	1.332 871	1.333 489	1.334 107
68	1.334 725	1.335 344	1.335 964	1.336 583	1.337 203	1.337 823	1.338 444	1.339 065	1.339 687	1.340 309
69	1.340 931	1.341 554	1.342 177	1.342 800	1.343 424	1.344 048	1.344 673	1.345 298	1.345 923	1.346 549
70	1.347 175	1.347 801	1.348 428	1.349 055	1.349 683	1.350 311	1.350 939	1.351 568	1.352 197	1.352 827
71	1.353 457	1.354 087	1.354 718	1.355 349	1.355 980	1.356 612	1.357 244	1.357 877	1.358 510	1.359 143
72	1.359 777	1.360 411	1.361 045	1.361 680	1.362 316	1.362 951	1.363 587	1.364 224	1.364 860	1.365 497
73	1.366 135	1.366 773	1.367 411	1.368 050	1.368 689	1.369 328	1.369 968	1.370 608	1.371 249	1.371 890
74	1.372 531	1.373 173	1.373 815	1.374 457	1.375 100	1.375 743	1.376 387	1.377 031	1.377 675	1.378 320
75	1.378 965	1.379 611	1.380 257	1.380 903	1.381 549	1.382 196	1.382 844	1.383 492	1.384 140	1.384 788
76	1.385 437	1.386 086	1.386 736	1.387 386	1.388 037	1.388 687	1.389 339	1.389 990	1.390 642	1.391 294
77	1.391 947	1.392 600	1.393 253	1.393 907	1.394 561	1.395 216	1.395 871	1.396 526	1.397 182	1.397 838
78	1.398 494	1.399 151	1.399 808	1.400 466	1.401 124	1.401 782	1.402 441	1.403 100	1.403 759	1.404 419
79	1.405 080	1.405 740	1.406 401	1.407 062	1.407 724	1.408 386	1.409 049	1.409 711	1.410 375	1.411 038
80	1.411 702	1.412 366	1.413 031	1.413 696	1.414 362	1.415 027	1.415 694	1.416 360	1.417 027	1.417 694
81	1.418 362	1.419 030	1.419 699	1.420 367	1.421 037	1.421 706	1.422 376	1.423 046	1.423 717	1.424 388
82	1.425 059	1.425 731	1.426 403	1.427 076	1.427 749	1.428 422	1.429 096	1.429 770	1.430 444	1.431 119

	0	1	2	3	4	5	6	7	8	9
83	1.431 794	1.432 469	1.433 145	1.433 821	1.434 498	1.435 175	1.435 852	1.436 530	1.437 208	1.437 886
84	1.438 565	1.439 244	1.439 924	1.440 604	1.441 284	1.441 965	1.442 646	1.443 327	1.444 009	1.444 691
85	1.445 374	1.446 056	1.446 740	1.447 423	1.448 107	1.448 792	1.449 476	1.450 161	1.450 847	1.451 532
86	1.452 219	1.452 905	1.453 592	1.454 279	1.454 967	1.455 655	1.456 343	1.457 032	1.457 721	1.458 411
87	1.459 100	1.459 791	1.460 481	1.461 172	1.461 863	1.462 555	1.463 247	1.463 939	1.464 632	1.465 325
88	1.466 019	1.466 712	1.467 407	1.468 101	1.468 796	1.469 491	1.470 187	1.470 883	1.471 579	1.472 276
89	1.472 973	1.473 670	1.474 368	1.475 066	1.475 765	1.476 464	1.477 163	1.477 863	1.478 563	1.479 263
90	1.479 964	1.480 665	1.481 366	1.482 068	1.482 770	1.483 472	1.484 175	1.484 878	1.485 582	1.486 286
91	1.486 990	1.487 695	1.488 400	1.489 105	1.489 811	1.490 517	1.491 223	1.491 930	1.492 637	1.493 345
92	1.494 052	1.494 761	1.495 469	1.496 178	1.496 887	1.497 597	1.498 307	1.499 017	1.499 728	1.500 439
93	1.501 150	1.501 862	1.502 574	1.503 287	1.503 999	1.504 713	1.505 426	1.506 140	1.506 854	1.507 569
94	1.508 284	1.508 999	1.509 715	1.510 431	1.511 147	1.511 864	1.512 581	1.513 298	1.514 016	1.514 734
95	1.515 452	1.516 171	1.516 890	1.517 610	1.518 329	1.519 050	1.519 770	1.520 491	1.521 212	1.521 934
96	1.522 656	1.523 378	1.524 101	1.524 823	1.525 547	1.526 270	1.526 994	1.527 719	1.528 443	1.529 168
97	1.529 894	1.530 620	1.531 346	1.532 072	1.532 799	1.533 526	1.534 253	1.534 981	1.535 709	1.536 438
98	1.537 167	1.537 896	1.538 625	1.539 355	1.540 085	1.540 816	1.541 547	1.542 278	1.543 010	1.543 741
99	1.544 474	1.545 206	1.545 939	1.546 673	1.547 406	1.548 140	1.548 874	1.549 609	1.550 344	1.551 079
100	1.551 022									

Table 7.13. Temperature corrections at different observed readings in °Brix (reference temperature 20 °C) (see [16]).

Temp. (°C)	Observed °Brix (percentage of sugar)													
	0	5	10	15	20	25	30	35	40	45	50	55	60	70
Subtract from the observed °Brix (percentage of sugar)														
0	0.30	0.49	0.65	0.77	0.89	0.99	1.08	1.16	1.24	1.31	1.37	1.41	1.44	1.49
5	0.36	0.47	0.56	0.65	0.73	0.80	0.86	0.91	0.97	1.01	1.05	1.08	1.10	1.14
10	0.32	0.38	0.43	0.48	0.52	0.57	0.60	0.64	0.67	0.70	0.72	0.74	0.75	0.77
11	0.31	0.35	0.40	0.44	0.48	0.51	0.55	0.58	0.60	0.63	0.65	0.66	0.68	0.70
12	0.29	0.32	0.36	0.40	0.43	0.46	0.50	0.52	0.54	0.56	0.58	0.59	0.60	0.62
13	0.26	0.29	0.32	0.35	0.38	0.41	0.44	0.46	0.48	0.49	0.51	0.52	0.53	0.55
14	0.24	0.26	0.29	0.31	0.34	0.36	0.38	0.40	0.41	0.42	0.44	0.45	0.46	0.47
15	0.20	0.22	0.24	0.26	0.28	0.30	0.32	0.33	0.34	0.36	0.36	0.37	0.38	0.39
16	0.17	0.18	0.20	0.22	0.23	0.25	0.26	0.27	0.28	0.28	0.29	0.30	0.31	0.32
17	0.13	0.14	0.15	0.16	0.18	0.19	0.20	0.20	0.21	0.21	0.22	0.23	0.23	0.24
17.5	0.11	0.12	0.12	0.14	0.15	0.16	0.16	0.17	0.17	0.18	0.18	0.19	0.19	0.20
18	0.09	0.10	0.10	0.11	0.12	0.13	0.13	0.14	0.14	0.14	0.15	0.15	0.15	0.16
19	0.05	0.05	0.05	0.06	0.06	0.06	0.07	0.07	0.07	0.07	0.08	0.08	0.08	0.08
Add to observed °Brix (percentage of sugar)														
21	0.04	0.05	0.06	0.06	0.06	0.07	0.07	0.07	0.07	0.08	0.08	0.08	0.08	0.09
22	0.10	0.10	0.11	0.12	0.12	0.13	0.14	0.14	0.15	0.15	0.16	0.16	0.16	0.16
23	0.16	0.16	0.17	0.17	0.19	0.20	0.21	0.21	0.22	0.23	0.24	0.24	0.24	0.24
24	0.21	0.22	0.23	0.24	0.26	0.27	0.28	0.29	0.30	0.31	0.32	0.32	0.32	0.32
25	0.27	0.28	0.30	0.31	0.32	0.34	0.35	0.36	0.38	0.38	0.39	0.39	0.40	0.40
26	0.33	0.34	0.36	0.37	0.40	0.40	0.42	0.44	0.46	0.47	0.47	0.48	0.48	0.48
27	0.40	0.41	0.42	0.44	0.46	0.48	0.50	0.52	0.54	0.54	0.55	0.56	0.56	0.56
28	0.46	0.47	0.49	0.51	0.54	0.56	0.58	0.60	0.61	0.62	0.63	0.64	0.64	0.64

29	0.54	0.55	0.56	0.59	0.61	0.63	0.66	0.68	0.70	0.70	0.71	0.72	0.72	0.72
30	0.61	0.62	0.63	0.66	0.68	0.71	0.73	0.76	0.78	0.78	0.79	0.80	0.80	0.81
35	0.99	1.01	1.02	1.06	1.10	1.13	1.16	1.18	1.20	1.21	1.22	1.22	1.23	1.22
40	1.42	1.45	1.47	1.51	1.54	1.57	1.60	1.62	1.64	1.65	1.65	1.65	1.66	1.65
45	1.91	1.94	1.96	2.00	2.03	2.05	2.07	2.09	2.10	2.10	2.10	2.10	2.10	2.08
50	2.46	2.48	2.50	2.53	2.56	2.57	2.58	2.59	2.59	2.58	2.58	2.57	2.56	2.52
55	3.05	3.07	3.09	3.12	3.12	3.12	3.12	3.11	3.10	3.08	3.07	3.05	3.03	2.97
60	3.69	3.72	3.73	3.73	3.72	3.70	3.67	3.65	3.62	3.60	3.57	3.54	3.50	3.43
65	4.4	4.4	4.4	4.4	4.4	4.4	4.3	4.2	4.2	4.1	4.1	4.0	4.0	3.9
70	5.1	5.1	5.1	5.0	5.0	5.0	4.9	4.8	4.8	4.7	4.7	4.6	4.6	4.4
75	6.1	6.0	6.0	5.9	5.8	5.8	5.7	5.6	5.5	5.4	5.4	5.3	5.2	5.0
80	7.1	7.0	7.0	6.9	6.8	6.7	6.6	6.4	6.3	6.2	6.1	6.0	5.9	5.6

Table 7.14. Density and surface tension at 20 °C of ethanol mixtures against alcoholic degree defined in terms of volume/volume. In this table the density is in kg m^{-3} and the surface tension in mN m^{-1}. * Includes an imaginary ethanol content of more than 100%. These figures are necessary for the adjustment of alcoholometers. It also represents the case of highly concentrated ethanol–water mixtures at temperatures between 20 and 40 °C, the density of which formally corresponds to ethanol content above 100%. Reference IS: 3608 (part 1) 1987.

Alcohol degree		0	1	2	3	4	5	6	7	8	9
0	Density	998.20	996.70	995.23	993.81	992.41	991.06	989.73	988.43	987.16	985.92
	ST	72.6	68.1	64.5	61.7	59.6	57.8	56.1	54.5	53.1	51.8
10	Density	984.71	983.52	982.35	981.21	980.08	978.97	977.87	976.79	975.71	974.63
	ST	50.5	49.4	48.3	47.2	46.3	45.4	44.5	43.7	42.9	42.1
20	Density	973.56	972.48	971.40	960.31	969.21	968.10	966.97	965.81	964.64	963.44
	ST	41.4	40.7	40.0	39.3	38.7	38.1	37.5	37.0	36.4	35.9
30	Density	962.21	960.95	959.66	958.34	956.98	955.59	954.15	952.69	951.18	949.63
	ST	35.4	35.0	34.5	34.1	33.7	33.3	32.9	32.6	32.3	31.9
40	Density	948.05	946.42	944.76	943.06	941.32	939.54	937.73	935.88	934.00	932.09
	ST	31.7	31.4	31.1	30.9	30.6	30.4	30.2	30.0	29.8	29.6
50	Density	930.14	928.16	926.16	924.12	922.06	919.96	917.84	915.70	913.53	911.33
	ST	29.4	29.3	29.1	28.9	28.8	28.6	28.5	28.3	28.2	28.1
60	Density	909.11	906.87	904.60	902.31	899.99	897.65	895.28	892.89	890.48	888.03
	ST	27.9	27.8	27.7	27.6	27.4	27.3	27.2	27.1	27.0	26.9
70	Density	885.56	883.06	880.54	877.99	875.40	872.79	870.15	867.48	864.78	862.04
	ST	26.7	26.6	26.5	26.4	26.3	26.2	26.1	25.9	25.8	25.7
80	Density	859.27	856.46	853.62	850.74	847.82	844.85	841.84	838.77	835.64	832.45
	ST	25.6	25.4	25.3	25.2	25.0	24.9	24.8	24.6	24.5	24.4
90	Density	829.18	825.83	822.39	818.85	815.18	811.38	807.42	803.27	798.90	794.25
	ST	24.2	24.1	23.9	23.8	23.6	23.4	23.3	23.1	22.9	22.6
100	Density	789.24	*783.75	*778.26	*772.77						
	ST	22.4	22.2	22.0	21.8						

Table 7.15. Density and surface tension at 20°C of ethanol mixtures against alcoholic degree defined in terms of mass/mass. In this table the density is in kg m^{-3} and the surface tension in mN m^{-1}. * Includes an imaginary ethanol content of more than 100%. These figures are necessary for the adjustment of alcoholometers. It also represents the case of highly concentrated ethanol–water mixtures at temperatures between 20 and 40 °C, the density of which formally corresponds to ethanol content above 100%. Reference: ISO 4805-1982(E).

Alcohol degree		0	1	2	3	4	5	6	7	8	9
0	Density	998.20	996.31	994.49	992.73	991.02	989.38	987.78	986.24	984.73	983.27
	ST	72.6	67.1	63.0	60.1	57.8	55.7	53.8	52.1	50.5	49.1
10	Density	981.85	980.46	979.10	977.76	976.44	975.13	973.83	972.54	971.24	969.93
	ST	47.8	46.6	45.5	44.4	43.4	42.5	41.6	40.7	39.9	39.1
20	Density	968.61	967.27	965.90	964.51	963.09	961.63	960.14	958.61	957.05	955.44
	ST	38.3	37.7	37.0	36.4	35.8	35.2	34.7	34.2	33.7	33.3
30	Density	953.78	952.09	950.36	948.58	946.77	944.92	943.03	941.11	939.15	937.16
	ST	32.8	32.5	32.1	31.8	31.4	31.1	30.9	30.6	30.3	30.1
40	Density	935.15	933.10	931.03	928.94	926.82	924.69	922.53	920.37	918.18	915.98
	ST	29.9	29.7	29.5	29.3	29.1	28.9	28.8	28.6	28.5	28.3
50	Density	913.77	911.55	909.31	907.07	904.81	902.55	900.28	897.99	895.70	893.40
	ST	28.2	28.1	28.0	27.8	27.7	27.6	27.5	27.3	27.2	27.1
60	Density	891.10	888.78	886.46	884.13	881.79	879.45	877.09	874.73	872.37	869.99
	ST	27.0	26.9	26.8	26.7	26.6	26.5	26.4	26.3	26.2	26.1
70	Density	867.61	865.22	862.83	860.43	858.02	855.60	853.17	850.74	848.30	845.85
	ST	26.0	25.8	25.7	25.6	25.5	25.4	25.3	25.2	25.1	25.0
80	Density	843.39	840.91	838.43	835.93	833.41	830.88	828.32	825.75	823.15	820.53
	ST	24.8	24.7	24.6	24.5	24.4	24.3	24.2	24.1	24.0	23.8
90	Density	817.88	815.21	812.49	809.75	806.97	804.14	801.27	798.36	795.38	792.35
	ST	23.7	23.6	23.5	23.4	23.2	23.1	23.0	22.8	22.7	22.6
100	Density	789.24	*786.13	*783.02	*779.91	*776.80	*773.69	*770.58			
	ST	22.4	22.3	22.2	22.1	22.0	21.9	21.8			

Table 7.16. Equipment used by IMGC, LNE and MIKES.

Equipment	IMGC, Italy	LNE, France	MIKES, Finland
Balance	Electronic 405 g capacity, readability & uncertainty ±0.1 mg	1.1 kg capacity, readability & uncertainty ±0.1 mg	300 g capacity, readability & uncertainty ±0.1 mg
Bath	Jar containing reference liquid inside a thermostatically controlled water of 200 l volume	Pyrex glass jar inside a thermostatic bath	Jacketed glass jar. Water in the jacket is thermostatically controlled
Temperature measurement	With Pt100 thermometer, uncertainty ±10 mK	With Pt25, thermometer uncertainty ±10 mK	With Pt100, thermometer uncertainty ±30 mK
Surface tension	A commercial tensiometer accuracy 0.5 mN m⁻¹	—	Surface tension is measured with a Pt ring
Suspension device for hydrometer	A special system to hang the hydrometer at the desired level	A special system to hang the hydrometer at the desired level	The bath is lowered/raised manually to adjust liquid level up to the desired mark of hydrometer

Commercially available instruments are used to measure air temperature, humidity and pressure

Photograph P7.3. System for calibrating hydrometers by hydrostatic weighing (courtesy MIKES, Finland).

The results from the calibration of density hydrometers by MIKES and IMGC were in good agreement. The MIKES results are, in general, 0.2 kg m^{-3} lower than those of IMGC. Similarly an inter-comparison of the calibration of an alcoholometer by IMGC and LNE showed good agreement although in this case the LNE results were 0.04 kg m^{-3} higher than those of the IMGC. However, the differences in the results are smaller than half the value of the smallest interval (least-scale division). Such exercises assess the compatibility and mutual confidence in measurements taken by the participating laboratories.

References

[1] Indian Standard Specification for density hydrometers, IS: 3104, 1965
[2] British Standard Specification for density hydrometers, BS: 718, 1979
[3] Glazebrook Sir R (ed) 1950 *A Dictionary of Applied Physics, Hydrometers* (New York: Smith) pp 431–44, reprinted with arrangement with Macmillan
[4] *NPL Notes on Applied Science, Hydrometry* 1964 (HMSO)

[5] Alcoholometry and Alcohol Hydrometers: OIML Recommendations IR-44, 1980

[6] Indian Standard Specification for Glass Alcoholometers: IS: 3608 (Part 1), 1987

[7] International Alcoholometric Tables: OIML Recommendations IR-22, 1975
Tables for Alcoholometry (by pyknometer method) IS: 3506, 1989

[8] Indian Standard Specification for Brix Hydrometers: IS: 7324, 1983

[9] Glazebrook Sir R (ed) 1950 *A Dictionary of Applied Physics, Saccharometry* (New York: Smith) pp 724–33 (Table III, p 728), reprinted by arrangement with Macmillan

[10] Jose and Jemy 1901/02 *Bull. Assoc. Sucr. Dist.* **XIX** 302

[11] Schonrock 1900 *Z. deut. Zuckerind.* **1** 419

[12] Masui R, Watanabe H and Iuzuka K 1978 Precision hydrostatic pyknometer *Japan. J. Phys.* **17** 755–6

[13] Cuckow F W 1949 A new method of high accuracy for the calibration of reference standard hydrometers *J. Chem. Industry* **68** 44–9

[14] Bowman H A and Gallagher H 1969 An improved method of high precision calibration procedure for reference standard hydrometers *J. Res. Natl Bur. Stand.* C **73** 57–65

[15] Gupta S V and Mohindernath 1984 A method for calibrating low density hydrometers using a standard hydrometer calibrated for higher densities *Bull. OIML* **94** 7–11

[16] Bates F J *et al* 1942 Polarimetry, saccharimetry and the sugars: C-44, National Bureau of Standards, Washington

[17] Lorefice S, Heinonen M and Madec T 2000 Bilateral comparisons of hydrometer calibrations between IMGC–LNE and the IMGC–MIKES *Metrologia* **37** 141–7

Chapter 8

Density of materials used in industry—liquids

Symbols

Δ	Density of standard weights
σ	Air density
γ	Cubic expansion coefficient of material of measures
σ_p	Density of air at pressure p
$\rho_{w(t,P)}$	Density of water at pressure P and temperature t
I_1, I_2	Indications by a damped balance
M, m or W	Mass
R_1, R_2	Rest points as calculated from the turning points of a freely oscillating balance
T	Reference temperature of the capacity measure
V	Volume
ρ_l	Density of liquid
P	Porosity.

8.1 Introduction

In earlier chapters, the measurement of density has been discussed in detail with respect to those materials which are essentially used as reference materials. For example, water and mercury are used as the reference medium in more than one way. Materials like silicon, ultra-low expansion glass, quartz and zerodur, which are transformed into solids with perfect geometrical shapes, have been considered from the point of view of achieving the objective of determining the density directly in terms of the base units of length and mass. These solid objects work as primary standards of density and are used to measure the density of reference materials. The special methods for density determination, which are quite often used for measuring the density of materials used in industry, have also been discussed. However, measurement of the density of materials used in industry requires more elaboration in terms of achieving quick results by comparative non-specialists and for the variety of materials used in industry. The density of a material reflects upon its other properties and is measured to guarantee that these required properties can be achieved. Therefore, more often than not, it is measured as part of a mutual agreement between the two parties. In such cases, therefore, the density should be determined according to certain specified national or international standards. Besides other national and international organizations, the American Society for Testing and Materials (ASTM) has prepared quite comprehensive methods for measuring the density of materials and products used equally by the public and industry for a variety of industrial materials. These standards, therefore, are often used here to help in narrating the methods and describing the apparatus used for measuring the density of these materials and products. Wherever such methods have been used, appropriate referencing is made.

8.2   Definitions

Terms relating to density and relative density have been defined elsewhere; however, it is advisable to repeat the definitions in terms of ASTM standards [1,2].

8.2.1   Density (of solids and liquids)

The density of a solid or liquid is the mass of unit volume of a material at a specific temperature. The SI unit of density is $kg\ m^{-3}$, but $g\ cm^{-3}$ or $g\ ml^{-1}$ may also be used. In the case of a solid, the volume shall be that of an impermeable portion. It should be expressed as $n\ kg\ m^{-3}$ at $x\ °C$. Here n and x are pure numbers.

As liquids and solids are very much less compressible, pressure is not mentioned in industrial measurement but when precision is demanded, pressure is also indicated. Normally all density values are given at atmospheric pressure, for example, water density is referred to a normal atmospheric pressure of 101 325 Pa unless qualified by some other value of pressure.

8.2.2   Density (of gases)

The density of a gas is the mass of unit volume of a gas at the stated temperature and pressure. Mostly the density of a gas is stated at normal pressure and temperature, which is 101 325 Pa and $0\ °C$ respectively in SI units. It should be expressed as $n\ kg\ m^{-3}$ at x Pa and $y\ °C$. Here also n, x and y are pure numbers. Instead of the pascal another unit of pressure like bar or mm Hg may also be used.

8.2.3   Apparent density (of solids and liquids)

The apparent density of a solid or a liquid is the apparent mass (weight) of a unit volume of a material at a specified temperature. In the case of a solid, the volume shall be that of an impermeable portion only. It should be expressed as n kg m^{-3} at x °C. Here n and x are pure numbers.

8.2.4   Bulk density (of solids)

The bulk density of a solid is the apparent mass (weight) of a unit volume of a permeable material (it includes both impermeable and permeable voids normal to the material) at a stated temperature. It should be expressed as n kg m^{-3} at x °C. Here n and x are pure numbers.

Note

In ASTM E 12-70 [1], the word 'weight' has been used instead of mass, though it has been explained that weight here means mass in air. No buoyancy correction is applied to the volume of solid or liquid, as the case may be, in the case of apparent mass. However, buoyancy corrections for the standards of mass, according to the author, should be applied, otherwise the apparent mass given by using standards of different materials will be different.

8.2.5   Relative density or specific gravity (of solids and liquids)

The relative density or specific gravity (of a solid or liquid) is the ratio of the mass of unit volume of a material at a stated temperature to the mass of the same volume of gas-free distilled water at a stated temperature. The two stated temperatures can be either the same or different. If the material is a solid, the volume shall be that of an impermeable portion. It should be expressed as n at x °C/y °C. Here n, x and y are pure numbers.

8.2.6   Relative density or specific gravity (of gases)

The relative density or specific gravity (of a gas) is the ratio of the mass of a certain volume of a gas under the observed temperature and pressure to the mass of the same volume of air containing 0.04% of CO_2 at the same temperature and pressure. It should be expressed as n at x °C/y °C. Here n, x and y are pure numbers.

   Here the relative humidity of air or gas is not mentioned. In the view of the author, the relative humidity of air should also be stated.

8.2.7   Apparent relative density or specific gravity (of solids and liquids)

The apparent relative density or specific gravity (of a solid or liquid) is the ratio of apparent mass (weight) of a certain volume of a material at stated temperature to the apparent mass (weight) of the same volume of gas-free distilled water at the stated temperature. If the material is a solid, the volume shall be that of an impermeable portion. It should be expressed as n at $x\,°C/y\,°C$. Here n, x and y are pure numbers.

8.2.8   Bulk relative density or specific gravity (of solids)

The bulk relative density or specific gravity (of a solid) is the ratio of the apparent mass (weight) of a certain volume of a permeable material (including both impermeable and permeable voids normal to the material) at the stated temperature to the apparent mass (weight) of the same volume of gas-free distilled water at the stated temperature. If the material is a solid, the volume shall be that of the impermeable and permeable portions. It should be expressed as n at $x\,°C/y\,°C$. Here n, x and y are pure numbers.

Notes

(1) The term 'distilled water' used in these definitions is vague, so instead of distilled water the term SMOW (Standard Mean Ocean Water) should be used.
(2) According to the guidelines of the International Union of Pure and Applied Chemistry (IUPAC: Compendium of Analytical Nomenclature—Definitions and Rules 1997), the term relative density is preferred in place of specific gravity. So the term relative density followed by specific gravity has been used.

8.2.9   Porosity

Porosity is the ratio of the volume of open pores (pervious volume) to that of the total volume of the solid. Generally porosity P is expressed in %.

8.3   Density of liquids

Different types of pyknometer, hydrometer and sinker of known volume by using hydrostatic weighing are employed for determining the density of petroleum liquids and products. The specific method to be chosen depends upon the viscosity and accuracy requirements. For pure liquids, hydrometers with the appropriate range and scale values are used. The method is quite convenient and easy. Hydrometers are also used to find the density of liquids at high pressures. The normal repeatability of the method is 0.0002 g cm^{-3}. For volatile or low viscous liquids Lipkin bicapillary pyknometers are used, while for very viscous liquids, a Bingham pyknometer is a good choice. Bingham pyknometers with a slightly different design are also used for petroleum liquids with a boiling point between 90 and 100 °C.

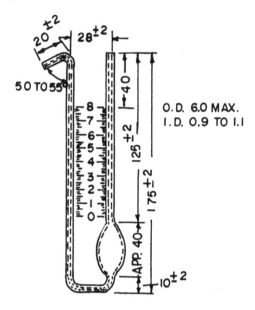

Figure 8.1. Pyknometer for low viscosity liquids.

8.4 Density of liquids using a pyknometer

8.4.1 Low viscosity liquids

The method given here [3] is suitable for liquids which do not have a vapour pressure greater than 600 mm Hg (0.8 times the atmospheric pressure) at room temperature and a viscosity no greater than 15 mm^2 s^{-1}.

Pyknometer

The pyknometer is U-shaped and has a bulb on one side, which holds the bulk of the liquid. The bulb culminates on both sides in capillary tubes of internal diameter 1 mm and is shown in figure 8.1. Beyond the bulb, the capillaries on either side are graduated. The zeros of the two scales are at the same level. The volume of the bulb, including the volumes of the two capillaries up to the zero of the scale on either side, is 45 ± 0.5 cm^3. The bulk volume is defined in between the zeros on the two scales via the bulb. The volume in between two consecutive graduations corresponds to 1 mm^3 approximately. The uniformity in the diameter of the capillary tube should be better than 10%, so that the error in the volume due to non-uniformity is less than 0.2 mm^3, which is about 4 ppm. The other dimensions are given in figure 8.1. The pyknometer is made from borosilicate glass.

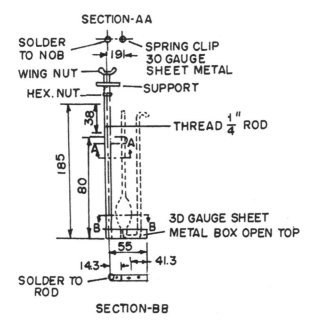

Figure 8.2. Pyknometer stand.

Pyknometer holder

A pyknometer holder of suitable dimensions so as to hold the pyknometer in a vertical position is used. A holder with basic dimensions is shown in figure 8.2.

Thermostat bath

A water bath of depth 300 mm in which the temperature can be maintained within ±0.02 °C at 20 or at 25 °C is suitable for this purpose.

Cleaning the pyknometer

The pyknometer is thoroughly cleaned with hot chromic acid. Keeping it submerged overnight in the hot chromic acid is advisable. The pyknometer is then rinsed with distilled water and is dried in a chamber for more than an hour. The temperature of the chamber is maintained between 105 and 110 °C. A filtered stream of hot air may also be passed through it if necessary. Whenever the draining water or liquid either in the bulb or capillary is not clean or the pyknometer is to be calibrated, it should be cleaned again. In between the tests it can be cleaned with isopropane or benzene and dried in vacuum. If acetone is used, the pyknometer should be rinsed with isopropane or benzene.

Thermometers

Thermometers which can read up to 5 mK should be used to measure the temperature.

Balance

A balance of suitable capacity, which can read up to 0.1 mg, should be used.

Standards of mass

Weights of F1 class from OIML will be appropriate. Each shall be calibrated within 0.05 mg. The same set of weights should be used for calibration of the pyknometer and density of liquids.

Calibration of pyknometer

The pyknometer should be calibrated against water.   Pure distilled water preferably with known isotopic ratio and free from dissolved air should be used. Water is held in a beaker and the tip of the siphon tube is dipped into the water. Water will rise due to surface tension (the diameter of the tube is only 1 mm) and will be able to reach the highest point and start flowing down. Let the pyknometer be filled by siphoning. The water will take about a minute to fill it up to the zero of the scale. Disconnect the hooked tip from the water and place the pyknometer in the water bath and leave it for 10 min so that thermal equilibrium between water, pyknometer and water bath is attained and the water acquires 20 or 25 °C as the case may be. Read the sum of the two readings of the scale. Calculate the capacity of the vessel from the following formula.

$$V = M(1 - \sigma/\Delta)/(\rho_w - \sigma) \tag{8.1}$$

where ρ_w is the density of water at 20 or 25 °C, σ is the density of air at the temperature, pressure and relative humidity of air at which the weighing was carried out and Δ is the density of weights used in the calibration. M is the difference in the mass of the standards used when weighing the pyknometer empty and when weighed with water. The calibration is repeated at least at three points of the scales.

The total volume V can be expressed as a sum of the bulb volume and the volume of water in the capillaries. If the sum of the scale readings is s then

$$V = V_0 + sa \tag{8.2}$$

where V_0 is the volume of the bulb between zeros of the two scales and a is the area of the cross section of the capillaries.

If we plot a graph between V and s it should be a straight line meeting the volume (y) axis at V_0. Then the three values of volume so obtained are plotted against the corresponding values of s on the x-axis. The straight line best fitted to the three points is drawn. Normally it should pass through all the three points but experimental error and the non-uniformity of the bore of the capillaries may cause some discrepancies. The vertical distance between the point and line should not increase by more than 0.0002 cm^3. However if this condition is not satisfied, repeat more observations and draw the straight line. Discard the pyknometer when a proper straight line cannot be drawn even after repeating the readings.

Procedure for density determination of liquids

Find the mass of the liquid following the same procedure as in the case of water; make sure that the pyknometer and its contents attain the bath temperature. Let the mass be W_1. From the value of the sum of the scale readings, find the volume from the graph. If it is V, then ρ, the density of the liquid at the temperature at which the bath is maintained, is given by

$$\rho = W_1/V. \tag{8.3}$$

8.4.2 Density of highly volatile liquids

For highly volatile liquids, i.e. those liquids which have components with a boiling point of 20 °C or less, cool the temperature of the liquid and pyknometer to around 5 °C and then fill it. The formation of dew on the pyknometer may be a problem. To prevent this, due care has to be taken. Either the air around it should be kept dry enough so that the dew point is below 5 °C or all dew should be properly removed with a lint-free cloth before weighing. Another problem, while wiping out dew so formed on the pyknometer or using dry air, is the accumulation of static charge on the glass pyknometer. Good care should be taken to discharge this static charge by earthing the pyknometer properly. It takes about 10 to 20 min for the charge to dissipate. The error due to static charge may be up to 1 mg.

This method is capable of giving a repeatability of 0.0001 g cm^{-3} and a reproducibility of 0.0002 g cm^{-3}.

8.4.3 Density of liquids with a viscosity of more than 15 mm^2 s^{-1}

The method described here is from [4] and is useful for liquids with a viscosity of more than 15 mm^2 s^{-1} at 20 °C and also for viscous oils and melted waxes at elevated temperatures but not for those liquids which have a vapour pressure of 13 kPa or more at room temperatures. The density at elevated temperatures such as 40 or 100 °C is required when converting kinematic viscosity to dynamic viscosity.

Pyknometer for viscous liquids

The pyknometer used for such liquids has only a few modifications to the Lipkin pyknometer described in section 8.4.1.

- The end of the capillary on the measuring bulb side ends in an open cup and is 2 cm above the end of the capillary on the other side.
- Prior to the two capillaries of internal diameter 1 ± 0.2 mm on which the scale is marked, two capillaries of internal diameter 2 ± 0.5 mm start immediately from the bulb.
- The other side of the capillary is well ground and polished so that a detachable bent side arm can be fitted on it. The bent side arm is also ground and polished from inside so that the side arm can be fitted to the pyknometer for filling it. The internal diameter of the capillary of the side arm is 2 mm.
- The scale is graduated in mm and every cm mark is numbered.
- The capacity of the bulb including the capacities of the two wider capillaries up to the zero of the scale on either side is 10 ± 0.1 cm^3.
- The uniformity in diameter of the capillary tube is better than 10%, so the error due to that non-uniformity is less than 0.2 mm^3, which makes the error in the volume due to this cause only about 0.002%.

A side-arm-type pyknometer conforming to these dimensions is shown in figure 8.3.

Rack

A rack which can be used in filling is shown in figure 8.4.

Constant temperature oven

Any hot air oven capable of maintaining a temperature around 100 °C and able to hold the rack is good enough. It is used in filling the pyknometer.

Constant temperature bath

A bath containing oil or a mixture of water and glycerine with a depth of 300 mm or more and capable of maintaining any temperature between 20 and 100 °C constant within 0.01 °C is good enough.

Bath thermometers

Calibrated mercury-in-glass thermometers with an uncertainty of 0.01 °C and graduated in 0.1 °C with a range of 17 to 22 °C are required.

    The rest of the equipment is more or less the same as that described in section 8.4.1.

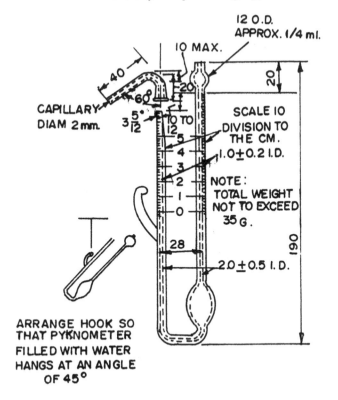

Figure 8.3. Pyknometer for viscous liquids.

Standards of mass

Weights of F1 class of OIML will be appropriate. Each should be calibrated with 0.05 mg. The same set of weights should be used to calibrate both the pyknometer and the density of the liquids.

Cleaning the pyknometer

The cleaning procedure for this pyknometer is the same as that previously given but isopentane instead of isopropane is used for rinsing and cleaning in between refills of the liquid.

Calibration of the pyknometer

Weigh the clean dry pyknometer without the side arm to the nearest 0.1 mg and record its mass. Fill the pyknometer with freshly boiled distilled water by placing its side arm in the beaker of water. Fill the pyknometer with a siphoning action.

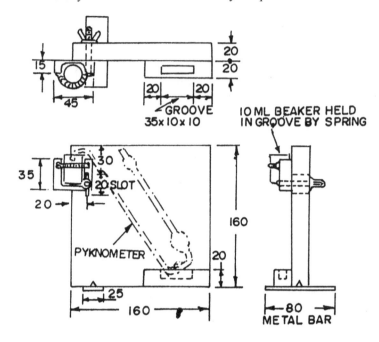

Figure 8.4. Rack for holding the pyknometer while filling.

Break the siphon by removing the side arm when the water reaches the 6 cm mark on the other scale.

Remove the excess water from the capillary tip by wiping it with a small filter paper. Place the pyknometer in the holder and the system in the constant temperature bath at *t* °C. When the water level on both scales reaches equilibrium, wait for another 15 min, before reading the scale to the nearest 0.2 mm. Wait 5 min, then read the scale again. If the difference between the sums of the two scale readings is more than ±0.04%, repeat the readings at 5 min intervals until the repeated sums of the readings are within ±0.04%.

Remove the pyknometer and clean it from outside with acetone, then with distilled water and dry it with chemically clean lint-free cloth slightly damped with water. Allow it to come to room temperature and weigh it. The apparent mass of water contained in the pyknometer is thus determined.

Similarly find the apparent mass of water, which fills up to three different positions on the scale. Obtain the apparent volume by dividing the mass of water by its density at the temperature of calibration. Plot the best-fit line for volume on the *y*-axis against the sum of the scale readings on the *x*-axis. The vertical distance of the point from the line of best fit in any case should not exceed 0.0002 cm^3. Calibration will be made at 20, 37.78 and 50 °C and the corresponding lines of best fit will be obtained.

Procedure for liquid density

Weigh the clean dry pyknometer without the side arm to the nearest 0.1 mg and record its mass. Place the sample beaker (capacity 10 cm^3) in the wooden rack. Before attaching the side arm to the pyknometer, drain a few drops of the liquid sample through the side arm to wet the inside surface to reduce the chances of trapping air bubbles in the capillary during filling. Fit the side arm onto the pyknometer and place the assembly on the rack with the side arm dipping into the liquid.

For very viscous oils or waxes with high melting points, place the whole assembly in a hot-air oven to facilitate filling. The temperature in the oven is maintained around 100 °C, which is enough for any viscous oil or wax.

Apply gentle suction to the bulb arm to start the siphoning action. The suction must be gentle to prevent bubbles forming. After the siphoning has started, let it continue until the liquid level in the bulb arm ceases to rise, but not above 6.4 cm. Stop siphoning by removing the side arm. Then remove the pyknometer from the rack and place it in a thermostatic bath in the same tilted position, until the level ceases to change. Place the pyknometer in an upright position and allow the upper liquid level in the bulb arm to reach the upper portion of the calibrated capillary.

After removing the side arm from the short arm of the pyknometer, wipe the tip and ground glass joint of the pyknometer. Adjust it in an upright position in the thermostatic bath. The bath liquid level shall be above the 6 cm mark on the pyknometer, but below the ground glass joint tip of the pyknometer.

Allow 15 min for equilibrium to be obtained. Read the meniscus levels in both arms of the pyknometer to the nearest one-fifth of the smallest graduation mark. If the sums of the readings at the two different times do not agree within ±0.04%, repeat at 5 min intervals until repeated readings within 0.04% are obtained. Record the sum of these readings together with the corresponding apparent volume from the calibration curve for the same temperature.

To avoid drainage error, the final level of the oil in the pyknometer should be less than 5 mm below the tip of the ground glass end of the pyknometer. The level in the bulb side of the pyknometer should be above the highest level which has ever been reached at any time during the procedure. This precaution is all the more important when very viscous liquid samples are used.

Remove the pyknometer from the bath and tilt it so that the liquid moves down in the short arm and up in the bulb arm. Clean and dry the outside of the pyknometer as described earlier and allow it to come to the room temperature of the balance. Weigh to the nearest 0.1 mg. Subtract the apparent mass of the empty pyknometer, without the side arm, to obtain the apparent mass of the sample.

Calculate the density of the sample, corrected to vacuum, from the following equation:

$$\text{Density in vacuum, } d_t = W/V + A \text{ g cm}^{-3} \qquad (8.4)$$

where W is the apparent mass of sample in air in g, V the apparent volume in cm^3

and A the air buoyancy correction given in table 8.2.

Calculate the relative density (specific gravity) of the sample at $(t_1/t_2\,°C)$ by dividing the density, as calculated from (8.4) by the density of water at the reference temperature t_2. The relative density at $t_1/15.5$ or $(t/60\,°F$ where t is expressed in degrees Fahrenheit) can be changed to the conventional $15.56/15.56\,°C$, i.e. $(60/60\,°F)$ relative density, from table 23 (the Petroleum Measurement Table) in ASTM D 1250.

8.4.4 Density of highly viscous liquids, viscosity up to 40 000 mm^2 s^{-1}

Bingham pyknometer [5]

This is a single capacity pyknometer. It has a cylindrical bulb, with a capacity which may take values from 25 to 2 cm^3. The bulb culminates in a thick capillary tube 7–8 mm in outer diameter but only 1–1.1 mm in internal diameter. The other end of the capillary widens into a 6 mm tube. The length of the tube may be around 30 mm. The end of the tube is properly ground so that a ground glass stopper can be fitted tightly in the pyknometer. A fine mark made either by a diamond point or etching goes all around the neck of the capillary, which defines its volume. The pyknometer is made from borosilicate glass. Pyknometers recommended in ASTM D 1480-81 [5] and ASTM D 1217-81 [6] are shown in figures 8.5(a) and 8.5(b). As the pyknometer has a closed-end capillary neck, it requires a special arrangement to either clean it or fill it. To fill it, a hypodermic syringe with a needle which has a smaller diameter than that of the neck of the pyknometer is used.

Cleaning arrangement

This essentially consists of an overflow tube with a ground-glass male socket on which the pyknometer can be fitted. The socket is connected to a two-way stop cock, which connects it either to a flask acting as a trap leading to a vacuum pump or to a beaker containing the cleaning agent. The arrangement is shown in figure 8.6.

Cleaning the pyknometer

The whole arrangement for cleaning is mounted firmly on a stand and the pyknometer is fixed on the socket upside down. The stopcock A is moved into a position so that pyknometer is connected to a vacuum pump via flask F. The end of the tube T is dipped in the hot chromic acid contained in the beaker B. When enough vacuum is generated, the stopcock is moved to another position so that hot chromic acid at 40–50 °C fills the pyknometer. Keep the pyknometer in this position for several hours. If necessary, the pyknometer may be emptied through the trap flask by evacuation and recharged again with hot chromic acid. Final cleaning is done by rinsing it with distilled water by repeating the process of

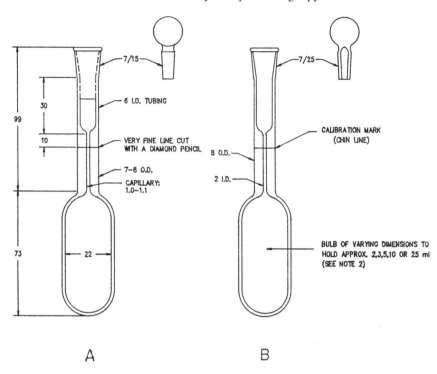

Figure 8.5. Two sketches of Bingham pyknometers.

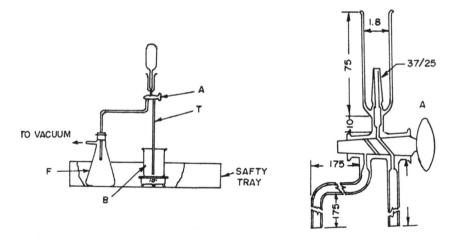

Figure 8.6. Glass cleaning assembly for a Bingham pyknometer.

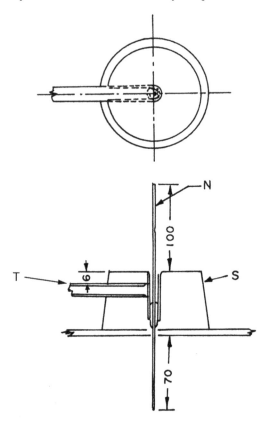

Figure 8.7. Cleaning assembly with a hypodermic needle.

evacuation followed by filling. Alternatively, the pyknometer may also be cleaned by a strong jet of solvent such as isopentane followed by distilled water through a hypodermic needle as shown in figure 8.7.

The pyknometer is fixed to the cork S and connected to a vacuum pump through tube T. The lower end of the needle is dipped in the isopentane contained in the beaker B (not shown). When the pressure is reduced, a jet of isopentane strikes the bottom of the pyknometer, not shown in the figure, cleaning the pyknometer and its walls. Finally the pyknometer is rinsed with distilled water and dried in the same manner by passing pure dry and clean air.

Such cleaning is necessary whenever the liquid sample is changed or it is observed that the liquid fails to drain cleanly either from the walls of the pyknometer or its capillary. In between the determination, it is cleaned with solvent, distilled water and dried by evacuation. The outside of the pyknometer is also cleaned with acetone and distilled water and dried finally with a moist lint-free cloth before any weighing.

Calibration of the pyknometer

The dried pyknometer is weighed in air on a balance to the nearest 0.1 mg. Glass has a tendency to pick up electrostatic charge from the air especially when the relative humidity is less than 50%. Therefore, proper care needs to be taken to discharge completely any electrostatic charge by earthing the balance properly and waiting for 10–15 min.

The pyknometer is filled with distilled water by using a hypodermic syringe until the water level is lower than the specified mark if the reference temperature is higher than the room temperature. However, if the reference temperature is lower than the room temperature, the water level is then kept a little higher than the specified mark. The pyknometer is placed in the constant temperature bath maintained at the reference temperature within ±0.01 °C. The temperature is read with a mercury-in-glass thermometer, which has been calibrated previously to the desired accuracy. The depth of the bath liquid must be such that the mark on the pyknometer is well below the level of the liquid in the bath. The thermostatic liquid may be water for a reference temperature of 20–25 °C but may be a water and glycerine mixture for higher reference temperatures. If the temperature of the bath is higher than that of the water inside the pyknometer, the water will expand and overshoot the mark. The excess water can be removed with the help of a syringe. To avoid drainage problems, the level of water should not be allowed to rise by more than 10 mm above the mark. The process is continued until there is no further rise or fall in the water level, indicating that thermal equilibrium between water inside the pyknometer and the thermostatic liquid of the bath has been reached. Keep the pyknometer in the bath for another 15 min and ensure that the specified mark is tangential to the water meniscus in the pyknometer. Take out the pyknometer, clean its outside with acetone and distilled water and dry it. Keep it near the balance until it acquires room temperature. Weigh it. The apparent mass of the water M_w is obtained by subtracting this value from the apparent mass of the empty pyknometer.

Filling the pyknometer with liquid sample

Warm the pyknometer, syringe and its needle to a convenient temperature in an oven. Draw the requisite amount of the specimen into the syringe and transfer the liquid into the pyknometer. Care should be taken that no air bubbles are occluded in the process. Fill the pyknometer until the specimen reaches about 10–20 mm above the specified mark. The exact distance above the mark will depend upon the contraction, which the specimen is likely to undergo when brought to the reference temperatures. Place the pyknometer in the constant temperature bath at the desired reference temperature and wait for 15 min after thermal equilibrium is reached. Take out the excess liquid with the help of a filter paper and make sure that the specimen stays up to the specified mark. Clean inside the pyknometer with a swab moist with a suitable solvent and dry it.

Filling the pyknometer with samples which require preheating

The specimen which needs to be melted, together with the pyknometer and syringe are kept in an oven. The oven is maintained at a temperature which is 2–3 °C above the melting point of the specimen. Allow all items to acquire the oven temperature. Charge the syringe and transfer the liquid specimen to the pyknometer. Fill the pyknometer above the specified mark to such a point that on cooling to the reference temperature, the specimen comes to the specified mark. Take out the excess liquid with a filter paper and ensure that the liquid level just touches the specified mark. Clean the pyknometer with a swab moist with a suitable solvent several times. When sure that it is clean, dry it with a dry swab. Leave the pyknometer in the constant temperature bath, which is being maintained at the desired reference temperature, until there is no change in the liquid level.

After proper filling, remove the pyknometer; clean it with a suitable solvent and acetone. Dry it and finally clean it with a lint-free slightly damp cloth. Weigh the pyknometer and find the mass or apparent mass whichever the case may be to calculate the density of the specimen at the reference temperature.

8.4.5   Determination of density of specimen at different temperatures

To find the density of the liquid at different reference temperatures, there are two methods. In the first one, the actual capacity of the pyknometer is calculated at different reference temperatures. In the second one the apparent relative density is found in terms of a factor which is the ratio of the density of water to its mass at the reference temperature. Air buoyancy correction terms are applied separately. Multiplication of the factor by the apparent mass of the same volume of liquid gives the apparent density of the liquid and applying the buoyancy correction gives the density of the liquid at that temperature.

The first method

If M_w is the mass of water filled up to the specified mark at temperature t °C, then a correction C is added to it to obtain the capacity of the pyknometer up to the graduation mark at a given reference temperature. From the same data, the values of correction C may be calculated at any other reference temperature. In this case it is not necessary to maintain the bath at all reference temperatures at which the pyknometer is required to function thus reducing the burden of calibrating the pyknometer at all reference temperatures. The values of C for borosilicate glass for reference temperatures from 20 to 80 °C in steps of 10 °C against temperature of water have been given in tables 8.1(a)–(n). Tables 8.1(a)–(g) are for standard weights of density 8.400 g cm^{-3} while tables 8.1(h)–(n) are for standard weights of density 8.000 g cm^{-3}. The temperature of the water for which the values of C have been tabulated ranges from 5 to 41 °C in steps of 0.1 °C. While calculating the values of C, the following have been taken into account:

- the buoyancy corrections due to the water being weighed in air against mass standards of different density,
- the water density at different temperatures and
- the cubic expansion coefficient of borosilicate glass.

The capacity, V_{tr}, of the measure at the reference temperature, t_r °C, is given by

$$V_{tr} = M_w + C.$$

If M_1 is the corrected mass of the standards, which equals the mass of the liquid filling the pyknometer to the specified mark at the same reference temperature T °C, then the density of the liquid ρ_1 is related:

$$M_1(1 - \sigma/\Delta) = V_{tr}(\rho_1 - \sigma) \tag{8.5}$$

giving

$$\rho_1 = [M_1(1 - \sigma/\Delta)/V_{tr}] + \sigma. \tag{8.6}$$

The second method

The second method is not to calculate the actual capacity of the pyknometer at the reference temperature, but to calculate instead a factor F, given by

$$F = \rho_w/M_w \tag{8.7}$$

where ρ_w is the density of water at the temperature of the bath, which should be the same as the reference temperature. If M_{la} is the apparent mass of the liquid with a volume equal to that of water at the same temperature, then the density, ρ_1, is given by

$$\rho_1 = M_{la}F + A \tag{8.8}$$

where A is an air buoyancy correction. In this method, all measurements are to be carried out at the desired reference temperature. The methods for arriving at the expressions for C and A are given in the next section.

The repeatability of the method is 0.000 05 g cm^{-3} while reproducibility is 0.000 14 g cm^{-3}.

8.4.6 Expression for C

Let M_w be the apparent mass of water contained in the pyknometer at temperature t °C. If σ, Δ are the density of air and that of the material of mass standards, then V_t, the volume of the pyknometer at t °C, is given by

$$M_w(1 - \sigma/\Delta) = V_t(\rho_w - \sigma). \tag{8.9}$$

If V_T is the capacity of the pyknometer at reference temperature t_r °C, then

$$V_t = V_{tr}\{1 + \gamma(t - T)\} \tag{8.10}$$

Table 8.1. (*a*) Value of correction C in mg for a 1 cm^3 borosilicate glass measure for $\gamma = 0.00000975$, $\sigma = 0.0012$ g cm^{-3}, $\Delta = 8.400$ g cm^{-3}, reference temperature $= 20\,°C$, where γ is the coefficient of expansion of the material of the measure, σ is the density of air (in g cm^{-3}) and Δ is the density of standard weights used (in g cm^{-3}).

T (°C)	0.0	0.1	0.2	0.3	0.4	0.5	0.6	0.7	0.8	0.9
5	1.2365	1.2372	1.2381	1.2391	1.2402	1.2416	1.2430	1.2446	1.2464	1.2483
6	1.2504	1.2526	1.2550	1.2575	1.2601	1.2629	1.2659	1.2690	1.2723	1.2757
7	1.2792	1.2829	1.2867	1.2907	1.2949	1.2991	1.3036	1.3081	1.3128	1.3177
8	1.3226	1.3278	1.3330	1.3385	1.3440	1.3497	1.3555	1.3615	1.3676	1.3739
9	1.3803	1.3868	1.3934	1.4002	1.4072	1.4143	1.4215	1.4288	1.4363	1.4439
10	1.4517	1.4595	1.4676	1.4757	1.4840	1.4924	1.5010	1.5097	1.5185	1.5274
11	1.5365	1.5457	1.5550	1.5645	1.5741	1.5838	1.5937	1.6037	1.6138	1.6240
12	1.6344	1.6449	1.6555	1.6662	1.6771	1.6881	1.6992	1.7105	1.7219	1.7334
13	1.7450	1.7567	1.7686	1.7806	1.7927	1.8050	1.8173	1.8298	1.8424	1.8552
14	1.8680	1.8810	1.8941	1.9073	1.9206	1.9341	1.9476	1.9613	1.9751	1.9891
15	2.0031	2.0173	2.0316	2.0460	2.0605	2.0751	2.0899	2.1047	2.1197	2.1348
16	2.1500	2.1654	2.1808	2.1964	2.2120	2.2278	2.2437	2.2597	2.2759	2.2921
17	2.3085	2.3249	2.3415	2.3582	2.3750	2.3919	2.4090	2.4261	2.4433	2.4607
18	2.4782	2.4957	2.5134	2.5312	2.5491	2.5672	2.5853	2.6035	2.6219	2.6403
19	2.6589	2.6775	2.6963	2.7152	2.7342	2.7533	2.7725	2.7918	2.8112	2.8307
20	2.8503	2.8701	2.8899	2.9098	2.9299	2.9500	2.9703	2.9906	3.0111	3.0317
21	3.0523	3.0731	3.0940	3.1149	3.1360	3.1572	3.1785	3.1999	3.2214	3.2430
22	3.2646	3.2864	3.3083	3.3303	3.3524	3.3746	3.3969	3.4193	3.4418	3.4644
23	3.4871	3.5099	3.5327	3.5557	3.5788	3.6020	3.6253	3.6487	3.6721	3.6957
24	3.7194	3.7432	3.7670	3.7910	3.8150	3.8392	3.8635	3.8878	3.9123	3.9368
25	3.9614	3.9862	4.0110	4.0359	4.0609	4.0860	4.1112	4.1365	4.1619	4.1874
26	4.2130	4.2387	4.2645	4.2903	4.3163	4.3423	4.3685	4.3947	4.4210	4.4474
27	4.4739	4.5005	4.5272	4.5540	4.5809	4.6079	4.6349	4.6621	4.6893	4.7167
28	4.7441	4.7716	4.7992	4.8269	4.8547	4.8825	4.9105	4.9385	4.9667	4.9949
29	5.0232	5.0516	5.0801	5.1087	5.1374	5.1661	5.1950	5.2239	5.2530	5.2821
30	5.3113	5.3406	5.3699	5.3994	5.4289	5.4586	5.4883	5.5181	5.5480	5.5780
31	5.6080	5.6382	5.6684	5.6987	5.7291	5.7596	5.7902	5.8209	5.8516	5.8825
32	5.9134	5.9444	5.9754	6.0066	6.0379	6.0692	6.1006	6.1321	6.1637	6.1954
33	6.2271	6.2590	6.2909	6.3229	6.3550	6.3871	6.4194	6.4517	6.4841	6.5166
34	6.5492	6.5819	6.6146	6.6474	6.6803	6.7133	6.7464	6.7795	6.8127	6.8461
35	6.8794	6.9129	6.9465	6.9801	7.0138	7.0476	7.0814	7.1154	7.1494	7.1835
36	7.2177	7.2520	7.2863	7.3207	7.3552	7.3898	7.4245	7.4592	7.4940	7.5289
37	7.5639	7.5989	7.6340	7.6692	7.7045	7.7398	7.7753	7.8108	7.8464	7.8820
38	7.9178	7.9536	7.9895	8.0254	8.0615	8.0976	8.1338	8.1701	8.2064	8.2428
39	8.2793	8.3159	8.3525	8.3893	8.4260	8.4629	8.4999	8.5369	8.5740	8.6111
40	8.6484	8.6857	8.7231	8.7605	8.7981	8.8357	8.8733	8.9111	8.9489	8.9868
41	9.0248	9.0628	9.1009	9.1391	9.1774	9.2157	9.2541	9.2926	9.3311	9.3697

Table 8.1. (*b*) Value of correction C in mg for a 1 cm^3 borosilicate glass measure for $\gamma = 0.000\,009\,75$, $\sigma = 0.0012$ g cm^{-3}, $\Delta = 8.400$ g cm^{-3}, reference temperature $= 30\,°C$, where γ is the coefficient of expansion of the material of the measure, σ is the density of air (in g cm^{-3}) and Δ is the density of standard weights used (in g cm^{-3}).

T (°C)	0.0	0.1	0.2	0.3	0.4	0.5	0.6	0.7	0.8	0.9
5	1.3339	1.3346	1.3355	1.3365	1.3376	1.3389	1.3404	1.3420	1.3438	1.3457
6	1.3478	1.3500	1.3523	1.3549	1.3575	1.3603	1.3633	1.3664	1.3697	1.3731
7	1.3766	1.3803	1.3841	1.3881	1.3922	1.3965	1.4009	1.4055	1.4102	1.4150
8	1.4200	1.4252	1.4304	1.4358	1.4414	1.4471	1.4529	1.4589	1.4650	1.4712
9	1.4776	1.4842	1.4908	1.4976	1.5046	1.5116	1.5188	1.5262	1.5337	1.5413
10	1.5490	1.5569	1.5649	1.5731	1.5814	1.5898	1.5983	1.6070	1.6158	1.6248
11	1.6338	1.6430	1.6524	1.6619	1.6714	1.6812	1.6910	1.7010	1.7111	1.7214
12	1.7317	1.7422	1.7528	1.7636	1.7745	1.7855	1.7966	1.8078	1.8192	1.8307
13	1.8423	1.8541	1.8659	1.8779	1.8901	1.9023	1.9147	1.9271	1.9397	1.9525
14	1.9653	1.9783	1.9914	2.0046	2.0179	2.0314	2.0450	2.0586	2.0725	2.0864
15	2.1004	2.1146	2.1289	2.1433	2.1578	2.1724	2.1872	2.2020	2.2170	2.2321
16	2.2473	2.2627	2.2781	2.2937	2.3093	2.3251	2.3410	2.3570	2.3732	2.3894
17	2.4057	2.4222	2.4388	2.4555	2.4723	2.4892	2.5062	2.5234	2.5406	2.5580
18	2.5754	2.5930	2.6107	2.6285	2.6464	2.6644	2.6825	2.7008	2.7191	2.7376
19	2.7561	2.7748	2.7935	2.8124	2.8314	2.8505	2.8697	2.8890	2.9084	2.9279
20	2.9476	2.9673	2.9871	3.0070	3.0271	3.0472	3.0675	3.0878	3.1083	3.1289
21	3.1495	3.1703	3.1912	3.2121	3.2332	3.2544	3.2757	3.2971	3.3185	3.3401
22	3.3618	3.3836	3.4055	3.4275	3.4496	3.4718	3.4941	3.5165	3.5389	3.5615
23	3.5842	3.6070	3.6299	3.6529	3.6760	3.6991	3.7224	3.7458	3.7693	3.7929
24	3.8165	3.8403	3.8642	3.8881	3.9122	3.9363	3.9606	3.9849	4.0094	4.0339
25	4.0585	4.0833	4.1081	4.1330	4.1580	4.1831	4.2083	4.2336	4.2590	4.2845
26	4.3101	4.3358	4.3615	4.3874	4.4133	4.4394	4.4655	4.4918	4.5181	4.5445
27	4.5710	4.5976	4.6243	4.6511	4.6779	4.7049	4.7320	4.7591	4.7864	4.8137
28	4.8411	4.8686	4.8962	4.9239	4.9517	4.9795	5.0075	5.0356	5.0637	5.0919
29	5.1202	5.1486	5.1771	5.2057	5.2344	5.2631	5.2920	5.3209	5.3499	5.3790
30	5.4082	5.4375	5.4669	5.4963	5.5259	5.5555	5.5852	5.6150	5.6449	5.6749
31	5.7050	5.7351	5.7653	5.7957	5.8261	5.8566	5.8871	5.9178	5.9485	5.9794
32	6.0103	6.0413	6.0724	6.1035	6.1348	6.1661	6.1975	6.2290	6.2606	6.2923
33	6.3240	6.3558	6.3878	6.4198	6.4518	6.4840	6.5163	6.5486	6.5810	6.6135
34	6.6461	6.6787	6.7114	6.7443	6.7772	6.8101	6.8432	6.8763	6.9096	6.9429
35	6.9763	7.0097	7.0433	7.0769	7.1106	7.1444	7.1782	7.2122	7.2462	7.2803
36	7.3145	7.3487	7.3831	7.4175	7.4520	7.4866	7.5212	7.5559	7.5907	7.6256
37	7.6606	7.6956	7.7308	7.7660	7.8012	7.8366	7.8720	7.9075	7.9431	7.9787
38	8.0145	8.0503	8.0862	8.1221	8.1582	8.1943	8.2305	8.2667	8.3031	8.3395
39	8.3760	8.4126	8.4492	8.4859	8.5227	8.5596	8.5965	8.6335	8.6706	8.7078
40	8.7450	8.7823	8.8197	8.8572	8.8947	8.9323	8.9700	9.0077	9.0455	9.0834
41	9.1214	9.1594	9.1975	9.2357	9.2740	9.3123	9.3507	9.3892	9.4277	9.4663

Table 8.1. (*c*) Value of correction *C* in mg for a 1 cm^3 borosilicate glass measure for $\gamma = 0.000\,009\,75$, $\sigma = 0.0012$ g cm^{-3}, $\Delta = 8.400$ g cm^{-3}, reference temperature $= 40\,°C$, where γ is the coefficient of expansion of the material of the measure, σ is the density of air (in g cm^{-3}) and Δ is the density of standard weights used (in g cm^{-3}).

T (°C)	0.0	0.1	0.2	0.3	0.4	0.5	0.6	0.7	0.8	0.9
5	1.4313	1.4320	1.4329	1.4339	1.4350	1.4363	1.4378	1.4394	1.4412	1.4431
6	1.4452	1.4474	1.4497	1.4523	1.4549	1.4577	1.4607	1.4638	1.4670	1.4704
7	1.4740	1.4777	1.4815	1.4855	1.4896	1.4939	1.4983	1.5029	1.5076	1.5124
8	1.5174	1.5225	1.5278	1.5332	1.5388	1.5445	1.5503	1.5563	1.5624	1.5686
9	1.5750	1.5815	1.5882	1.5950	1.6019	1.6090	1.6162	1.6236	1.6310	1.6387
10	1.6464	1.6543	1.6623	1.6704	1.6787	1.6871	1.6957	1.7044	1.7132	1.7221
11	1.7312	1.7404	1.7497	1.7592	1.7688	1.7785	1.7884	1.7984	1.8085	1.8187
12	1.8291	1.8396	1.8502	1.8609	1.8718	1.8828	1.8939	1.9052	1.9165	1.9280
13	1.9397	1.9514	1.9633	1.9753	1.9874	1.9996	2.0120	2.0245	2.0371	2.0498
14	2.0626	2.0756	2.0887	2.1019	2.1152	2.1287	2.1423	2.1560	2.1698	2.1837
15	2.1977	2.2119	2.2262	2.2406	2.2551	2.2697	2.2845	2.2993	2.3143	2.3294
16	2.3446	2.3599	2.3754	2.3909	2.4066	2.4224	2.4383	2.4543	2.4704	2.4867
17	2.5030	2.5195	2.5361	2.5528	2.5696	2.5865	2.6035	2.6206	2.6379	2.6552
18	2.6727	2.6903	2.7079	2.7257	2.7436	2.7617	2.7798	2.7980	2.8163	2.8348
19	2.8533	2.8720	2.8908	2.9097	2.9286	2.9477	2.9669	2.9862	3.0056	3.0252
20	3.0448	3.0645	3.0843	3.1043	3.1243	3.1444	3.1647	3.1851	3.2055	3.2261
21	3.2467	3.2675	3.2884	3.3093	3.3304	3.3516	3.3729	3.3942	3.4157	3.4373
22	3.4590	3.4808	3.5027	3.5247	3.5467	3.5689	3.5912	3.6136	3.6361	3.6587
23	3.6814	3.7042	3.7270	3.7500	3.7731	3.7963	3.8196	3.8429	3.8664	3.8900
24	3.9137	3.9374	3.9613	3.9852	4.0093	4.0334	4.0577	4.0820	4.1065	4.1310
25	4.1556	4.1804	4.2052	4.2301	4.2551	4.2802	4.3054	4.3307	4.3561	4.3816
26	4.4072	4.4328	4.4586	4.4845	4.5104	4.5365	4.5626	4.5888	4.6151	4.6416
27	4.6681	4.6947	4.7213	4.7481	4.7750	4.8020	4.8290	4.8562	4.8834	4.9107
28	4.9381	4.9656	4.9932	5.0209	5.0487	5.0766	5.1045	5.1326	5.1607	5.1889
29	5.2172	5.2456	5.2741	5.3027	5.3314	5.3601	5.3890	5.4179	5.4469	5.4760
30	5.5052	5.5345	5.5639	5.5933	5.6229	5.6525	5.6822	5.7120	5.7419	5.7719
31	5.8019	5.8321	5.8623	5.8926	5.9230	5.9535	5.9841	6.0147	6.0455	6.0763
32	6.1072	6.1382	6.1693	6.2004	6.2317	6.2630	6.2944	6.3259	6.3575	6.3892
33	6.4209	6.4527	6.4846	6.5166	6.5487	6.5809	6.6131	6.6454	6.6778	6.7103
34	6.7429	6.7756	6.8083	6.8411	6.8740	6.9070	6.9400	6.9732	7.0064	7.0397
35	7.0731	7.1065	7.1401	7.1737	7.2074	7.2412	7.2750	7.3090	7.3430	7.3771
36	7.4113	7.4455	7.4798	7.5143	7.5488	7.5833	7.6180	7.6527	7.6875	7.7224
37	7.7573	7.7924	7.8275	7.8627	7.8980	7.9333	7.9687	8.0042	8.0398	8.0755
38	8.1112	8.1470	8.1829	8.2188	8.2549	8.2910	8.3272	8.3634	8.3998	8.4362
39	8.4727	8.5092	8.5459	8.5826	8.6194	8.6562	8.6932	8.7302	8.7673	8.8044
40	8.8416	8.8789	8.9163	8.9538	8.9913	9.0289	9.0666	9.1043	9.1421	9.1800
41	9.2180	9.2560	9.2941	9.3323	9.3705	9.4089	9.4473	9.4857	9.5243	9.5629

Table 8.1. (*d*) Value of correction C in mg for a 1 cm^3 borosilicate glass measure for $\gamma = 0.000\,009\,75$, $\sigma = 0.0012$ g cm^{-3}, $\Delta = 8.400$ g cm^{-3}, reference temperature $= 50\,°C$, where γ is the coefficient of expansion of the material of the measure, σ is the density of air (in g cm^{-3}) and Δ is the density of standard weights used (in g cm^{-3}).

T (°C)	0.0	0.1	0.2	0.3	0.4	0.5	0.6	0.7	0.8	0.9
5	1.5287	1.5294	1.5303	1.5313	1.5324	1.5337	1.5352	1.5368	1.5386	1.5405
6	1.5425	1.5448	1.5471	1.5496	1.5523	1.5551	1.5581	1.5612	1.5644	1.5678
7	1.5714	1.5751	1.5789	1.5829	1.5870	1.5913	1.5957	1.6003	1.6050	1.6098
8	1.6148	1.6199	1.6252	1.6306	1.6362	1.6418	1.6477	1.6536	1.6598	1.6660
9	1.6724	1.6789	1.6856	1.6924	1.6993	1.7064	1.7136	1.7209	1.7284	1.7360
10	1.7438	1.7516	1.7597	1.7678	1.7761	1.7845	1.7931	1.8017	1.8105	1.8195
11	1.8286	1.8378	1.8471	1.8566	1.8662	1.8759	1.8857	1.8957	1.9058	1.9161
12	1.9264	1.9369	1.9475	1.9583	1.9691	1.9801	1.9913	2.0025	2.0139	2.0254
13	2.0370	2.0487	2.0606	2.0726	2.0847	2.0970	2.1093	2.1218	2.1344	2.1471
14	2.1600	2.1729	2.1860	2.1992	2.2126	2.2260	2.2396	2.2533	2.2671	2.2810
15	2.2950	2.3092	2.3235	2.3379	2.3524	2.3670	2.3818	2.3966	2.4116	2.4267
16	2.4419	2.4572	2.4727	2.4882	2.5039	2.5197	2.5356	2.5516	2.5677	2.5840
17	2.6003	2.6168	2.6333	2.6500	2.6668	2.6837	2.7008	2.7179	2.7351	2.7525
18	2.7699	2.7875	2.8052	2.8230	2.8409	2.8589	2.8770	2.8953	2.9136	2.9320
19	2.9506	2.9692	2.9880	3.0069	3.0259	3.0450	3.0642	3.0835	3.1029	3.1224
20	3.1420	3.1617	3.1815	3.2015	3.2215	3.2417	3.2619	3.2823	3.3027	3.3233
21	3.3439	3.3647	3.3856	3.4065	3.4276	3.4488	3.4701	3.4914	3.5129	3.5345
22	3.5562	3.5780	3.5998	3.6218	3.6439	3.6661	3.6884	3.7108	3.7333	3.7558
23	3.7785	3.8013	3.8242	3.8472	3.8703	3.8934	3.9167	3.9401	3.9636	3.9871
24	4.0108	4.0346	4.0584	4.0824	4.1064	4.1306	4.1548	4.1792	4.2036	4.2281
25	4.2528	4.2775	4.3023	4.3272	4.3522	4.3773	4.4025	4.4278	4.4532	4.4787
26	4.5043	4.5299	4.5557	4.5815	4.6075	4.6335	4.6597	4.6859	4.7122	4.7386
27	4.7651	4.7917	4.8184	4.8452	4.8720	4.8990	4.9261	4.9532	4.9804	5.0077
28	5.0352	5.0627	5.0903	5.1179	5.1457	5.1736	5.2015	5.2296	5.2577	5.2859
29	5.3142	5.3426	5.3711	5.3997	5.4284	5.4571	5.4859	5.5149	5.5439	5.5730
30	5.6022	5.6315	5.6608	5.6903	5.7198	5.7494	5.7791	5.8089	5.8388	5.8688
31	5.8989	5.9290	5.9592	5.9895	6.0199	6.0504	6.0810	6.1116	6.1424	6.1732
32	6.2041	6.2351	6.2662	6.2973	6.3286	6.3599	6.3913	6.4228	6.4544	6.4860
33	6.5178	6.5496	6.5815	6.6135	6.6456	6.6777	6.7100	6.7423	6.7747	6.8072
34	6.8398	6.8724	6.9051	6.9379	6.9708	7.0038	7.0369	7.0700	7.1032	7.1365
35	7.1699	7.2033	7.2369	7.2705	7.3042	7.3380	7.3718	7.4058	7.4398	7.4739
36	7.5080	7.5423	7.5766	7.6110	7.6455	7.6801	7.7147	7.7495	7.7843	7.8191
37	7.8541	7.8891	7.9242	7.9594	7.9947	8.0300	8.0655	8.1010	8.1365	8.1722
38	8.2079	8.2437	8.2796	8.3155	8.3516	8.3877	8.4239	8.4601	8.4965	8.5329
39	8.5693	8.6059	8.6425	8.6792	8.7160	8.7529	8.7898	8.8268	8.8639	8.9011
40	8.9383	8.9756	9.0130	9.0504	9.0879	9.1255	9.1632	9.2009	9.2387	9.2766
41	9.3146	9.3526	9.3907	9.4289	9.4671	9.5055	9.5438	9.5823	9.6208	9.6594

Table 8.1. (*e*) Value of correction C in mg for a 1 cm³ borosilicate glass measure for $\gamma = 0.00000975$, $\sigma = 0.0012$ g cm⁻³, $\Delta = 8.400$ g cm⁻³, reference temperature $= 60\,°C$, where γ is the coefficient of expansion of the material of the measure, σ is the density of air (in g cm⁻³) and Δ is the density of standard weights used (in g cm⁻³).

T (°C)	0.0	0.1	0.2	0.3	0.4	0.5	0.6	0.7	0.8	0.9
5	1.6261	1.6268	1.6277	1.6287	1.6298	1.6311	1.6326	1.6342	1.6360	1.6379
6	1.6399	1.6422	1.6445	1.6470	1.6497	1.6525	1.6555	1.6586	1.6618	1.6652
7	1.6688	1.6725	1.6763	1.6803	1.6844	1.6887	1.6931	1.6976	1.7023	1.7072
8	1.7122	1.7173	1.7226	1.7280	1.7335	1.7392	1.7451	1.7510	1.7571	1.7634
9	1.7698	1.7763	1.7829	1.7897	1.7967	1.8037	1.8110	1.8183	1.8258	1.8334
10	1.8411	1.8490	1.8570	1.8652	1.8735	1.8819	1.8904	1.8991	1.9079	1.9168
11	1.9259	1.9351	1.9445	1.9539	1.9635	1.9732	1.9831	1.9931	2.0032	2.0134
12	2.0238	2.0343	2.0449	2.0556	2.0665	2.0775	2.0886	2.0998	2.1112	2.1227
13	2.1343	2.1461	2.1579	2.1699	2.1820	2.1943	2.2066	2.2191	2.2317	2.2445
14	2.2573	2.2703	2.2833	2.2966	2.3099	2.3233	2.3369	2.3506	2.3644	2.3783
15	2.3924	2.4065	2.4208	2.4352	2.4497	2.4643	2.4791	2.4939	2.5089	2.5240
16	2.5392	2.5545	2.5700	2.5855	2.6012	2.6170	2.6329	2.6489	2.6650	2.6812
17	2.6976	2.7140	2.7306	2.7473	2.7641	2.7810	2.7980	2.8152	2.8324	2.8497
18	2.8672	2.8848	2.9025	2.9202	2.9381	2.9562	2.9743	2.9925	3.0108	3.0293
19	3.0478	3.0665	3.0853	3.1041	3.1231	3.1422	3.1614	3.1807	3.2001	3.2196
20	3.2392	3.2589	3.2788	3.2987	3.3187	3.3389	3.3591	3.3795	3.3999	3.4205
21	3.4411	3.4619	3.4828	3.5037	3.5248	3.5460	3.5672	3.5886	3.6101	3.6317
22	3.6534	3.6751	3.6970	3.7190	3.7411	3.7633	3.7856	3.8079	3.8304	3.8530
23	3.8757	3.8985	3.9213	3.9443	3.9674	3.9906	4.0139	4.0372	4.0607	4.0843
24	4.1079	4.1317	4.1555	4.1795	4.2035	4.2277	4.2519	4.2763	4.3007	4.3252
25	4.3499	4.3746	4.3994	4.4243	4.4493	4.4744	4.4996	4.5249	4.5503	4.5758
26	4.6013	4.6270	4.6528	4.6786	4.7046	4.7306	4.7567	4.7830	4.8093	4.8357
27	4.8622	4.8888	4.9154	4.9422	4.9691	4.9960	5.0231	5.0502	5.0775	5.1048
28	5.1322	5.1597	5.1873	5.2150	5.2427	5.2706	5.2985	5.3266	5.3547	5.3829
29	5.4112	5.4396	5.4681	5.4967	5.5253	5.5541	5.5829	5.6119	5.6409	5.6700
30	5.6992	5.7284	5.7578	5.7872	5.8168	5.8464	5.8761	5.9059	5.9358	5.9657
31	5.9958	6.0259	6.0562	6.0865	6.1169	6.1473	6.1779	6.2086	6.2393	6.2701
32	6.3010	6.3320	6.3631	6.3942	6.4255	6.4568	6.4882	6.5197	6.5513	6.5829
33	6.6147	6.6465	6.6784	6.7104	6.7424	6.7746	6.8068	6.8392	6.8716	6.9040
34	6.9366	6.9692	7.0020	7.0348	7.0677	7.1006	7.1337	7.1668	7.2000	7.2333
35	7.2667	7.3002	7.3337	7.3673	7.4010	7.4348	7.4686	7.5025	7.5366	7.5707
36	7.6048	7.6391	7.6734	7.7078	7.7423	7.7769	7.8115	7.8462	7.8810	7.9159
37	7.9508	7.9859	8.0210	8.0562	8.0914	8.1268	8.1622	8.1977	8.2332	8.2689
38	8.3046	8.3404	8.3763	8.4122	8.4483	8.4844	8.5205	8.5568	8.5931	8.6295
39	8.6660	8.7026	8.7392	8.7759	8.8127	8.8495	8.8865	8.9235	8.9605	8.9977
40	9.0349	9.0722	9.1096	9.1470	9.1845	9.2221	9.2598	9.2975	9.3353	9.3732
41	9.4112	9.4492	9.4873	9.5255	9.5637	9.6020	9.6404	9.6789	9.7174	9.7560

Table 8.1. (*f*) Value of correction C in mg for a 1 cm^3 borosilicate glass measure for $\gamma = 0.000\,009\,75$, $\sigma = 0.0012$ g cm^{-3}, $\Delta = 8.400$ g cm^{-3}, reference temperature $= 70\,^\circ$C, where γ is the coefficient of expansion of the material of the measure, σ is the density of air (in g cm^{-3}) and Δ is the density of standard weights used (in g cm^{-3}).

T (°C)	0.0	0.1	0.2	0.3	0.4	0.5	0.6	0.7	0.8	0.9
5	1.7235	1.7242	1.7251	1.7261	1.7272	1.7285	1.7300	1.7316	1.7334	1.7353
6	1.7373	1.7395	1.7419	1.7444	1.7471	1.7499	1.7529	1.7560	1.7592	1.7626
7	1.7662	1.7698	1.7737	1.7777	1.7818	1.7861	1.7905	1.7950	1.7997	1.8046
8	1.8096	1.8147	1.8200	1.8254	1.8309	1.8366	1.8424	1.8484	1.8545	1.8608
9	1.8671	1.8737	1.8803	1.8871	1.8941	1.9011	1.9083	1.9157	1.9231	1.9308
10	1.9385	1.9464	1.9544	1.9625	1.9708	1.9792	1.9878	1.9965	2.0053	2.0142
11	2.0233	2.0325	2.0418	2.0513	2.0609	2.0706	2.0804	2.0904	2.1005	2.1108
12	2.1211	2.1316	2.1422	2.1530	2.1638	2.1748	2.1859	2.1972	2.2086	2.2200
13	2.2317	2.2434	2.2553	2.2673	2.2794	2.2916	2.3040	2.3165	2.3291	2.3418
14	2.3546	2.3676	2.3807	2.3939	2.4072	2.4206	2.4342	2.4479	2.4617	2.4756
15	2.4897	2.5038	2.5181	2.5325	2.5470	2.5616	2.5764	2.5912	2.6062	2.6213
16	2.6365	2.6518	2.6673	2.6828	2.6985	2.7143	2.7301	2.7462	2.7623	2.7785
17	2.7949	2.8113	2.8279	2.8446	2.8614	2.8783	2.8953	2.9124	2.9297	2.9470
18	2.9645	2.9820	2.9997	3.0175	3.0354	3.0534	3.0715	3.0898	3.1081	3.1265
19	3.1451	3.1637	3.1825	3.2014	3.2203	3.2394	3.2586	3.2779	3.2973	3.3168
20	3.3364	3.3562	3.3760	3.3959	3.4159	3.4361	3.4563	3.4767	3.4971	3.5177
21	3.5383	3.5591	3.5800	3.6009	3.6220	3.6432	3.6644	3.6858	3.7073	3.7289
22	3.7505	3.7723	3.7942	3.8162	3.8383	3.8604	3.8827	3.9051	3.9276	3.9502
23	3.9728	3.9956	4.0185	4.0415	4.0646	4.0877	4.1110	4.1344	4.1578	4.1814
24	4.2051	4.2288	4.2527	4.2766	4.3007	4.3248	4.3491	4.3734	4.3978	4.4224
25	4.4470	4.4717	4.4965	4.5214	4.5464	4.5715	4.5967	4.6220	4.6474	4.6729
26	4.6984	4.7241	4.7498	4.7757	4.8016	4.8277	4.8538	4.8800	4.9063	4.9327
27	4.9592	4.9858	5.0125	5.0393	5.0661	5.0931	5.1201	5.1473	5.1745	5.2018
28	5.2292	5.2567	5.2843	5.3120	5.3398	5.3676	5.3956	5.4236	5.4517	5.4799
29	5.5082	5.5366	5.5651	5.5937	5.6223	5.6511	5.6799	5.7088	5.7378	5.7669
30	5.7961	5.8254	5.8548	5.8842	5.9137	5.9434	5.9731	6.0029	6.0327	6.0627
31	6.0927	6.1229	6.1531	6.1834	6.2138	6.2443	6.2748	6.3055	6.3362	6.3670
32	6.3979	6.4289	6.4600	6.4911	6.5224	6.5537	6.5851	6.6166	6.6481	6.6798
33	6.7115	6.7434	6.7753	6.8072	6.8393	6.8715	6.9037	6.9360	6.9684	7.0009
34	7.0334	7.0661	7.0988	7.1316	7.1645	7.1975	7.2305	7.2636	7.2969	7.3301
35	7.3635	7.3970	7.4305	7.4641	7.4978	7.5316	7.5654	7.5993	7.6333	7.6674
36	7.7016	7.7358	7.7702	7.8046	7.8391	7.8736	7.9083	7.9430	7.9778	8.0126
37	8.0476	8.0826	8.1177	8.1529	8.1882	8.2235	8.2589	8.2944	8.3300	8.3656
38	8.4013	8.4371	8.4730	8.5089	8.5450	8.5811	8.6172	8.6535	8.6898	8.7262
39	8.7627	8.7992	8.8359	8.8726	8.9093	8.9462	8.9831	9.0201	9.0572	9.0943
40	9.1316	9.1689	9.2062	9.2437	9.2812	9.3188	9.3564	9.3941	9.4320	9.4698
41	9.5078	9.5458	9.5839	9.6221	9.6603	9.6986	9.7370	9.7754	9.8140	9.8526

Table 8.1. (*g*) Value of correction C in mg for a 1 cm^3 borosilicate glass measure for $\gamma = 0.000\,009\,75$, $\sigma = 0.0012$ g cm^{-3}, $\Delta = 8.400$ g cm^{-3}, reference temperature $= 80\,°C$, where γ is the coefficient of expansion of the material of the measure, σ is the density of air (in g cm^{-3}) and Δ is the density of standard weights used (in g cm^{-3}).

T (°C)	0.0	0.1	0.2	0.3	0.4	0.5	0.6	0.7	0.8	0.9
5	1.8209	1.8216	1.8224	1.8235	1.8246	1.8259	1.8274	1.8290	1.8307	1.8327
6	1.8347	1.8369	1.8393	1.8418	1.8445	1.8473	1.8502	1.8533	1.8566	1.8600
7	1.8635	1.8672	1.8711	1.8750	1.8792	1.8834	1.8879	1.8924	1.8971	1.9020
8	1.9069	1.9121	1.9173	1.9227	1.9283	1.9340	1.9398	1.9458	1.9519	1.9581
9	1.9645	1.9710	1.9777	1.9845	1.9914	1.9985	2.0057	2.0130	2.0205	2.0281
10	2.0359	2.0438	2.0518	2.0599	2.0682	2.0766	2.0851	2.0938	2.1026	2.1116
11	2.1206	2.1298	2.1392	2.1486	2.1582	2.1679	2.1778	2.1878	2.1979	2.2081
12	2.2185	2.2289	2.2396	2.2503	2.2612	2.2722	2.2833	2.2945	2.3059	2.3174
13	2.3290	2.3407	2.3526	2.3646	2.3767	2.3889	2.4013	2.4138	2.4264	2.4391
14	2.4519	2.4649	2.4780	2.4912	2.5045	2.5180	2.5315	2.5452	2.5590	2.5729
15	2.5870	2.6011	2.6154	2.6298	2.6443	2.6589	2.6737	2.6885	2.7035	2.7186
16	2.7338	2.7491	2.7645	2.7801	2.7958	2.8115	2.8274	2.8434	2.8596	2.8758
17	2.8921	2.9086	2.9252	2.9418	2.9586	2.9755	2.9926	3.0097	3.0269	3.0443
18	3.0617	3.0793	3.0970	3.1148	3.1327	3.1507	3.1688	3.1870	3.2053	3.2238
19	3.2423	3.2610	3.2797	3.2986	3.3176	3.3367	3.3558	3.3751	3.3945	3.4140
20	3.4337	3.4534	3.4732	3.4931	3.5132	3.5333	3.5535	3.5739	3.5943	3.6149
21	3.6355	3.6563	3.6772	3.6981	3.7192	3.7404	3.7616	3.7830	3.8045	3.8260
22	3.8477	3.8695	3.8914	3.9134	3.9354	3.9576	3.9799	4.0023	4.0247	4.0473
23	4.0700	4.0928	4.1157	4.1386	4.1617	4.1849	4.2081	4.2315	4.2550	4.2785
24	4.3022	4.3259	4.3498	4.3737	4.3978	4.4219	4.4462	4.4705	4.4949	4.5195
25	4.5441	4.5688	4.5936	4.6185	4.6435	4.6686	4.6938	4.7191	4.7445	4.7699
26	4.7955	4.8212	4.8469	4.8728	4.8987	4.9247	4.9509	4.9771	5.0034	5.0298
27	5.0563	5.0829	5.1095	5.1363	5.1632	5.1901	5.2172	5.2443	5.2715	5.2988
28	5.3263	5.3537	5.3813	5.4090	5.4368	5.4646	5.4926	5.5206	5.5487	5.5769
29	5.6052	5.6336	5.6621	5.6907	5.7193	5.7481	5.7769	5.8058	5.8348	5.8639
30	5.8931	5.9224	5.9517	5.9812	6.0107	6.0403	6.0700	6.0998	6.1297	6.1596
31	6.1897	6.2198	6.2500	6.2803	6.3107	6.3412	6.3718	6.4024	6.4331	6.4639
32	6.4948	6.5258	6.5569	6.5880	6.6193	6.6506	6.6820	6.7135	6.7450	6.7767
33	6.8084	6.8402	6.8721	6.9041	6.9362	6.9683	7.0006	7.0329	7.0653	7.0977
34	7.1303	7.1629	7.1957	7.2285	7.2613	7.2943	7.3273	7.3605	7.3937	7.4270
35	7.4603	7.4938	7.5273	7.5609	7.5946	7.6284	7.6622	7.6961	7.7301	7.7642
36	7.7984	7.8326	7.8669	7.9013	7.9358	7.9704	8.0050	8.0397	8.0745	8.1094
37	8.1443	8.1794	8.2145	8.2496	8.2849	8.3202	8.3556	8.3911	8.4267	8.4623
38	8.4980	8.5338	8.5697	8.6056	8.6417	8.6778	8.7139	8.7502	8.7865	8.8229
39	8.8594	8.8959	8.9325	8.9692	9.0060	9.0429	9.0798	9.1168	9.1538	9.1910
40	9.2282	9.2655	9.3028	9.3403	9.3778	9.4154	9.4530	9.4908	9.5286	9.5664
41	9.6044	9.6424	9.6805	9.7187	9.7569	9.7952	9.8336	9.8720	9.9105	9.9491

Table 8.1. (*h*) Value of correction C in mg for a 1 cm^3 borosilicate glass measure for $\gamma = 0.000\,009\,75$, $\sigma = 0.0012$ g cm^{-3}, $\Delta = 8.000$ g cm^{-3}, reference temperature $= 20\,^\circ$C, where γ is the coefficient of expansion of the material of the measure, σ is the density of air (in g cm^{-3}) and Δ is the density of standard weights used (in g cm^{-3}).

T (°C)	0.0	0.1	0.2	0.3	0.4	0.5	0.6	0.7	0.8	0.9
5	1.2294	1.2301	1.2310	1.2320	1.2331	1.2344	1.2359	1.2375	1.2393	1.2412
6	1.2432	1.2455	1.2478	1.2503	1.2530	1.2558	1.2588	1.2619	1.2651	1.2685
7	1.2721	1.2758	1.2796	1.2836	1.2877	1.2920	1.2964	1.3010	1.3057	1.3105
8	1.3155	1.3206	1.3259	1.3313	1.3369	1.3426	1.3484	1.3544	1.3605	1.3667
9	1.3731	1.3797	1.3863	1.3931	1.4001	1.4071	1.4143	1.4217	1.4292	1.4368
10	1.4445	1.4524	1.4604	1.4686	1.4769	1.4853	1.4938	1.5025	1.5113	1.5203
11	1.5294	1.5386	1.5479	1.5574	1.5670	1.5767	1.5865	1.5965	1.6066	1.6169
12	1.6272	1.6377	1.6484	1.6591	1.6700	1.6810	1.6921	1.7034	1.7147	1.7262
13	1.7379	1.7496	1.7615	1.7735	1.7856	1.7978	1.8102	1.8227	1.8353	1.8480
14	1.8609	1.8738	1.8869	1.9001	1.9135	1.9269	1.9405	1.9542	1.9680	1.9819
15	1.9960	2.0101	2.0244	2.0388	2.0534	2.0680	2.0827	2.0976	2.1126	2.1277
16	2.1429	2.1582	2.1737	2.1892	2.2049	2.2207	2.2366	2.2526	2.2687	2.2850
17	2.3013	2.3178	2.3344	2.3511	2.3679	2.3848	2.4018	2.4190	2.4362	2.4536
18	2.4710	2.4886	2.5063	2.5241	2.5420	2.5600	2.5782	2.5964	2.6147	2.6332
19	2.6517	2.6704	2.6892	2.7081	2.7270	2.7461	2.7653	2.7846	2.8041	2.8236
20	2.8432	2.8629	2.8828	2.9027	2.9228	2.9429	2.9632	2.9835	3.0040	3.0245
21	3.0452	3.0660	3.0869	3.1078	3.1289	3.1501	3.1714	3.1928	3.2142	3.2358
22	3.2575	3.2793	3.3012	3.3232	3.3453	3.3675	3.3898	3.4122	3.4347	3.4573
23	3.4799	3.5027	3.5256	3.5486	3.5717	3.5949	3.6182	3.6415	3.6650	3.6886
24	3.7123	3.7360	3.7599	3.7839	3.8079	3.8321	3.8563	3.8807	3.9051	3.9297
25	3.9543	3.9790	4.0039	4.0288	4.0538	4.0789	4.1041	4.1294	4.1548	4.1803
26	4.2059	4.2316	4.2573	4.2832	4.3092	4.3352	4.3613	4.3876	4.4139	4.4403
27	4.4668	4.4934	4.5201	4.5469	4.5738	4.6008	4.6278	4.6550	4.6822	4.7095
28	4.7370	4.7645	4.7921	4.8198	4.8476	4.8754	4.9034	4.9314	4.9596	4.9878
29	5.0161	5.0445	5.0730	5.1016	5.1303	5.1590	5.1879	5.2168	5.2459	5.2750
30	5.3042	5.3334	5.3628	5.3923	5.4218	5.4515	5.4812	5.5110	5.5409	5.5709
31	5.6009	5.6311	5.6613	5.6916	5.7220	5.7525	5.7831	5.8138	5.8445	5.8753
32	5.9063	5.9373	5.9683	5.9995	6.0308	6.0621	6.0935	6.1250	6.1566	6.1883
33	6.2200	6.2519	6.2838	6.3158	6.3479	6.3800	6.4123	6.4446	6.4770	6.5095
34	6.5421	6.5748	6.6075	6.6403	6.6732	6.7062	6.7393	6.7724	6.8056	6.8390
35	6.8723	6.9058	6.9394	6.9730	7.0067	7.0405	7.0743	7.1083	7.1423	7.1764
36	7.2106	7.2449	7.2792	7.3136	7.3481	7.3827	7.4174	7.4521	7.4869	7.5218
37	7.5568	7.5918	7.6269	7.6621	7.6974	7.7328	7.7682	7.8037	7.8393	7.8749
38	7.9107	7.9465	7.9824	8.0184	8.0544	8.0905	8.1267	8.1630	8.1993	8.2357
39	8.2722	8.3088	8.3454	8.3822	8.4190	8.4558	8.4928	8.5298	8.5669	8.6040
40	8.6413	8.6786	8.7160	8.7534	8.7910	8.8286	8.8663	8.9040	8.9418	8.9797
41	9.0177	9.0557	9.0939	9.1320	9.1703	9.2086	9.2470	9.2855	9.3240	9.3627

Table 8.1. (*i*) Value of correction C in mg for a 1 cm^3 borosilicate glass measure for $\gamma = 0.000\,009\,75$, $\sigma = 0.0012$ g cm^{-3}, $\Delta = 8.000$ g cm^{-3}, reference temperature $= 30\,°C$, where γ is the coefficient of expansion of the material of the measure, σ is the density of air (in g cm^{-3}) and Δ is the density of standard weights used (in g cm^{-3}).

T (°C)	0.0	0.1	0.2	0.3	0.4	0.5	0.6	0.7	0.8	0.9
5	1.3268	1.3275	1.3283	1.3294	1.3305	1.3318	1.3333	1.3349	1.3367	1.3386
6	1.3406	1.3428	1.3452	1.3477	1.3504	1.3532	1.3562	1.3593	1.3625	1.3659
7	1.3695	1.3732	1.3770	1.3810	1.3851	1.3894	1.3938	1.3984	1.4031	1.4079
8	1.4129	1.4180	1.4233	1.4287	1.4343	1.4400	1.4458	1.4518	1.4579	1.4641
9	1.4705	1.4770	1.4837	1.4905	1.4974	1.5045	1.5117	1.5191	1.5265	1.5341
10	1.5419	1.5498	1.5578	1.5659	1.5742	1.5827	1.5912	1.5999	1.6087	1.6176
11	1.6267	1.6359	1.6453	1.6547	1.6643	1.6740	1.6839	1.6939	1.7040	1.7142
12	1.7246	1.7351	1.7457	1.7564	1.7673	1.7783	1.7894	1.8007	1.8121	1.8236
13	1.8352	1.8469	1.8588	1.8708	1.8829	1.8952	1.9075	1.9200	1.9326	1.9453
14	1.9582	1.9712	1.9843	1.9975	2.0108	2.0243	2.0378	2.0515	2.0653	2.0792
15	2.0933	2.1075	2.1217	2.1361	2.1507	2.1653	2.1800	2.1949	2.2099	2.2250
16	2.2402	2.2555	2.2710	2.2865	2.3022	2.3180	2.3339	2.3499	2.3660	2.3823
17	2.3986	2.4151	2.4317	2.4484	2.4652	2.4821	2.4991	2.5162	2.5335	2.5508
18	2.5683	2.5859	2.6036	2.6214	2.6393	2.6573	2.6754	2.6936	2.7120	2.7304
19	2.7490	2.7676	2.7864	2.8053	2.8243	2.8434	2.8626	2.8819	2.9013	2.9208
20	2.9404	2.9602	2.9800	2.9999	3.0200	3.0401	3.0604	3.0807	3.1012	3.1217
21	3.1424	3.1632	3.1840	3.2050	3.2261	3.2473	3.2686	3.2899	3.3114	3.3330
22	3.3547	3.3765	3.3984	3.4204	3.4425	3.4647	3.4869	3.5093	3.5318	3.5544
23	3.5771	3.5999	3.6228	3.6458	3.6688	3.6920	3.7153	3.7387	3.7622	3.7857
24	3.8094	3.8332	3.8570	3.8810	3.9051	3.9292	3.9535	3.9778	4.0023	4.0268
25	4.0514	4.0762	4.1010	4.1259	4.1509	4.1760	4.2012	4.2265	4.2519	4.2774
26	4.3030	4.3287	4.3544	4.3803	4.4062	4.4323	4.4584	4.4846	4.5110	4.5374
27	4.5639	4.5905	4.6172	4.6440	4.6708	4.6978	4.7249	4.7520	4.7792	4.8066
28	4.8340	4.8615	4.8891	4.9168	4.9446	4.9724	5.0004	5.0284	5.0566	5.0848
29	5.1131	5.1415	5.1700	5.1986	5.2273	5.2560	5.2849	5.3138	5.3428	5.3719
30	5.4011	5.4304	5.4598	5.4892	5.5188	5.5484	5.5781	5.6079	5.6378	5.6678
31	5.6979	5.7280	5.7582	5.7886	5.8190	5.8495	5.8800	5.9107	5.9414	5.9723
32	6.0032	6.0342	6.0653	6.0964	6.1277	6.1590	6.1904	6.2219	6.2535	6.2852
33	6.3169	6.3488	6.3807	6.4127	6.4447	6.4769	6.5092	6.5415	6.5739	6.6064
34	6.6390	6.6716	6.7043	6.7372	6.7701	6.8030	6.8361	6.8692	6.9025	6.9358
35	6.9692	7.0026	7.0362	7.0698	7.1035	7.1373	7.1711	7.2051	7.2391	7.2732
36	7.3074	7.3416	7.3760	7.4104	7.4449	7.4795	7.5141	7.5489	7.5837	7.6185
37	7.6535	7.6886	7.7237	7.7589	7.7941	7.8295	7.8649	7.9004	7.9360	7.9717
38	8.0074	8.0432	8.0791	8.1151	8.1511	8.1872	8.2234	8.2597	8.2960	8.3324
39	8.3689	8.4055	8.4421	8.4788	8.5156	8.5525	8.5894	8.6264	8.6635	8.7007
40	8.7379	8.7752	8.8126	8.8501	8.8876	8.9252	8.9629	9.0006	9.0384	9.0763
41	9.1143	9.1523	9.1904	9.2286	9.2669	9.3052	9.3436	9.3821	9.4206	9.4592

Table 8.1. (*j*) Value of correction C in mg for a 1 cm^3 borosilicate glass measure for $\gamma = 0.000\,009\,75$, $\sigma = 0.0012$ g cm^{-3}, $\Delta = 8.000$ g cm^{-3}, reference temperature $= 40\,°\text{C}$, where γ is the coefficient of expansion of the material of the measure, σ is the density of air (in g cm^{-3}) and Δ is the density of standard weights used (in g cm^{-3}).

T (°C)	0.0	0.1	0.2	0.3	0.4	0.5	0.6	0.7	0.8	0.9
5	1.4242	1.4249	1.4257	1.4267	1.4279	1.4292	1.4307	1.4323	1.4340	1.4360
6	1.4380	1.4402	1.4426	1.4451	1.4478	1.4506	1.4535	1.4567	1.4599	1.4633
7	1.4669	1.4706	1.4744	1.4784	1.4825	1.4868	1.4912	1.4957	1.5004	1.5053
8	1.5103	1.5154	1.5207	1.5261	1.5316	1.5373	1.5432	1.5491	1.5552	1.5615
9	1.5679	1.5744	1.5811	1.5879	1.5948	1.6019	1.6091	1.6164	1.6239	1.6315
10	1.6393	1.6471	1.6552	1.6633	1.6716	1.6800	1.6886	1.6972	1.7061	1.7150
11	1.7241	1.7333	1.7426	1.7521	1.7617	1.7714	1.7812	1.7912	1.8013	1.8116
12	1.8219	1.8324	1.8430	1.8538	1.8647	1.8757	1.8868	1.8980	1.9094	1.9209
13	1.9325	1.9443	1.9561	1.9681	1.9803	1.9925	2.0049	2.0173	2.0299	2.0427
14	2.0555	2.0685	2.0816	2.0948	2.1081	2.1216	2.1351	2.1488	2.1626	2.1766
15	2.1906	2.2048	2.2190	2.2334	2.2480	2.2626	2.2773	2.2922	2.3072	2.3223
16	2.3375	2.3528	2.3683	2.3838	2.3995	2.4153	2.4312	2.4472	2.4633	2.4795
17	2.4959	2.5124	2.5289	2.5456	2.5624	2.5793	2.5964	2.6135	2.6307	2.6481
18	2.6656	2.6831	2.7008	2.7186	2.7365	2.7545	2.7727	2.7909	2.8092	2.8277
19	2.8462	2.8649	2.8837	2.9025	2.9215	2.9406	2.9598	2.9791	2.9985	3.0180
20	3.0377	3.0574	3.0772	3.0971	3.1172	3.1373	3.1576	3.1779	3.1984	3.2189
21	3.2396	3.2604	3.2812	3.3022	3.3233	3.3445	3.3657	3.3871	3.4086	3.4302
22	3.4519	3.4737	3.4956	3.5175	3.5396	3.5618	3.5841	3.6065	3.6290	3.6516
23	3.6743	3.6970	3.7199	3.7429	3.7660	3.7892	3.8124	3.8358	3.8593	3.8829
24	3.9065	3.9303	3.9542	3.9781	4.0022	4.0263	4.0506	4.0749	4.0994	4.1239
25	4.1485	4.1733	4.1981	4.2230	4.2480	4.2731	4.2983	4.3236	4.3490	4.3745
26	4.4001	4.4257	4.4515	4.4774	4.5033	4.5293	4.5555	4.5817	4.6080	4.6344
27	4.6609	4.6875	4.7142	4.7410	4.7679	4.7948	4.8219	4.8490	4.8763	4.9036
28	4.9310	4.9585	4.9861	5.0138	5.0416	5.0695	5.0974	5.1255	5.1536	5.1818
29	5.2101	5.2385	5.2670	5.2956	5.3243	5.3530	5.3819	5.4108	5.4398	5.4689
30	5.4981	5.5274	5.5568	5.5862	5.6157	5.6454	5.6751	5.7049	5.7348	5.7648
31	5.7948	5.8250	5.8552	5.8855	5.9159	5.9464	5.9770	6.0076	6.0384	6.0692
32	6.1001	6.1311	6.1622	6.1933	6.2246	6.2559	6.2873	6.3188	6.3504	6.3821
33	6.4138	6.4456	6.4775	6.5095	6.5416	6.5738	6.6060	6.6383	6.6707	6.7032
34	6.7358	6.7685	6.8012	6.8340	6.8669	6.8999	6.9329	6.9661	6.9993	7.0326
35	7.0660	7.0994	7.1330	7.1666	7.2003	7.2341	7.2679	7.3019	7.3359	7.3700
36	7.4042	7.4384	7.4728	7.5072	7.5417	7.5762	7.6109	7.6456	7.6804	7.7153
37	7.7503	7.7853	7.8204	7.8556	7.8909	7.9262	7.9616	7.9971	8.0327	8.0684
38	8.1041	8.1399	8.1758	8.2118	8.2478	8.2839	8.3201	8.3563	8.3927	8.4291
39	8.4656	8.5021	8.5388	8.5755	8.6123	8.6491	8.6861	8.7231	8.7602	8.7973
40	8.8346	8.8719	8.9092	8.9467	8.9842	9.0218	9.0595	9.0972	9.1351	9.1729
41	9.2109	9.2489	9.2870	9.3252	9.3635	9.4018	9.4402	9.4787	9.5172	9.5558

Table 8.1. (k) Value of correction C in mg for a 1 cm^3 borosilicate glass measure for $\gamma = 0.000\,009\,75$, $\sigma = 0.0012$ g cm^{-3}, $\Delta = 8.000$ g cm^{-3}, reference temperature $= 50\,°C$, where γ is the coefficient of expansion of the material of the measure, σ is the density of air (in g cm^{-3}) and Δ is the density of standard weights used (in g cm^{-3}).

T (°C)	0.0	0.1	0.2	0.3	0.4	0.5	0.6	0.7	0.8	0.9
5	1.5216	1.5223	1.5231	1.5241	1.5253	1.5266	1.5281	1.5297	1.5314	1.5333
6	1.5354	1.5376	1.5400	1.5425	1.5452	1.5480	1.5509	1.5540	1.5573	1.5607
7	1.5642	1.5679	1.5718	1.5758	1.5799	1.5842	1.5886	1.5931	1.5978	1.6027
8	1.6077	1.6128	1.6181	1.6235	1.6290	1.6347	1.6405	1.6465	1.6526	1.6589
9	1.6653	1.6718	1.6784	1.6852	1.6922	1.6992	1.7065	1.7138	1.7213	1.7289
10	1.7366	1.7445	1.7525	1.7607	1.7690	1.7774	1.7859	1.7946	1.8034	1.8124
11	1.8214	1.8306	1.8400	1.8494	1.8590	1.8687	1.8786	1.8886	1.8987	1.9089
12	1.9193	1.9298	1.9404	1.9511	1.9620	1.9730	1.9841	1.9954	2.0067	2.0182
13	2.0299	2.0416	2.0535	2.0655	2.0776	2.0898	2.1022	2.1147	2.1273	2.1400
14	2.1528	2.1658	2.1789	2.1921	2.2054	2.2189	2.2325	2.2461	2.2599	2.2739
15	2.2879	2.3021	2.3164	2.3307	2.3453	2.3599	2.3746	2.3895	2.4045	2.4196
16	2.4348	2.4501	2.4655	2.4811	2.4968	2.5126	2.5285	2.5445	2.5606	2.5768
17	2.5932	2.6096	2.6262	2.6429	2.6597	2.6766	2.6936	2.7108	2.7280	2.7454
18	2.7628	2.7804	2.7981	2.8159	2.8338	2.8518	2.8699	2.8881	2.9065	2.9249
19	2.9435	2.9621	2.9809	2.9998	3.0188	3.0378	3.0570	3.0763	3.0957	3.1153
20	3.1349	3.1546	3.1744	3.1944	3.2144	3.2345	3.2548	3.2751	3.2956	3.3162
21	3.3368	3.3576	3.3784	3.3994	3.4205	3.4417	3.4629	3.4843	3.5058	3.5274
22	3.5491	3.5708	3.5927	3.6147	3.6368	3.6590	3.6813	3.7037	3.7261	3.7487
23	3.7714	3.7942	3.8171	3.8401	3.8631	3.8863	3.9096	3.9330	3.9564	3.9800
24	4.0037	4.0274	4.0513	4.0753	4.0993	4.1235	4.1477	4.1720	4.1965	4.2210
25	4.2456	4.2704	4.2952	4.3201	4.3451	4.3702	4.3954	4.4207	4.4461	4.4716
26	4.4971	4.5228	4.5486	4.5744	4.6004	4.6264	4.6525	4.6788	4.7051	4.7315
27	4.7580	4.7846	4.8113	4.8381	4.8649	4.8919	4.9189	4.9461	4.9733	5.0006
28	5.0281	5.0556	5.0832	5.1108	5.1386	5.1665	5.1944	5.2225	5.2506	5.2788
29	5.3071	5.3355	5.3640	5.3926	5.4213	5.4500	5.4788	5.5078	5.5368	5.5659
30	5.5951	5.6244	5.6537	5.6832	5.7127	5.7423	5.7720	5.8018	5.8317	5.8617
31	5.8918	5.9219	5.9521	5.9824	6.0128	6.0433	6.0739	6.1045	6.1353	6.1661
32	6.1970	6.2280	6.2591	6.2902	6.3215	6.3528	6.3842	6.4157	6.4473	6.4789
33	6.5107	6.5425	6.5744	6.6064	6.6385	6.6706	6.7029	6.7352	6.7676	6.8001
34	6.8327	6.8653	6.8980	6.9308	6.9637	6.9967	7.0298	7.0629	7.0961	7.1294
35	7.1628	7.1962	7.2298	7.2634	7.2971	7.3309	7.3647	7.3987	7.4327	7.4668
36	7.5009	7.5352	7.5695	7.6039	7.6384	7.6730	7.7076	7.7424	7.7772	7.8120
37	7.8470	7.8820	7.9171	7.9523	7.9876	8.0229	8.0584	8.0939	8.1294	8.1651
38	8.2008	8.2366	8.2725	8.3085	8.3445	8.3806	8.4168	8.4530	8.4894	8.5258
39	8.5623	8.5988	8.6355	8.6722	8.7089	8.7458	8.7827	8.8197	8.8568	8.8940
40	8.9312	8.9685	9.0059	9.0433	9.0808	9.1184	9.1561	9.1938	9.2317	9.2695
41	9.3075	9.3455	9.3836	9.4218	9.4601	9.4984	9.5368	9.5752	9.6138	9.6524

Table 8.1. (*l*) Value of correction C in mg for a 1 cm^3 borosilicate glass measure for $\gamma = 0.000\,009\,75$, $\sigma = 0.0012$ g cm^{-3}, $\Delta = 8.000$ g cm^{-3}, reference temperature $= 60\,°C$, where γ is the coefficient of expansion of the material of the measure, σ is the density of air (in g cm^{-3}) and Δ is the density of standard weights used (in g cm^{-3}).

T (°C)	0.0	0.1	0.2	0.3	0.4	0.5	0.6	0.7	0.8	0.9
5	1.6190	1.6197	1.6205	1.6215	1.6227	1.6240	1.6255	1.6271	1.6288	1.6307
6	1.6328	1.6350	1.6374	1.6399	1.6426	1.6454	1.6483	1.6514	1.6547	1.6581
7	1.6616	1.6653	1.6692	1.6731	1.6773	1.6815	1.6860	1.6905	1.6952	1.7001
8	1.7050	1.7102	1.7154	1.7209	1.7264	1.7321	1.7379	1.7439	1.7500	1.7562
9	1.7626	1.7692	1.7758	1.7826	1.7895	1.7966	1.8038	1.8112	1.8186	1.8263
10	1.8340	1.8419	1.8499	1.8580	1.8663	1.8747	1.8833	1.8920	1.9008	1.9097
11	1.9188	1.9280	1.9373	1.9468	1.9564	1.9661	1.9760	1.9859	1.9960	2.0063
12	2.0166	2.0271	2.0377	2.0485	2.0594	2.0704	2.0815	2.0927	2.1041	2.1156
13	2.1272	2.1389	2.1508	2.1628	2.1749	2.1872	2.1995	2.2120	2.2246	2.2373
14	2.2502	2.2631	2.2762	2.2894	2.3028	2.3162	2.3298	2.3435	2.3573	2.3712
15	2.3852	2.3994	2.4137	2.4281	2.4426	2.4572	2.4719	2.4868	2.5018	2.5169
16	2.5321	2.5474	2.5628	2.5784	2.5941	2.6098	2.6257	2.6417	2.6579	2.6741
17	2.6905	2.7069	2.7235	2.7402	2.7570	2.7739	2.7909	2.8080	2.8253	2.8426
18	2.8601	2.8777	2.8953	2.9131	2.9310	2.9490	2.9672	2.9854	3.0037	3.0222
19	3.0407	3.0594	3.0781	3.0970	3.1160	3.1351	3.1543	3.1736	3.1930	3.2125
20	3.2321	3.2518	3.2716	3.2916	3.3116	3.3318	3.3520	3.3723	3.3928	3.4134
21	3.4340	3.4548	3.4756	3.4966	3.5177	3.5389	3.5601	3.5815	3.6030	3.6246
22	3.6462	3.6680	3.6899	3.7119	3.7340	3.7562	3.7784	3.8008	3.8233	3.8459
23	3.8686	3.8914	3.9142	3.9372	3.9603	3.9835	4.0067	4.0301	4.0536	4.0771
24	4.1008	4.1246	4.1484	4.1724	4.1964	4.2206	4.2448	4.2692	4.2936	4.3181
25	4.3428	4.3675	4.3923	4.4172	4.4422	4.4673	4.4925	4.5178	4.5432	4.5687
26	4.5942	4.6199	4.6457	4.6715	4.6974	4.7235	4.7496	4.7758	4.8022	4.8286
27	4.8551	4.8817	4.9083	4.9351	4.9620	4.9889	5.0160	5.0431	5.0704	5.0977
28	5.1251	5.1526	5.1802	5.2079	5.2356	5.2635	5.2914	5.3195	5.3476	5.3758
29	5.4041	5.4325	5.4610	5.4896	5.5182	5.5470	5.5758	5.6048	5.6338	5.6629
30	5.6921	5.7213	5.7507	5.7801	5.8097	5.8393	5.8690	5.8988	5.9287	5.9586
31	5.9887	6.0188	6.0491	6.0794	6.1098	6.1402	6.1708	6.2015	6.2322	6.2630
32	6.2939	6.3249	6.3560	6.3871	6.4184	6.4497	6.4811	6.5126	6.5442	6.5758
33	6.6076	6.6394	6.6713	6.7033	6.7354	6.7675	6.7997	6.8321	6.8645	6.8969
34	6.9295	6.9621	6.9949	7.0277	7.0606	7.0935	7.1266	7.1597	7.1929	7.2262
35	7.2596	7.2931	7.3266	7.3602	7.3939	7.4277	7.4615	7.4955	7.5295	7.5636
36	7.5977	7.6320	7.6663	7.7007	7.7352	7.7698	7.8044	7.8391	7.8739	7.9088
37	7.9438	7.9788	8.0139	8.0491	8.0843	8.1197	8.1551	8.1906	8.2262	8.2618
38	8.2975	8.3333	8.3692	8.4052	8.4412	8.4773	8.5135	8.5497	8.5861	8.6225
39	8.6589	8.6955	8.7321	8.7688	8.8056	8.8425	8.8794	8.9164	8.9535	8.9906
40	9.0278	9.0651	9.1025	9.1400	9.1775	9.2151	9.2527	9.2905	9.3283	9.3662
41	9.4041	9.4421	9.4802	9.5184	9.5566	9.5950	9.6333	9.6718	9.7103	9.7489

Table 8.1. (m) Value of correction C in mg for a 1 cm^3 borosilicate glass measure for γ = 0.000 009 75, σ = 0.0012 g cm^{-3}, Δ = 8.000 g cm^{-3}, reference temperature = 70 °C, where γ is the coefficient of expansion of the material of the measure, σ is the density of air (in g cm^{-3}) and Δ is the density of standard weights used (in g cm^{-3}).

T (°C)	0.0	0.1	0.2	0.3	0.4	0.5	0.6	0.7	0.8	0.9
5	1.7164	1.7171	1.7179	1.7189	1.7201	1.7214	1.7228	1.7245	1.7262	1.7281
6	1.7302	1.7324	1.7348	1.7373	1.7400	1.7428	1.7457	1.7488	1.7521	1.7555
7	1.7590	1.7627	1.7666	1.7705	1.7747	1.7789	1.7833	1.7879	1.7926	1.7974
8	1.8024	1.8076	1.8128	1.8182	1.8238	1.8295	1.8353	1.8413	1.8474	1.8536
9	1.8600	1.8665	1.8732	1.8800	1.8869	1.8940	1.9012	1.9085	1.9160	1.9236
10	1.9314	1.9393	1.9473	1.9554	1.9637	1.9721	1.9807	1.9893	1.9981	2.0071
11	2.0161	2.0254	2.0347	2.0441	2.0537	2.0635	2.0733	2.0833	2.0934	2.1036
12	2.1140	2.1245	2.1351	2.1458	2.1567	2.1677	2.1788	2.1901	2.2014	2.2129
13	2.2245	2.2363	2.2481	2.2601	2.2723	2.2845	2.2968	2.3093	2.3219	2.3346
14	2.3475	2.3605	2.3735	2.3867	2.4001	2.4135	2.4271	2.4408	2.4546	2.4685
15	2.4825	2.4967	2.5110	2.5254	2.5399	2.5545	2.5692	2.5841	2.5991	2.6142
16	2.6294	2.6447	2.6601	2.6757	2.6914	2.7071	2.7230	2.7390	2.7552	2.7714
17	2.7877	2.8042	2.8208	2.8374	2.8542	2.8711	2.8882	2.9053	2.9225	2.9399
18	2.9573	2.9749	2.9926	3.0104	3.0283	3.0463	3.0644	3.0826	3.1010	3.1194
19	3.1380	3.1566	3.1754	3.1942	3.2132	3.2323	3.2515	3.2708	3.2902	3.3097
20	3.3293	3.3490	3.3689	3.3888	3.4088	3.4290	3.4492	3.4696	3.4900	3.5106
21	3.5312	3.5520	3.5728	3.5938	3.6149	3.6360	3.6573	3.6787	3.7002	3.7217
22	3.7434	3.7652	3.7871	3.8091	3.8311	3.8533	3.8756	3.8980	3.9205	3.9431
23	3.9657	3.9885	4.0114	4.0344	4.0574	4.0806	4.1039	4.1273	4.1507	4.1743
24	4.1979	4.2217	4.2456	4.2695	4.2936	4.3177	4.3419	4.3663	4.3907	4.4152
25	4.4399	4.4646	4.4894	4.5143	4.5393	4.5644	4.5896	4.6149	4.6403	4.6657
26	4.6913	4.7170	4.7427	4.7686	4.7945	4.8206	4.8467	4.8729	4.8992	4.9256
27	4.9521	4.9787	5.0054	5.0322	5.0590	5.0860	5.1130	5.1402	5.1674	5.1947
28	5.2221	5.2496	5.2772	5.3049	5.3326	5.3605	5.3885	5.4165	5.4446	5.4728
29	5.5011	5.5295	5.5580	5.5866	5.6152	5.6440	5.6728	5.7017	5.7307	5.7598
30	5.7890	5.8183	5.8477	5.8771	5.9066	5.9363	5.9660	5.9957	6.0256	6.0556
31	6.0856	6.1158	6.1460	6.1763	6.2067	6.2372	6.2677	6.2984	6.3291	6.3599
32	6.3908	6.4218	6.4529	6.4840	6.5153	6.5466	6.5780	6.6095	6.6411	6.6727
33	6.7044	6.7363	6.7682	6.8002	6.8322	6.8644	6.8966	6.9289	6.9613	6.9938
34	7.0264	7.0590	7.0917	7.1245	7.1574	7.1904	7.2234	7.2566	7.2898	7.3231
35	7.3564	7.3899	7.4234	7.4570	7.4907	7.5245	7.5583	7.5923	7.6263	7.6603
36	7.6945	7.7288	7.7631	7.7975	7.8320	7.8665	7.9012	7.9359	7.9707	8.0055
37	8.0405	8.0755	8.1106	8.1458	8.1811	8.2164	8.2518	8.2873	8.3229	8.3585
38	8.3942	8.4300	8.4659	8.5019	8.5379	8.5740	8.6102	8.6464	8.6827	8.7191
39	8.7556	8.7922	8.8288	8.8655	8.9023	8.9391	8.9760	9.0130	9.0501	9.0873
40	9.1245	9.1618	9.1991	9.2366	9.2741	9.3117	9.3493	9.3871	9.4249	9.4628
41	9.5007	9.5387	9.5768	9.6150	9.6532	9.6915	9.7299	9.7684	9.8069	9.8455

Table 8.1. (*n*) Value of correction C in mg for a 1 cm^3 borosilicate glass measure for $\gamma = 0.000\,009\,75$, $\sigma = 0.0012$ g cm^{-3}, $\Delta = 8.000$ g cm^{-3}, reference temperature $= 80\,°C$, where γ is the coefficient of expansion of the material of the measure, σ is the density of air (in g cm^{-3}) and Δ is the density of standard weights used (in g cm^{-3}).

T (°C)	0.0	0.1	0.2	0.3	0.4	0.5	0.6	0.7	0.8	0.9
5	1.8138	1.8145	1.8153	1.8163	1.8175	1.8188	1.8202	1.8219	1.8236	1.8255
6	1.8276	1.8298	1.8322	1.8347	1.8373	1.8402	1.8431	1.8462	1.8495	1.8529
7	1.8564	1.8601	1.8639	1.8679	1.8720	1.8763	1.8807	1.8853	1.8900	1.8948
8	1.8998	1.9049	1.9102	1.9156	1.9212	1.9269	1.9327	1.9386	1.9448	1.9510
9	1.9574	1.9639	1.9706	1.9774	1.9843	1.9914	1.9986	2.0059	2.0134	2.0210
10	2.0287	2.0366	2.0446	2.0528	2.0611	2.0695	2.0780	2.0867	2.0955	2.1044
11	2.1135	2.1227	2.1320	2.1415	2.1511	2.1608	2.1707	2.1806	2.1907	2.2010
12	2.2113	2.2218	2.2324	2.2432	2.2540	2.2650	2.2762	2.2874	2.2988	2.3103
13	2.3219	2.3336	2.3455	2.3575	2.3696	2.3818	2.3942	2.4067	2.4193	2.4320
14	2.4448	2.4578	2.4709	2.4841	2.4974	2.5108	2.5244	2.5381	2.5519	2.5658
15	2.5798	2.5940	2.6083	2.6227	2.6372	2.6518	2.6665	2.6814	2.6964	2.7115
16	2.7267	2.7420	2.7574	2.7730	2.7886	2.8044	2.8203	2.8363	2.8524	2.8687
17	2.8850	2.9015	2.9180	2.9347	2.9515	2.9684	2.9854	3.0026	3.0198	3.0371
18	3.0546	3.0722	3.0898	3.1076	3.1255	3.1435	3.1617	3.1799	3.1982	3.2166
19	3.2352	3.2538	3.2726	3.2915	3.3105	3.3295	3.3487	3.3680	3.3874	3.4069
20	3.4265	3.4463	3.4661	3.4860	3.5060	3.5262	3.5464	3.5668	3.5872	3.6078
21	3.6284	3.6492	3.6700	3.6910	3.7121	3.7332	3.7545	3.7759	3.7974	3.8189
22	3.8406	3.8624	3.8843	3.9062	3.9283	3.9505	3.9728	3.9952	4.0176	4.0402
23	4.0629	4.0857	4.1085	4.1315	4.1546	4.1778	4.2010	4.2244	4.2479	4.2714
24	4.2951	4.3188	4.3427	4.3666	4.3907	4.4148	4.4391	4.4634	4.4878	4.5124
25	4.5370	4.5617	4.5865	4.6114	4.6364	4.6615	4.6867	4.7120	4.7374	4.7628
26	4.7884	4.8141	4.8398	4.8657	4.8916	4.9176	4.9438	4.9700	4.9963	5.0227
27	5.0492	5.0758	5.1024	5.1292	5.1561	5.1830	5.2101	5.2372	5.2644	5.2917
28	5.3191	5.3466	5.3742	5.4019	5.4297	5.4575	5.4855	5.5135	5.5416	5.5698
29	5.5981	5.6265	5.6550	5.6836	5.7122	5.7410	5.7698	5.7987	5.8277	5.8568
30	5.8860	5.9153	5.9446	5.9741	6.0036	6.0332	6.0629	6.0927	6.1226	6.1525
31	6.1826	6.2127	6.2429	6.2732	6.3036	6.3341	6.3647	6.3953	6.4260	6.4568
32	6.4877	6.5187	6.5498	6.5809	6.6122	6.6435	6.6749	6.7064	6.7379	6.7696
33	6.8013	6.8331	6.8650	6.8970	6.9291	6.9612	6.9935	7.0258	7.0582	7.0906
34	7.1232	7.1558	7.1886	7.2214	7.2542	7.2872	7.3203	7.3534	7.3866	7.4199
35	7.4532	7.4867	7.5202	7.5538	7.5875	7.6213	7.6551	7.6890	7.7230	7.7571
36	7.7913	7.8255	7.8599	7.8943	7.9287	7.9633	7.9979	8.0326	8.0674	8.1023
37	8.1372	8.1723	8.2074	8.2425	8.2778	8.3131	8.3485	8.3840	8.4196	8.4552
38	8.4910	8.5267	8.5626	8.5986	8.6346	8.6707	8.7068	8.7431	8.7794	8.8158
39	8.8523	8.8888	8.9255	8.9622	8.9989	9.0358	9.0727	9.1097	9.1468	9.1839
40	9.2211	9.2584	9.2958	9.3332	9.3707	9.4083	9.4460	9.4837	9.5215	9.5594
41	9.5973	9.6353	9.6734	9.7116	9.7498	9.7881	9.8265	9.8649	9.9035	9.9421

Table 8.1. (*o*) Value of correction C in mg for a 1 cm^3 soda glass measure for $\gamma = 0.000\,030$, $\sigma = 0.0012$ g cm^{-3}, $\Delta = 8.000$ g cm^{-3}, reference temperature $= 20\,°$C, where γ is the coefficient of expansion of the material of the measure, σ is the density of air (in g cm^{-3}) and Δ is the density of standard weights used (in g cm^{-3}).

T (°C)	0.0	0.1	0.2	0.3	0.4	0.5	0.6	0.7	0.8	0.9
5	1.5328	1.5315	1.5303	1.5293	1.5284	1.5277	1.5272	1.5268	1.5265	1.5264
6	1.5264	1.5266	1.5270	1.5274	1.5281	1.5289	1.5298	1.5309	1.5321	1.5335
7	1.5350	1.5367	1.5385	1.5405	1.5426	1.5448	1.5472	1.5498	1.5524	1.5553
8	1.5582	1.5613	1.5646	1.5680	1.5715	1.5752	1.5790	1.5829	1.5870	1.5912
9	1.5956	1.6001	1.6047	1.6095	1.6144	1.6195	1.6247	1.6300	1.6354	1.6410
10	1.6468	1.6526	1.6586	1.6647	1.6710	1.6774	1.6839	1.6906	1.6974	1.7043
11	1.7113	1.7185	1.7258	1.7333	1.7409	1.7486	1.7564	1.7643	1.7724	1.7806
12	1.7890	1.7975	1.8061	1.8148	1.8236	1.8326	1.8417	1.8509	1.8603	1.8698
13	1.8794	1.8891	1.8989	1.9089	1.9190	1.9292	1.9396	1.9500	1.9606	1.9713
14	1.9822	1.9931	2.0042	2.0154	2.0267	2.0381	2.0496	2.0613	2.0731	2.0850
15	2.0970	2.1092	2.1214	2.1338	2.1463	2.1589	2.1717	2.1845	2.1975	2.2105
16	2.2237	2.2370	2.2505	2.2640	2.2776	2.2914	2.3053	2.3193	2.3334	2.3476
17	2.3620	2.3764	2.3910	2.4056	2.4204	2.4353	2.4503	2.4654	2.4807	2.4960
18	2.5114	2.5270	2.5427	2.5584	2.5743	2.5903	2.6064	2.6226	2.6390	2.6554
19	2.6719	2.6886	2.7053	2.7222	2.7392	2.7562	2.7734	2.7907	2.8081	2.8256
20	2.8432	2.8609	2.8787	2.8967	2.9147	2.9328	2.9510	2.9694	2.9878	3.0064
21	3.0250	3.0438	3.0626	3.0816	3.1006	3.1198	3.1391	3.1584	3.1779	3.1975
22	3.2172	3.2369	3.2568	3.2768	3.2969	3.3170	3.3373	3.3577	3.3782	3.3987
23	3.4194	3.4402	3.4611	3.4820	3.5031	3.5243	3.5455	3.5669	3.5884	3.6099
24	3.6316	3.6533	3.6752	3.6971	3.7192	3.7413	3.7636	3.7859	3.8083	3.8308
25	3.8535	3.8762	3.8990	3.9219	3.9449	3.9680	3.9912	4.0145	4.0379	4.0613
26	4.0849	4.1086	4.1323	4.1562	4.1801	4.2042	4.2283	4.2525	4.2768	4.3012
27	4.3257	4.3503	4.3750	4.3998	4.4246	4.4496	4.4746	4.4998	4.5250	4.5503
28	4.5757	4.6012	4.6268	4.6525	4.6783	4.7042	4.7301	4.7561	4.7823	4.8085
29	4.8348	4.8612	4.8877	4.9143	4.9409	4.9677	4.9945	5.0214	5.0485	5.0756
30	5.1028	5.1300	5.1574	5.1848	5.2124	5.2400	5.2677	5.2955	5.3234	5.3514
31	5.3794	5.4076	5.4358	5.4641	5.4925	5.5210	5.5496	5.5782	5.6070	5.6358
32	5.6647	5.6937	5.7228	5.7520	5.7812	5.8105	5.8400	5.8695	5.8990	5.9287
33	5.9585	5.9883	6.0182	6.0482	6.0783	6.1084	6.1387	6.1690	6.1994	6.2299
34	6.2605	6.2912	6.3219	6.3527	6.3836	6.4146	6.4457	6.4768	6.5080	6.5393
35	6.5707	6.6022	6.6337	6.6654	6.6971	6.7289	6.7607	6.7927	6.8247	6.8568
36	6.8890	6.9213	6.9536	6.9860	7.0185	7.0511	7.0838	7.1165	7.1493	7.1822
37	7.2152	7.2482	7.2813	7.3145	7.3478	7.3812	7.4146	7.4481	7.4817	7.5154
38	7.5491	7.5829	7.6168	7.6508	7.6849	7.7190	7.7532	7.7875	7.8218	7.8562
39	7.8907	7.9253	7.9600	7.9947	8.0295	8.0644	8.0993	8.1343	8.1694	8.2046
40	8.2399	8.2752	8.3106	8.3460	8.3816	8.4172	8.4529	8.4887	8.5245	8.5604
41	8.5964	8.6324	8.6685	8.7047	8.7410	8.7774	8.8138	8.8503	8.8868	8.9234

Table 8.1. (p) Value of correction C in mg for a 1 cm^3 soda glass measure for $\gamma = 0.000\,030$, $\sigma = 0.0012$ g cm^{-3}, $\Delta = 8.400$ g cm^{-3}, reference temperature $= 20\,°C$, where γ is the coefficient of expansion of the material of the measure, σ is the density of air (in g cm^{-3}) and Δ is the density of standard weights used (in g cm^{-3}).

T (°C)	0.0	0.1	0.2	0.3	0.4	0.5	0.6	0.7	0.8	0.9
5	1.5400	1.5386	1.5375	1.5364	1.5356	1.5349	1.5343	1.5339	1.5336	1.5335
6	1.5336	1.5337	1.5341	1.5346	1.5352	1.5360	1.5369	1.5380	1.5393	1.5406
7	1.5422	1.5438	1.5456	1.5476	1.5497	1.5520	1.5544	1.5569	1.5596	1.5624
8	1.5654	1.5685	1.5717	1.5751	1.5786	1.5823	1.5861	1.5900	1.5941	1.5984
9	1.6027	1.6072	1.6119	1.6166	1.6216	1.6266	1.6318	1.6371	1.6426	1.6482
10	1.6539	1.6597	1.6657	1.6719	1.6781	1.6845	1.6911	1.6977	1.7045	1.7114
11	1.7185	1.7257	1.7330	1.7404	1.7480	1.7557	1.7635	1.7715	1.7796	1.7878
12	1.7961	1.8046	1.8132	1.8219	1.8308	1.8397	1.8488	1.8581	1.8674	1.8769
13	1.8865	1.8962	1.9061	1.9160	1.9261	1.9364	1.9467	1.9572	1.9677	1.9785
14	1.9893	2.0002	2.0113	2.0225	2.0338	2.0452	2.0568	2.0684	2.0802	2.0921
15	2.1042	2.1163	2.1286	2.1409	2.1534	2.1661	2.1788	2.1916	2.2046	2.2177
16	2.2309	2.2442	2.2576	2.2711	2.2848	2.2985	2.3124	2.3264	2.3405	2.3547
17	2.3691	2.3835	2.3981	2.4128	2.4275	2.4424	2.4574	2.4726	2.4878	2.5031
18	2.5186	2.5341	2.5498	2.5656	2.5815	2.5975	2.6136	2.6298	2.6461	2.6625
19	2.6791	2.6957	2.7125	2.7293	2.7463	2.7634	2.7805	2.7978	2.8152	2.8327
20	2.8503	2.8680	2.8859	2.9038	2.9218	2.9399	2.9582	2.9765	2.9949	3.0135
21	3.0321	3.0509	3.0697	3.0887	3.1078	3.1269	3.1462	3.1656	3.1850	3.2046
22	3.2243	3.2440	3.2639	3.2839	3.3040	3.3241	3.3444	3.3648	3.3853	3.4058
23	3.4265	3.4473	3.4682	3.4891	3.5102	3.5314	3.5526	3.5740	3.5955	3.6170
24	3.6387	3.6604	3.6823	3.7042	3.7263	3.7484	3.7707	3.7930	3.8154	3.8380
25	3.8606	3.8833	3.9061	3.9290	3.9520	3.9751	3.9983	4.0216	4.0450	4.0685
26	4.0920	4.1157	4.1394	4.1633	4.1872	4.2113	4.2354	4.2596	4.2839	4.3083
27	4.3328	4.3574	4.3821	4.4069	4.4317	4.4567	4.4818	4.5069	4.5321	4.5574
28	4.5829	4.6084	4.6339	4.6596	4.6854	4.7113	4.7372	4.7633	4.7894	4.8156
29	4.8419	4.8683	4.8948	4.9214	4.9480	4.9748	5.0016	5.0286	5.0556	5.0827
30	5.1099	5.1371	5.1645	5.1920	5.2195	5.2471	5.2748	5.3026	5.3305	5.3585
31	5.3865	5.4147	5.4429	5.4712	5.4996	5.5281	5.5567	5.5854	5.6141	5.6429
32	5.6718	5.7008	5.7299	5.7591	5.7883	5.8176	5.8471	5.8766	5.9061	5.9358
33	5.9656	5.9954	6.0253	6.0553	6.0854	6.1155	6.1458	6.1761	6.2065	6.2370
34	6.2676	6.2983	6.3290	6.3598	6.3907	6.4217	6.4528	6.4839	6.5151	6.5464
35	6.5778	6.6093	6.6408	6.6725	6.7042	6.7360	6.7678	6.7998	6.8318	6.8639
36	6.8961	6.9283	6.9607	6.9931	7.0256	7.0582	7.0908	7.1236	7.1564	7.1893
37	7.2223	7.2553	7.2884	7.3216	7.3549	7.3883	7.4217	7.4552	7.4888	7.5225
38	7.5562	7.5900	7.6239	7.6579	7.6920	7.7261	7.7603	7.7945	7.8289	7.8633
39	7.8978	7.9324	7.9671	8.0018	8.0366	8.0715	8.1064	8.1414	8.1765	8.2117
40	8.2469	8.2823	8.3177	8.3531	8.3887	8.4243	8.4600	8.4957	8.5316	8.5675
41	8.6035	8.6395	8.6756	8.7118	8.7481	8.7844	8.8208	8.8573	8.8939	8.9305

Table 8.1. (*q*) Value of correction C in mg for a 1 cm^3 soda glass measure for $\gamma = 0.000\,030$, $\sigma = 0.001\,1685$ g cm^{-3}, $\Delta = 8.000$ g cm^{-3}, reference temperature $= 27\,°C$, where γ is the coefficient of expansion of the material of the measure, σ is the density of air (in g cm^{-3}) and Δ is the density of standard weights used (in g cm^{-3}).

T (°C)	0.0	0.1	0.2	0.3	0.4	0.5	0.6	0.7	0.8	0.9
6	1.7151	1.7137	1.7125	1.7115	1.7107	1.7099	1.7094	1.7090	1.7087	1.7086
7	1.7086	1.7088	1.7092	1.7097	1.7103	1.7111	1.7120	1.7131	1.7143	1.7157
8	1.7172	1.7189	1.7207	1.7227	1.7248	1.7270	1.7294	1.7320	1.7346	1.7375
9	1.7404	1.7435	1.7468	1.7502	1.7537	1.7573	1.7612	1.7651	1.7692	1.7734
10	1.7778	1.7823	1.7869	1.7917	1.7966	1.8016	1.8068	1.8121	1.8176	1.8232
11	1.8289	1.8348	1.8408	1.8469	1.8532	1.8595	1.8661	1.8727	1.8795	1.8864
12	1.8935	1.9007	1.9080	1.9154	1.9230	1.9307	1.9385	1.9465	1.9545	1.9628
13	1.9711	1.9796	1.9882	1.9969	2.0057	2.0147	2.0238	2.0330	2.0424	2.0519
14	2.0615	2.0712	2.0810	2.0910	2.1011	2.1113	2.1216	2.1321	2.1427	2.1534
15	2.1642	2.1752	2.1862	2.1974	2.2087	2.2201	2.2317	2.2434	2.2551	2.2670
16	2.2791	2.2912	2.3035	2.3158	2.3283	2.3409	2.3537	2.3665	2.3795	2.3925
17	2.4057	2.4190	2.4324	2.4460	2.4596	2.4734	2.4873	2.5013	2.5154	2.5296
18	2.5439	2.5583	2.5729	2.5876	2.6023	2.6172	2.6322	2.6474	2.6626	2.6779
19	2.6934	2.7089	2.7246	2.7403	2.7562	2.7722	2.7883	2.8045	2.8208	2.8373
20	2.8538	2.8705	2.8872	2.9041	2.9210	2.9381	2.9553	2.9726	2.9899	3.0074
21	3.0250	3.0427	3.0606	3.0785	3.0965	3.1146	3.1328	3.1512	3.1696	3.1882
22	3.2068	3.2255	3.2444	3.2634	3.2824	3.3016	3.3208	3.3402	3.3597	3.3792
23	3.3989	3.4187	3.4385	3.4585	3.4786	3.4987	3.5190	3.5394	3.5599	3.5804
24	3.6011	3.6219	3.6427	3.6637	3.6848	3.7059	3.7272	3.7485	3.7700	3.7916
25	3.8132	3.8350	3.8568	3.8787	3.9008	3.9229	3.9452	3.9675	3.9899	4.0124
26	4.0350	4.0578	4.0806	4.1035	4.1265	4.1496	4.1727	4.1960	4.2194	4.2429
27	4.2664	4.2901	4.3138	4.3377	4.3616	4.3856	4.4098	4.4340	4.4583	4.4827
28	4.5072	4.5318	4.5564	4.5812	4.6061	4.6310	4.6561	4.6812	4.7064	4.7317
29	4.7571	4.7826	4.8082	4.8339	4.8597	4.8855	4.9115	4.9375	4.9636	4.9898
30	5.0161	5.0425	5.0690	5.0956	5.1222	5.1490	5.1758	5.2027	5.2297	5.2568
31	5.2840	5.3113	5.3387	5.3661	5.3936	5.4213	5.4490	5.4768	5.5046	5.5326
32	5.5607	5.5888	5.6170	5.6453	5.6737	5.7022	5.7308	5.7594	5.7881	5.8170
33	5.8459	5.8749	5.9039	5.9331	5.9623	5.9916	6.0211	6.0505	6.0801	6.1098
34	6.1395	6.1694	6.1993	6.2293	6.2593	6.2895	6.3197	6.3500	6.3804	6.4109
35	6.4415	6.4721	6.5029	6.5337	6.5646	6.5956	6.6266	6.6578	6.6890	6.7203
36	6.7516	6.7831	6.8146	6.8463	6.8780	6.9097	6.9416	6.9735	7.0056	7.0377
37	7.0698	7.1021	7.1344	7.1668	7.1993	7.2319	7.2646	7.2973	7.3301	7.3630
38	7.3959	7.4290	7.4621	7.4953	7.5286	7.5619	7.5953	7.6289	7.6624	7.6961
39	7.7298	7.7636	7.7975	7.8315	7.8655	7.8996	7.9338	7.9681	8.0024	8.0369
40	8.0714	8.1059	8.1406	8.1753	8.2101	8.2449	8.2799	8.3149	8.3500	8.3852
41	8.4204	8.4557	8.4911	8.5266	8.5621	8.5977	8.6334	8.6691	8.7050	8.7409
42	8.7768	8.8129	8.8490	8.8852	8.9214	8.9578	8.9942	9.0306	9.0672	9.1038

Table 8.1. (r) Value of correction C in mg for a 1 cm^3 soda glass measure for $\gamma = 0.000\,030$, $\sigma = 0.001\,1685$ g cm^{-3}, $\Delta = 8.400$ g cm^{-3}, reference temperature $= 27\,°C$, where γ is the coefficient of expansion of the material of the measure, σ is the density of air (in g cm^{-3}) and Δ is the density of standard weights used (in g cm^{-3}).

T (°C)	0.0	0.1	0.2	0.3	0.4	0.5	0.6	0.7	0.8	0.9
6	1.7220	1.7207	1.7195	1.7185	1.7176	1.7169	1.7163	1.7159	1.7157	1.7155
7	1.7156	1.7158	1.7161	1.7166	1.7172	1.7180	1.7190	1.7200	1.7213	1.7227
8	1.7242	1.7258	1.7277	1.7296	1.7317	1.7340	1.7364	1.7389	1.7416	1.7444
9	1.7474	1.7505	1.7537	1.7571	1.7606	1.7643	1.7681	1.7720	1.7761	1.7804
10	1.7847	1.7892	1.7939	1.7986	1.8035	1.8086	1.8138	1.8191	1.8245	1.8301
11	1.8359	1.8417	1.8477	1.8538	1.8601	1.8665	1.8730	1.8797	1.8865	1.8934
12	1.9004	1.9076	1.9149	1.9224	1.9299	1.9376	1.9455	1.9534	1.9615	1.9697
13	1.9780	1.9865	1.9951	2.0038	2.0127	2.0217	2.0308	2.0400	2.0493	2.0588
14	2.0684	2.0781	2.0880	2.0979	2.1080	2.1182	2.1286	2.1390	2.1496	2.1603
15	2.1712	2.1821	2.1932	2.2044	2.2157	2.2271	2.2386	2.2503	2.2621	2.2740
16	2.2860	2.2981	2.3104	2.3228	2.3353	2.3479	2.3606	2.3734	2.3864	2.3995
17	2.4127	2.4260	2.4394	2.4529	2.4666	2.4803	2.4942	2.5082	2.5223	2.5365
18	2.5508	2.5653	2.5798	2.5945	2.6093	2.6242	2.6392	2.6543	2.6695	2.6848
19	2.7003	2.7158	2.7315	2.7473	2.7632	2.7792	2.7953	2.8115	2.8278	2.8442
20	2.8607	2.8774	2.8941	2.9110	2.9280	2.9450	2.9622	2.9795	2.9969	3.0144
21	3.0320	3.0497	3.0675	3.0854	3.1034	3.1215	3.1398	3.1581	3.1765	3.1951
22	3.2137	3.2325	3.2513	3.2703	3.2893	3.3085	3.3278	3.3471	3.3666	3.3862
23	3.4058	3.4256	3.4455	3.4654	3.4855	3.5057	3.5259	3.5463	3.5668	3.5874
24	3.6080	3.6288	3.6497	3.6706	3.6917	3.7129	3.7341	3.7555	3.7769	3.7985
25	3.8201	3.8419	3.8637	3.8857	3.9077	3.9299	3.9521	3.9744	3.9968	4.0194
26	4.0420	4.0647	4.0875	4.1104	4.1334	4.1565	4.1797	4.2030	4.2263	4.2498
27	4.2734	4.2970	4.3208	4.3446	4.3685	4.3926	4.4167	4.4409	4.4652	4.4896
28	4.5141	4.5387	4.5634	4.5881	4.6130	4.6379	4.6630	4.6881	4.7133	4.7387
29	4.7641	4.7896	4.8151	4.8408	4.8666	4.8924	4.9184	4.9444	4.9705	4.9968
30	5.0231	5.0495	5.0759	5.1025	5.1292	5.1559	5.1827	5.2097	5.2367	5.2638
31	5.2909	5.3182	5.3456	5.3730	5.4006	5.4282	5.4559	5.4837	5.5116	5.5395
32	5.5676	5.5957	5.6239	5.6522	5.6806	5.7091	5.7377	5.7663	5.7951	5.8239
33	5.8528	5.8818	5.9108	5.9400	5.9692	5.9986	6.0280	6.0575	6.0870	6.1167
34	6.1464	6.1763	6.2062	6.2362	6.2662	6.2964	6.3266	6.3570	6.3874	6.4178
35	6.4484	6.4791	6.5098	6.5406	6.5715	6.6025	6.6335	6.6647	6.6959	6.7272
36	6.7586	6.7900	6.8216	6.8532	6.8849	6.9167	6.9485	6.9805	7.0125	7.0446
37	7.0767	7.1090	7.1413	7.1738	7.2062	7.2388	7.2715	7.3042	7.3370	7.3699
38	7.4028	7.4359	7.4690	7.5022	7.5355	7.5688	7.6023	7.6358	7.6693	7.7030
39	7.7367	7.7705	7.8044	7.8384	7.8724	7.9065	7.9407	7.9750	8.0093	8.0438
40	8.0783	8.1128	8.1475	8.1822	8.2170	8.2518	8.2868	8.3218	8.3569	8.3921
41	8.4273	8.4626	8.4980	8.5335	8.5690	8.6046	8.6403	8.6760	8.7118	8.7477
42	8.7837	8.8198	8.8559	8.8921	8.9283	8.9647	9.0011	9.0375	9.0741	9.1107

where γ is the coefficient of cubic thermal expansion. This gives

$$M_w = V_{tr}\{1 + \gamma(t - t_r)\}(\rho_w - \sigma)/(1 - \sigma/\Delta). \tag{8.11}$$

We define C such that

$$M_w + C = V_{tr}. \tag{8.12}$$

Taking the value of M_w from (8.11) and solving for C, we get

$$C = V_{tr}[1 - \{1 + \gamma(t - t_r)\}(\rho_w - \sigma)/(1 - \sigma/\Delta)]. \tag{8.13}$$

The value of the air density has been taken to be constant, as the error due to a variation in the density of air for capacities up to 25 cm^3 will be negligible.

The values of C in mg, for t_r, the reference temperatures, from 20 to 80 °C in steps of 10 °C, have been calculated for water temperatures from 5 to 40 °C in steps of 0.1 °C. These have been tabulated in tables 8.1(a)–(n). The nominal capacity has been taken as 1 cm^3. The values of C are proportional to the nominal capacity. Hence by multiplying the given value of C by a factor equal to the nominal capacity of the pyknometer, one can calculate the value of C for any other capacity.

8.4.7   Expression for A

Let M_w be the corrected mass of standard weights of density Δ and σ the density of the mass. If V is the capacity of the pyknometer and ρ_w the density of water at temperature t_r °C, then

$$M_w(1 - \sigma/\Delta) = V(\rho_w - \sigma). \tag{8.14}$$

Similarly if M_1 is the apparent mass of the liquid with the same volume and temperature as those of water, then

$$M_1(1 - \sigma/\Delta) = V(\rho_1 - \sigma). \tag{8.15}$$

Dividing (8.14) by (8.15), we get

$$M_w/M_1 = (\rho_w - \sigma)/(\rho_1 - \sigma) \tag{8.16}$$

which gives

$$(\rho_1 - \sigma) = (\rho_w - \sigma)M_1/M_w \tag{8.17}$$

giving

$$\rho_1 = (\rho_w/M_w)M_1\{(1 - \sigma/\rho_w)/(1 - \sigma/\rho_1)\}$$
$$\rho_1 = (\rho_w/M_w)M_1\{(1 - \sigma/\rho_w + \sigma/\rho_1)\}. \tag{8.18}$$

Equation (8.18) can be expressed as

$$\rho_1 = (\rho_w/M_w)M_1 + (\rho_w/M_w)M_1\sigma(1/\rho_1 - 1/\rho_w). \tag{8.19}$$

Table 8.2. Values of $A \times 10^3$ due to buoyancy correction for different values for the density of air and observed values of density of liquids.

Density of liquid	Air density $\times 10^3$ (g cm^{-3})					
	1.14	1.16	1.18	1.20	1.22	1.24
0.60	0.4548	0.4627	0.4707	0.4787	0.4867	0.4947
0.65	0.3977	0.4046	0.4116	0.4186	0.4256	0.4326
0.70	0.3406	0.3465	0.3525	0.3585	0.3645	0.3704
0.75	0.2835	0.2884	0.2934	0.2984	0.3034	0.3083
0.80	0.2264	0.2303	0.2343	0.2383	0.2422	0.2462
0.85	0.1693	0.1722	0.1752	0.1782	0.1811	0.1841
0.90	0.1122	0.1141	0.1161	0.1181	0.1200	0.1220
0.95	0.0551	0.0560	0.0570	0.0580	0.0589	0.0599
	Subtract from apparent density					
1.00	0.0020	0.0021	0.0021	0.0022	0.0022	0.0022
1.05	0.0592	0.0602	0.0612	0.0623	0.0633	0.0643
1.10	0.1163	0.1183	0.1203	0.1224	0.1244	0.1265
1.15	0.1734	0.1764	0.1794	0.1825	0.1855	0.1886
1.20	0.2305	0.2345	0.2385	0.2426	0.2466	0.2507
1.25	0.2876	0.2926	0.2976	0.3027	0.3077	0.3128
1.30	0.3447	0.3507	0.3568	0.3628	0.3688	0.3749
1.35	0.4018	0.4088	0.4159	0.4229	0.4300	0.4370
1.40	0.4589	0.4669	0.4750	0.4830	0.4911	0.4991
1.45	0.5160	0.5250	0.5341	0.5431	0.5522	0.5612
1.50	0.5731	0.5831	0.5932	0.6032	0.6133	0.6233
1.55	0.6302	0.6412	0.6523	0.6633	0.6744	0.6855
1.60	0.6873	0.6993	0.7114	0.7234	0.7355	0.7476

As ρ_l is nearly equal to $(\rho_w/M_w)M_l$ to the first approximation, ρ_l may be substituted for $M_l(\rho_w/M_w)$ in the second term of (8.19) as this term is itself quite small, giving us

$$\rho_l = (\rho_w/M_w)M_l + \sigma(1 - \rho_l/\rho_w). \tag{8.20}$$

This may be expressed as

$$\rho_l = FM_w + A \tag{8.21}$$

where

$$A = \sigma(1 - \rho_l/\rho_w). \tag{8.22}$$

The values of A are given in table 8.2 for different values of σ and ρ_l. The value of ρ_w, at $20\,°C$ has been taken as $0.998\,2072$ g cm^{-3}, which is the most recent value for air-free SMOW. The density values for SMOW at other temperatures are given in table 4.4.

8.4.8   Relative advantages and disadvantages of the two methods

In the first method the pyknometer need not be calibrated with water every time
the density of a sample of liquid is to be measured. The water density table refers
to the density of SMOW. The water sample used in the laboratory may not match
SMOW but a user laboratory is in no position to apply the necessary correction for
the use of a different sample of water. However, if necessary, a good calibration
laboratory may apply the correction. Once the pyknometer has been calibrated,
it may remain good for a couple of years. Therefore, it will take half the time
to determine the density of liquids. The disadvantage is that the cubic expansion
coefficient for glass has not been measured so any error in the assumed value
will be reflected in the determinations of the densities of liquids. Instead of the
apparent mass of the liquid its true mass should be determined, which involves
some calculation.

Special shields are used to heat the highly viscous petroleum liquids as given
in [6].

8.4.9   Density determinations of liquids at elevated pressure [7]

The density of the liquid at high pressures may be determined with a density
hydrometer.

Procedure for density determination with a hydrometer

The liquid sample whose density is to be determined is placed in a cylindrical jar
along with the hydrometer. The jar has an air-tight lid and can be connected to a
pressure-generating device. The whole system is kept in a thermostatic bath with
its temperature maintained at a given reference temperature. The air pressure
inside the cylinder is increased to the desired value. Sufficient time is given to
attain thermal equilibrium. The hydrometer is read and the necessary corrections
applied as explained in section 8.4.10. With a hydrometer which has lower
and upper density values of, respectively 1.000 and 1.600 g cm^{-3}, on average
0.009 g cm^{-3} is subtracted from the observed reading. So the repeatability of this
method is only about 0.01 g cm^{-3}.

However, much better repeatability may be obtained from the use of
hydrometers, provided due account is taken care of the influencing factors by
applying the appropriate corrections.   The correction applicable at elevated
pressure depends upon many parameters of the hydrometer. To get the exact value
of the correction, the hydrometer should be calibrated with pure distilled air-free
water when pressurized to the desired value. The density of the air-free water at a
high pressure P may be determined by using the following relations.

$$\rho_w(t, p) = \rho_{wt}(1 + k_t(P \text{ kPa}^{-1}))$$

where P is the pressure in kPa and k_t is given by

$$k_t = 5.088\,496 \times 10^{-8} + 6.163\,813 \times 10^{-9}(t\,°C^{-1})$$
$$+ 1.459\,187 \times 10^{-11}(t\,°C^{-1})^2 + 20.084\,438 \times 10^{-14}(t\,°C)^3$$
$$- 5.847\,727 \times 10^{-16}(t\,°C^{-1})^4 + 4.104\,110 \times 10^{-18}(t\,°C^{-1})^5. \quad (8.23)$$

The density of water at temperature $t\,°C$ at a normal atmospheric pressure of 101 325 Pa is given by

$$(1 - \rho/\rho_{max}) \times 10^6 = A_1((t\,°C^{-1}) - 3.983\,035)$$
$$+ A_2((t\,°C^{-1}) - 3.983\,035)^2$$
$$+ A_3((t\,°C^{-1}) - 3.983\,035)^3$$
$$+ A_4((t\,°C^{-1}) - 3.983\,035)^4$$
$$+ A_5((t\,°C^{-1}) - 3.983\,035)^5 \quad (8.24)$$

where

$$A_1 = -2.381\,848 \times 10^{-2}$$
$$A_2 = 7.969\,992\,983$$
$$A_3 = -7.999\,081 \times 10^{-2} \quad (8.25)$$
$$A_4 = 8.842\,680 \times 10^{-4}$$
$$A_5 = -5.446\,145 \times 10^{-6}.$$

The value of ρ_{max} has been taken to be $999.974\,950 \pm 0.000\,84$ kg m^{-3}.

At other points, the hydrometer is calibrated at room temperature and pressure. Depending upon the range of the hydrometer the appropriate liquids and their mixtures are used. For example for a hydrometer in the range 1.000–1.600 g cm^{-3} liquids such as trichloro-monofluoro-methane and carbon tetrachloride may be used. Once we know the correction at high pressure on the lowest density, we can find the necessary corrections at other points as explained in section 8.4.10.

Hydrometers can normally withstand a pressure of 2.4 MPa (about 25 times atmospheric pressure). However, the cylindrical jar should be tested for the required high pressure especially for any leaks etc.

8.4.10   Corrections to hydrometer indications calibrated at atmospheric pressure but used at high pressures

When a hydrometer is calibrated at normal atmospheric pressure by the comparison method, the upthrust of air on the exposed portion on the stems of the two hydrometers is not taken into account. But if the hydrometer so calibrated is used at elevated pressure say up to 25 bar, the upward thrust due to air on the exposed portion of the hydrometer cannot be neglected, as in this case the density of air is pretty high and differs from that at which the hydrometer was originally calibrated. There will, therefore, be some error in the hydrometer indication and the appropriate correction should be applied.

Let the hydrometer of mass M be calibrated at atmospheric pressure with air of density σ and let it float up to a certain mark on the hydrometer in a liquid of density ρ. Then the equilibrium equation is

$$M = (V + v_1)\rho + (v - v_1 + E)\sigma. \tag{8.26}$$

If the hydrometer reads the same density at an elevated pressure when the density of air is σ_p, then the equilibrium equation will be

$$M = (V + v_1)\rho_p + (v - v_1 + E)\sigma_p \tag{8.27}$$

where E is the volume of the stem above the highest graduation mark. When comparing (8.26) and (8.27), it should be noted that as σ_p will be larger than σ, ρ_p will be smaller than ρ. The difference depends upon the values of V, v and v_1 of the hydrometer. But these values depend upon the range, scale interval and class of the hydrometer. To get the value of the correction, let us proceed with a hydrometer of density range ρ_1–ρ_2, where ρ_2 is greater than ρ_1.

Under normal atmospheric pressure with air density σ, the equilibrium equation is

$$M = V\rho_2 + (v + E)\sigma. \tag{8.28}$$

Similarly when floating in a liquid of density ρ_1,

$$M = (V + v)\rho_1 + E\sigma. \tag{8.29}$$

Subtracting (8.29) from (8.28),

$$V(\rho_2 - \rho_1) - v(\rho_1 - \sigma) = 0 \tag{8.30}$$

or

$$V = v(\rho_1 - \sigma)/(\rho_2 - \rho_1). \tag{8.31}$$

If ρ_{p2} is the density of the pressurized liquid in which the hydrometer floats at the lowest graduated mark (ρ_2) and the air density is σ_p then

$$M = V\rho_{p2} + (v + E)\sigma_p. \tag{8.32}$$

Subtracting (8.28) from (8.32), we get

$$\rho_{p2} = \rho_2 - (v + E)(\sigma_p - \sigma)/V. \tag{8.33}$$

In some hydrometers, we can neglect E in comparison with v and use (8.31), which will give us

$$\rho_{p2} = \rho_2 - (\rho_2 - \rho_1)(\sigma_p - \sigma)/(\rho_1 - \sigma). \tag{8.34}$$

Hence the correction C is equal to $-(\rho_2 - \rho_1)(\sigma_p - \sigma)/(\rho_1 - \sigma)$. However if E is not negligible in comparison with v, we proceed as follows.

The hydrometer is calibrated at the highest graduation mark both at elevated pressure as well as at room pressure. If the density of air is σ_p and σ respectively, then we get the following equations:

$$M = (V + v)\rho_{p1} + E\sigma_p \qquad \text{at elevated pressure} \qquad (8.35)$$
$$M = (V + v)\rho_1 + E\sigma \qquad \text{at room pressure} \qquad (8.36)$$

giving

$$(V + v)(\rho_{p1} - \rho_1) = -E(\sigma_p - \sigma). \qquad (8.37)$$

Putting v in terms of V from (8.31),

$$V[1 + (\rho_2 - \rho_1)/(\rho_1 - \sigma)](\rho_1 - \rho_{p1}) = E(\sigma_p - \sigma)$$

giving

$$E = V[(\rho_2 - \sigma)/(\rho_1 - \sigma)][(\rho_1 - \rho_{p1})/(\sigma_p - \sigma)]. \qquad (8.38)$$

Therefore the value of E from (8.38) can be substituted into (8.33) to give the value of ρ_{p2}.

Let the hydrometer float in the pressurized liquid to the mark representing the density ρ (the volume of the stem above the lowest mark is v_1), then

$$M = V\rho_p + (v - v_1 + E)\sigma_p. \qquad (8.39)$$

This gives

$$v_1 = V(\rho_{p2} - \rho_p)/(\rho_{p2} - \sigma_p). \qquad (8.40)$$

In normal conditions

$$v_1 = V(\rho - \rho_2)/(\rho - \sigma). \qquad (8.41)$$

Dividing (8.41) by (8.42), we get

$$(\rho_{p2} - \rho_p)/(\rho_p - \sigma_p) = (\rho_2 - \rho)/(\rho - \sigma)$$

which on simplification becomes

$$\rho_p(\rho_2 - \sigma) = \rho_{p2}(\rho - \sigma) - \sigma_p(\rho - \rho_2). \qquad (8.42)$$

Substituting the value of ρ_{p2} from (8.34), we get the corrected value of ρ_p to the first approximation:

$$\rho_p(\rho_2 - \sigma) = \{\rho_2 - (\rho_2 - \rho_1)(\sigma_p - \sigma)/(\rho_1 - \sigma)\}(\rho - \sigma) - \sigma_p(\rho - \rho_2). \quad (8.43)$$

Otherwise, substituting the value of E from (8.38) in (8.33) and using the value ρ_{p2}, we can calculate the correct value of ρ_p.

It should be noted that σ_p is the density of air plus the liquid vapours above the surface of the liquid and σ_p will not follow the normal gas equation because of the high pressure and the presence of liquid vapours. In some industrial processes at high pressure, no air is present, thus measuring the density of liquids by this method is not always ideal.

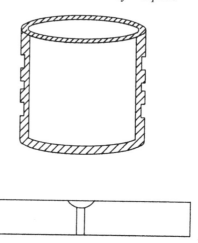

Figure 8.8. Capacity measure and its striking glass to define the capacity.

8.5   Density of liquid adhesives [8]

This method essentially consists of finding the mass of adhesive required to completely fill a pre-calibrated measure. Then the density of the adhesive is the mass of adhesive divided by the capacity of the measure.

Capacity measure

The measure used for this purpose is a single capacity one and is cylindrical in shape. To define its capacity it is preferable to have it with a striking glass. The rim of the measure is plane and smooth. Then its capacity is defined by the volume of water contained, at a reference temperature, in between the measure and the striking glass placed on its open end so that there are no air bubbles in between the water surface and the surface of the glass. The striking glass should be of appropriate diameter with a small cavity at the centre and a small hole that goes all the way through it. The capacity measure and its striking glass are shown in figure 8.8.

8.5.1   Calibration of a capacity measure

The gravimetric method for determining the capacity of a measure is used.

Gravimetric method of determining capacity of a volumetric measure

This consists of finding the mass of water required to completely fill the measure. The mass of water so obtained is changed into the capacity by adding a correction

C, which will account for the density of water, air buoyancy and expansion of the measure to state its capacity at certain reference temperature.

Purity of water

The density of the water naturally depends upon its purity. Therefore, triple-distilled air-free water is used in calibrating such a measure.

Cleanliness and surface condition of the measure

The second factor which affects the precision of the measurement is the cleanliness and condition of the surface of the measure. It should be emphasized here that both the outer and inner surfaces of the measure will matter in the capacity determination. If the outer surface is not clean, a change in humidity will cause the mass of the thin film of water remaining in contact with it to vary. A dirty surface will also catch more dust particles during the weighing process. The striking glass provided with the measure should also be properly cleaned on both sides.

Determination of mass of water

The mass of water required to completely fill the measure is determined as follows.

Step 1. The measure under test with its striking glass and another measure with a similar outer surface, if possible, are taken.

Step 2. On the right-hand pan of the balance, place the measure under test with its striking glass and standard weights at the rate of 1 g per cm^3 (ml). Meanwhile on the left pan the similar measure, if available, and sufficient weights to counterpoise it are placed so that the beam swings within the scale. In a single-pan balance, the capacity measure with standard weights at least at the rate of 1 g per cm^3 are placed in the pan and equilibrium is restored by manipulating the built-in weights.

Step 3. Record the scale readings and calculate the rest point (R_1) or indication of the balance I_1.

Step 4. Take out the measure and fill it with triple-distilled water. The water is kept in the same room overnight so that it acquires the room temperature. Add the water. To fill the measure let the water flow along the walls of the measure which is done with the help of a rod. No splashing or entrapping of air should take place. Continue to fill it until the level of water is slightly higher than the rim of the surface. (This is possible because of surface tension.)

Step 5. Remove all air bubbles sticking to the wall and bottom of the measure with a glass rod.

Step 6. Determine the temperature of the water with the help of a calibrated (mercury-in-glass) thermometer with a scale interval of 0.1 °C. Let the temperature be T_1 °C.

Step 7. Slide the striking glass horizontally to remove any excess water. Ensure that there is no air bubble between the surface of the water and the striking glass. The presence of air bubbles indicates that more water is needed. So add water in the spherical cavity of the striking glass and press it, air will escape and be replaced by water. If not, remove the striking glass, add more water and repeat the process.

Step 8. Clean the measure from all sides with ash-less filter paper. Special attention should be paid to the bottom of the measure and the sides of the rings if provided to strengthen the measure. The top of the striking glass should also be properly cleaned. Ensure that there are no traces of water on any side especially on the bottom or on the striking glass. The handling of the measure should be minimal, as handling changes the temperature of the measure and air bubbles will appear. Prolonged handling may also change the water temperature and then excess water will start coming out.

Step 9. Put the measure in the pan, remove the necessary weights, so that the pointer swings within the scale or the indication is within the scale. This way, water has substituted the standard weights. Take observations and calculate the rest point. Let this be R_2. Alternatively the indication of the single-pan balance is I_2 and mass of weights removed is Ms.

Then the apparent mass of water M_w is given by

$$M_w = Ms - (R_2 - R_1)SR \qquad \text{or} \qquad M_w = Ms + (I_2 - I_1)$$

for a single-pan balance. SR is the sensitivity reciprocal of the balance or the mass value between successive graduations.

Step 10. Take out the measure, remove the striking glass by sliding and take the temperature of the water. Let this be T_2.

Step 11. Take the mean of T_1 and T_2. Let this be T.

Step 12. Knowing the mean temperature, T, of the water and the reference temperature, look up the appropriate part of the table—the intersection of a row and column corresponding to the mean temperature, T, gives the correction C. For a stainless steel cylindrical measure, table 8.1(*o*) or 8.1(*p*) is sufficient, as soda glass has a volume expansion of $30 \times 10^{-6}\,°\text{C}^{-1}$ while that of steel is from 30 to $33 \times 10^{-6}\,°\text{C}^{-1}$. If the capacity of the measure differs from unity then multiply C from the table by the nominal capacity of the measure. Then the capacity of the measure in cm^3 at $20\,°\text{C}$ is given by

$$V_{20} = M + C \qquad\qquad (8.44)$$

where C is in mg.

Determination of the mass of the adhesive and its density

Step 13. Take the calibrated measure and pour the liquid adhesive into it. Then follow all the steps from 1 to 11 and find the apparent mass of the adhesive required to fill the measure completely. The mass divided by the volume obtained from equation (8.44) gives the apparent density of the adhesive at the temperature of measurement. For the true density calculate the true mass by applying buoyancy corrections.

8.6 Density of waxlike materials in both solid and liquid phase [9]

This method is suitable for determining the density of wax and similar materials in both the solid and liquid phases. It comprises the following steps:

(I)   calibration of the capacity of a special pyknometer at several reference temperatures, say from 10 to 80 °C in steps of 10 °C;
(II) determination of the density of glycerine at the same reference temperatures; and finally
(III) determination of the combined volume of the known masses of wax and glycerine at different temperatures. Knowing the density of glycerine and its volume, the volume of wax can be calculated. The mass of wax divided by its volume gives the density of the wax.

8.6.1 Pyknometer

The pyknometer is made from borosilicate glass. It consists of a wide mouth bottle with a ground-glass neck to which a ground-glass stopper may be fixed. The stopper has a stopcock and a small tube of about 1 mm internal diameter. The face of the tube is made square to its axis. From the bottom it is connected to a long capillary tube of uniform bore and bent once at right angles. The length of the tube may be around 20 cm or more. The tube is graduated at intervals of 1 mm and has an internal diameter of 2.8 mm. The pyknometer is shown in figure 8.9. The volume of the bottle is around 20 cm^3.

8.6.2 Calibration of the pyknometer

The capacity of the pyknometer is calibrated at several reference temperatures against water as the standard of density and using the gravimetric method as explained in section 8.4.5 (first method and using tables 8.1(a)–(n)). The capacity of the pyknometer is calibrated at least four points at each reference temperature. However the mass of water at each of the four points will only be determined once at any known temperature. A plot of the capacity (y-axis) against the position of the water level meniscus on the scale (x-axis) is drawn. If the capillary has a uniform bore, it will be a straight line and meet the y-axis at a point which will give the capacity of the bottle including that of the capillary up to the point in the level of the top of the stopper. A separate graph for each reference temperature is to be drawn to give the capacity up to any point of the scale. The reference

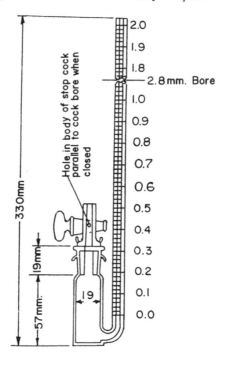

Figure 8.9. Special pyknometer.

temperatures should cover a temperature range much below the melting point and slightly below the boiling points of wax or similar materials.

8.6.3 Determination of the density of glycerine

Take the clean and dried pyknometer and weigh it with its stopper. Remove the stopper and fill it with glycerine applying a little suction at the end of the capillary tube. Holding the suction, put in the stopper and open the stopcock. As soon as the suction is removed, glycerine will fill the capillary of the stopper and stopcock and will start to come out. Close the stopcock. The initial filling with the glycerine should be such that the level of the glycerine is at the 1.8 mark (almost the highest point) of the scale. Clean the pyknometer from outside and weigh it. The necessary precautions should be taken while weighing the dried pyknometer. These have already been described in sections 8.4.1 and 8.4.2. The pyknometer is placed in a thermo-controlled bath maintained at the lowest reference temperature. Constancy in the reading of the level in the capillary indicates that thermal equilibrium has been attained. Readings on the scale are taken at 5 min intervals. Note the reading of the scale. From the calibration curve of capacity versus scale reading, at the reference temperature, the volume of the glycerine is determined. Therefore the density of glycerine can be calculated. The same exercise is repeated at all other reference temperatures to calculate the density of glycerine at various temperatures. This way we can calculate the density of glycerine at any other temperature by the interpolation method.

8.6.4 Preparation of the sample, procedure for determination of density

Preparation of the sample

Heat the wax to $17\,^\circ$C above melting point and pour it into a special rectangular mould. Let it cool down slowly and cut pieces $50 \times 5 \times 3$ mm^3, polish the cut sides of the pieces to eliminate all voids. A mould is made from a wooden rectangular frame of size 75×125 mm^2. Two cellulose films are stretched on each side of the frame and the distance between the two films is kept at 3 mm. This way, a cake 3 mm thick is formed when molten wax or any other material is poured into it.

Place three or four pieces of wax in the pre-dried and weighed pyknometer. Put in the stopper and weigh it. Let the mass of wax be M_1. Fill the pyknometer with glycerine by applying a little suction on the capillary side until the level reaches near the 1.5 mark. Put in the stopper and open the stopcock. On releasing the suction, glycerine will fill the capillary of the stopper and the stopcock. Allow some of the glycerine to ooze out from the stopcock and then close the stopcock. All parts of the stopper and stopcock should remain full with glycerine. Clean the pyknometer carefully so that no glycerine is sucked out from the capillary of the stopcock. Weigh the pyknometer. The mass of glycerine so filled is calculated. The pyknometer is then placed in the bath maintained at about $10\,^\circ$C below the melting point of the wax; note the level of the glycerine in the capillary when thermal equilibrium is attained. In this condition three consecutive readings taken at intervals of 5 min will be the same. Note the volume from the calibration curve; let this be V_t. Raise the temperature in steps of $2\,^\circ$C and calculate the volumes at different temperatures. Proceed in this way even beyond the melting point of wax but below its boiling point. The density of the wax is calculated at a particular temperature as follows.

Calculations

Let V_{1t}, V_{2t} be the volumes of the wax and glycerine respectively at any temperature $t\,^\circ$C. Then

$$V_t = V_{1t} + V_{2t}$$

but $V_{2t} = M_2/d_{2t}$, where M_2 and d_{2t} are the mass and density of glycerine. Therefore

$$V_{1t} = V_t - M_2/d_{2t}$$

giving us d_{1t} the density of the wax at temperature $t\,^\circ$C as $M_1/(V_t - M_2/d_{2t})$.

Thus the density at all measured temperatures can be calculated and, if necessary, the expansion coefficient or density can be determined for waxlike materials in both the solid and liquid phases.

References

[1] ASTM E 12-70; Standard definitions of terms relating to density and specific gravity of solids, liquids and gases
[2] ASTM C 904-81; Definitions of chemical resistant non-metallic materials
[3] ASTM D 941-55 (Re-approved 1978); Density and relative density (specific gravity) of liquids by Lipkin's bi-capillary pyknometer
[4] ASTM D 1481-62 (Re-approved 1976); Density and relative density (specific gravity) of viscous materials by Lipkin's bi-capillary pyknometer
[5] ASTM D 1480-81; Density and relative density (specific gravity) of viscous materials by Bingham pyknometer
[6] ASTM D 1217-81; Density and relative density (specific gravity) of liquids by Bingham pyknometer
[7] ASTM D 3096-79; Liquid density of pressurized products
[8] ASTM D 1875-69 (Re-approved 1980); Density of adhesive in liquid form
[9] ASTM D 1168-61 (Re-approved 1977); Hydrocarbon wax used for electrical insulation

Chapter 9

Density of materials used in industry—solids

Symbols

d_s	Density of solids
B	Bulk density of solids
D	Constant dry mass
C_ρ	Change of density with respect to temperature
α	Coefficient of linear expansion
h	Height of a point inside a liquid from an arbitrary reference line
P	Porosity
V_{imp}	Volume of impervious portion of the solid
S	Suspended weight in water
W	Saturated weight.

9.1 Introduction

Solid materials, from the point of view of density measurement, can be broadly classified into two categories, namely those in the form of powders and those in the form of any geometrical shape. In turn, solid materials in other forms may be grouped into materials which may be transformed into bodies of regular geometrical shape, have smooth impervious surfaces and are able to acquire a good degree of polish, such as quartz, zerodur or silicon. There are some other materials the densities of which are to be determined after articles have been manufactured from them such as articles manufactured from carbon and graphite. The densities of such materials change when pressed, moulded or machined and in such cases, the density of the processed article is measured.

We have already discussed in earlier chapters the density of solids, such as quartz, glass, zerodur and silicon, transformed into regular geometrical shapes.

For other solid materials, the methods for measuring their densities may be grouped into the following categories:

- By finding the density of liquid mixtures, the density of which has been made equal to that of the solid. In this state, the solid body remains suspended wherever it is displaced in the liquid, i.e. the solid remains in a state of neutral equilibrium. Furthermore, any slight change in the density of the liquid mixture causes the body to move either upwards or downwards. This may be accomplished either by establishing a vertical density gradient or by having a solution of uniform density equal to that of the solid.
- By hydrostatic weighing in which the mass of water displaced by the body is determined by finding the upthrust of water. The method also includes those methods in which the water is replaced by some other liquid such as toluene, wax or mercury etc.
- By the displacement method, in which the mass of the liquid of known density displaced by the solid body is measured by weighing. This method is commonly known as the specific gravity bottle method.
- By finding the volume of the body by measuring its dimensions. Normally this method is used in industry when only an ordinary accuracy of about 1% is required.

For powders there are several methods but each one essentially consists of allowing the powder to fall through a funnel with an orifice of specified diameter into a calibrated volumetric measure. The methods differ only in the way in which the powder is received; the volumetric measure is tapped, rotated or vibrated. In one case, the powder falls first into a vibrating pan and then into the volumetric measure. The mass of the powder in the measure is determined. The apparent density is finally calculated from knowledge of the volume of the measure.

9.2   Density of ceramic-like materials (by making the density of the liquid equal to that of the solid by mixing two liquids with different densities)

This method consists of making a mixture of suitable liquids, whose density is made equal to that of the specimen (ceramic) and measuring the density of the liquid mixture hydrostatically using a glass plummet. (The method described here is based on ASTM [1]).

9.2.1   Materials and apparatus required

The following materials and apparatus are necessary:

(1) high-density liquids such as thallium malonate–formate solutions,
(2) air-free distilled water,
(3) a glass plummet,

(4) a cylinder and
(5) a balance with an appropriate capacity reading up to 1 mg and capable of
     weighing under the pan.

Caution: Thallium malonate–formate solution is highly toxic and must be handled
with care and only by competent persons who know the implications of handling
such solutions.

9.2.2 Procedure

Finding the volume of the plummet

Hang the plummet from the pan of the balance and counterpoise it with
appropriate weights. Fill a cylinder with freshly distilled water and bring it under
the pan. Raise it until the plummet is submerged 2 cm below the water surface.
Place additional weights of known mass on the pan from which the plummet
is hanging, until equilibrium is reached. The mass of placed weights will be
equivalent to the loss in apparent mass of the plummet when immersed in water.
Note down the temperature T_w.

Preparing a liquid mixture of appropriate density

Clean the ceramic specimen whose density is to be measured and dry it properly.
Put about 50 cm^3 of thallium malonate–formate solution of maximum density into
a beaker. The solution should be sufficient for the specimen to float freely. Using
tweezers, immerse the specimen in the solution, which will float with partial
immersion. Lower the density of the solution by adding water, drop by drop,
until the density of the solution equals that of the specimen. Each drop should
not lower the density by more than 0.003 g cm^{-3}. In this condition the specimen
will lose the tendency to regain its position on slight displacement. Measure the
temperature of the solution; let this be T_s. Take out the specimen and thoroughly
clean it before putting it back.

Finding the density of the solution

Pour the solution into a cylinder and find the loss in weight of the plummet when
immersed in the solution. The ratio of the loss in weights in the solution and the
water will give the relative density of the solution. If W_1, W_2 are the losses in
weight of the plummet when immersed in water and the solution respectively and
d_w is the density of the water at the temperature T_w, then the density d_s of the
specimen is given by

$$d_s = d_w W_2 / W_1. \qquad (9.1)$$

It has been assumed that $T_s = T_w$.
       If this condition is not fulfilled, then the volume of the plummet may be
derived from the measured one and the coefficient of expansion of the material of

the plummet. As the plummet is made from borosilicate glass, the error due to the difference in temperature will be quite small and it may be ignored.

9.3 Density of glasslike materials (by making the density of the solution equal to that of the solid by temperature adjustment)

The method described here is based on ASTM [2].

9.3.1 Principle

If the density of the solid and that of the liquid are nearly same, then by changing the temperature of the liquid its density can be made equal to that of the solid. In this case the solid will float in a stationary position. This has been discussed in detail in section 6.6. Here the liquid mixture is chosen so that its density at 35 °C is within ± 0.02 g cm^{-3} of that of the solid at 25 °C whose density is to be measured. A glass standard of known density such that its density at 35 °C is almost same as that of the liquid mixture at 35 °C is taken. Therefore, at 25 °C, the density of the liquid mixture will be higher than that of either solid, so these will float with partially submerged volumes or, if made to submerge in the liquid mixture and left alone, they will move upwards. Therefore, these are kept fully submerged in the liquid with the help of a wire mesh cage (figure 9.1(b)), i.e. the specimen and the glass standard will rest against the roof of the cage. The temperature of the liquid is raised slowly, which will decrease the density of the liquid. Therefore, a temperature will be reached at which the density of the liquid will just equal or be very slightly less than that of the denser solid. In this situation the particular solid will leave the roof of the cage and start to sink or it will remain suspended in the liquid. Note the temperature. Similarly a stage will come at which the density of the liquid will become equal to or be slightly less than that of the lighter solid. That temperature should also be noted. By knowing the expansion properties of the liquid mixture and the linear coefficients of the two solids, one can determine the density of the solid in terms of that of the other.

9.3.2 Apparatus

A single-tube comparator is shown diagrammatically in figure 9.1(a). It has the following components.

Water bath

The circulating water bath contains a 4 litre beaker B1, a cover plate C with holes and rings for holding the tubes and thermometers. A copper cooling coil CC, through which pre-cooled water is circulated, an immersion heater with a rheostat and an electric stirrer to stir the water are placed inside the beaker. An external hot

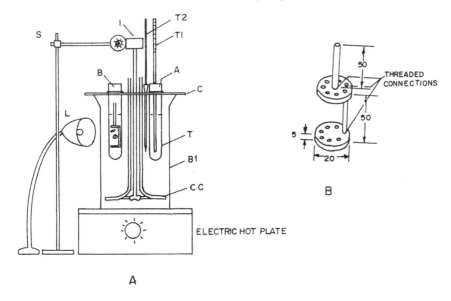

Figure 9.1. (*a*) Tubes for observing the suspended state of the solid piece. (*b*) Cage for stopping the solid pieces.

plate for controlling the heating rate may also be used instead of the immersion heater.

Tubes

Two similar tubes A and B contain the same liquid mixture. In tube A, only a thermometer T1 with 0.1 °C graduation is fitted while tube B contains a platinum wire mesh cage in which several samples and a reference standard of known density are held. This is the measuring tube. In fact it is assumed that the liquid in tube B will have the same temperature as the liquid in tube A. This assumption is valid as the liquid is stirred continuously and the tubes are similar and have the same thermal capacity.

Reference standard of density

This is a solid piece of glass with a volume between 0.1 and 0.15 cm^3 and may be in any form but preferably in the form of a cylinder or sphere. The ratio of the major to minor dimensions should not exceed 2. Its surface should be smooth, free from seeds, cords and cracks. Its density, at 25 °C, should be known within ± 0.0001 g cm^{-3}.

Liquids

The following organic liquids can be used to make a suitable liquid mixture:

(1) isopropyl salicylate, with a density at 25 °C of about 1.10 g cm^{-3},
(2) alpha bromo naphthalene, with a density at 25 °C of about 1.49 g cm^{-3},
(3) sym-tetra bromoethane, with a density at 25 °C of about 2.96 g cm^{-3} or
(4) methyl iodide, with a density at 25 °C of about 3.32 g cm^{-3}.

It should be noted that liquids (2)–(4) are light sensitive so they should be stored in the dark. Some more liquids were given in chapter 6.

9.3.3 Procedure

Prepare the specimen by cutting pieces comparable in size to that of the standard. The test specimen should be smooth and free from seeds, cords and cracks. Make a distinguishing mark on each sample piece to identify it from the others. Clean all specimens with pure alcohol or acetone and dry them. Place them along with the reference standard into the cage, which should then be hung in the liquid mixture. The temperature of the liquid mixture should be about 25 ± 3 °C.

Place all the tubes, thermometers, stoppers etc in the bath and rapidly heat the bath (1–2 °C min^{-1}) noting the temperature of the liquid mixture at which the specimen or the standard begins to be suspended freely in the liquid mixture. This is to obtain a rough estimate.

Adjust the bath temperature about 2–4 °C below the temperature at which the specimen or the standard became freely suspended. Allow the bath and liquid mixture to reach equilibrium for 10 min. Then raise the temperature at a slower rate, which may be 0.1–0.2 °C min^{-1}.

Note the temperature of the liquid mixture at which either the standard or specimen starts to become freely suspended halfway between the upper and lower parts of the cage. The temperature of the liquid mixture and the bath should agree within 0.4 °C. Record the temperature of the liquid mixture and correct it for any difference in temperatures of bath and liquid mixture.

9.3.4 To prepare a liquid mixture of density ρ_s

If V_1, the volume of liquid A of density ρ_1, is mixed with V_2, the volume of liquid B of density ρ_2, to make a liquid mixture of density ρ_s and volume $V_1 + V_2$, then by equating the masses of the two liquids to that of the mixture, we get

$$\rho_s(V_1 + V_2) = V_1\rho_1 + V_2\rho_2 \qquad \text{or} \qquad \rho_s = (V_1\rho_1 + V_2\rho_2)/(V_1 + V_2). \quad (9.2)$$

Note

Quite often, the volume of the mixture of two liquids is not the algebraic sum of the volumes of its components. In this case, the difference in the actual volume of

Table 9.1. The volumes of component liquids to give 300 cm^3 of the liquid mixture.

d_s (g cm^{-3} at 35 °C)	Isopropyl salicylate	Sym. tetra-bromo-ethane	Methyl iodide
2.103	135	165	—
2.136	127	173	—
2.190	120	180	—
2.222	115	185	—
2.236	113	187	—
2.257	109	191	—
2.291	104	196	—
2.315	100	200	—
2.335	95	205	—
2.365	92	208	—
2.403	85	215	—
2.434	80	220	—
2.448	78	222	—
2.473	74	226	—
2.495	70	230	—
2.511	68	232	—
2.529	65	235	—
2.560	60	240	—
2.589	56	244	—
2.596	54	246	—
2.619	50	250	—
2.633	48	252	—
2.669	42	258	—
2.702	37	263	—
2.728	33	267	—
2.757	28	272	—
2.812	19	281	—
2.847	13	287	—
2.863	10	290	—
2.893	6	294	—
2.933	—	300	1
2.960	—	277	23
2.999	—	248	52
3.035	—	214	86
3.054	—	198	102
3.096	—	168	132

the mixture from the sum of the volumes of its components is taken on the basis of molar fractions.

The volumes of various liquids which will give a liquid mixture of volume

300 cm^3 of density ρ_s are given in table 9.1. It should be noted that ρ_s is the density of the reference standard at $35\,^{\circ}\text{C}$.

Mix two liquids in the volumes given in table 9.1 in a beaker. Place the beaker on the hot plate and slowly raise its temperature. Place the reference density standard in the solution and adjust the density of the mixture by adding a few drops of the required liquid so that the density of the standard becomes equal to that of the mixture at $35 \pm 0.2\,^{\circ}\text{C}$.

9.3.5   Preparing a density–temperature table

Measure the density of the liquid mixture with the help of a pyknometer at two temperatures, one at t_1 about $25\,^{\circ}\text{C}$ and the other t_2 at $45\,^{\circ}\text{C}$. If ρ_1 and ρ_2 are the densities at temperature t_1 and t_2 respectively then C_ρ, the density coefficient of the liquid in g cm$^{-3}\,^{\circ}\text{C}^{-1}$, is given by

$$C_\rho = (\rho_1 - \rho_2)/(t_1 - t_2). \tag{9.3}$$

Let d_s be the density of the reference standard at t_s, which also equals the density of the liquid at t_s. Then the density d_t of the solid at the equilibrium temperature t is

$$d_t = d_s + C_\rho(t - t_s). \tag{9.4}$$

It should be noted that t_s is close to $35\,^{\circ}\text{C}$.

9.3.6   Calculating the density of the specimen at $25\,^{\circ}\text{C}$

If the coefficients of linear expansion of the standard and specimen are equal, then the density of specimen d_{25} is given in terms of the density of the reference standard d_{25s} at $25\,^{\circ}\text{C}$ and the coefficients of linear expansion of the solids and the coefficient of the density of the liquid mixture as follows.

$$d_{25} = d_{25s} + [C_\rho + 3\alpha d_s](t - t_s). \tag{9.5}$$

Although t_s is close to $35\,^{\circ}\text{C}$, it is not exactly equal to it, therefore some corresponding corrections have to be applied to the equilibrium temperature t for the specimen as follows.

$$t_c = t + (35\,^{\circ}\text{C} - t_s). \tag{9.6}$$

Using this value of t we get

$$d_{25} = d_{25s} + [C_\rho + 3\alpha d_s](t_c - 35\,^{\circ}\text{C}). \tag{9.7}$$

When the coefficients of linear expansion of the reference and specimen are not equal, a difference of $2 \times 10^{-7}\,^{\circ}\text{C}^{-1}$ in their coefficients of expansion will give an error in the determined density of 0.0001 g cm^{-3}. The error is positive if the equilibrium temperature is above $35\,^{\circ}\text{C}$ and negative if the equilibrium temperature for the specimen is below $35\,^{\circ}\text{C}$. Using (9.7), the density of the specimen versus the corrected temperature at which its density becomes equal to that of the liquid mixture has been calculated in table 9.2.

Table 9.2. Density of solid (g cm^{-3} at 25 °C) versus equilibrium temperature t_c for solid. t_c ranges from 25 to 45 °C in steps of 0.1 °C. The density coefficient is 0.001 925 g cm^{-3} °C in the table. The standard is made from soda lime glass.

t_c	ρ_{25}	t_c	ρ_{25}	t_c	ρ_{25}	t_c	ρ_{25}	t_c	ρ_{25}
25.0	2.5305	29.0	2.5228	33.0	2.5151	37.0	2.5074	41.0	2.4997
25.1	2.5303	29.1	2.5226	33.1	2.5149	37.1	2.5072	41.1	2.4995
25.2	2.5301	29.2	2.5224	33.2	2.5148	37.2	2.5071	41.2	2.4993
25.3	2.5299	29.3	2.5222	33.3	2.5146	37.3	2.5069	41.3	2.4991
25.4	2.5297	29.4	2.5221	33.4	2.5144	37.4	2.5067	41.4	2.4989
25.5	2.5295	29.5	2.5219	33.5	2.5142	37.5	2.5065	41.5	2.4988
25.6	2.5293	29.6	2.5217	33.6	2.5140	37.6	2.5063	41.6	2.4986
25.7	2.5292	29.7	2.5215	33.7	2.5138	37.7	2.5061	41.7	2.4984
25.8	2.5290	29.8	2.5213	33.8	2.5136	37.8	2.5059	41.8	2.4982
25.9	2.5288	29.9	2.5211	33.9	2.5134	37.9	2.5057	41.9	2.4980
26.0	2.5286	30.0	2.5209	34.0	2.5132	38.0	2.5055	42.0	2.4978
26.1	2.5284	30.1	2.5207	34.1	2.5130	38.1	2.5053	42.1	2.4976
26.2	2.5282	30.2	2.5205	34.2	2.5128	38.2	2.5051	42.2	2.4974
26.3	2.5280	30.3	2.5203	34.3	2.5126	38.3	2.5049	42.3	2.4972
26.4	2.5278	30.4	2.5201	34.4	2.5125	38.4	2.5047	42.4	2.4970
26.5	2.5276	30.5	2.5199	34.5	2.5123	38.5	2.5045	42.5	2.4968
26.6	2.5274	30.6	2.5197	34.6	2.5121	38.6	2.5044	42.6	2.4966
26.7	2.5272	30.7	2.5196	34.7	2.5119	38.7	2.5042	42.7	2.4964
26.8	2.5270	30.8	2.5194	34.8	2.5117	38.8	2.5040	42.8	2.4962
26.9	2.5269	30.9	2.5192	34.9	2.5115	38.9	2.5038	42.9	2.4961
27.0	2.5267	31.0	2.5190	35.0	2.5113	39.0	2.5036	43.0	2.4959
27.1	2.5265	31.1	2.5188	35.1	2.5111	39.1	2.5034	43.1	2.4957
27.2	2.5263	31.2	2.5186	35.2	2.5109	39.2	2.5032	43.2	2.4955
27.3	2.5261	31.3	2.5184	35.3	2.5107	39.3	2.5030	43.3	2.4953
27.4	2.5259	31.4	2.5182	35.4	2.5105	39.4	2.5028	43.4	2.4951
27.5	2.5257	31.5	2.5180	35.5	2.5103	39.5	2.5206	43.5	2.4949
27.6	2.5255	31.6	2.5178	35.6	2.5101	39.6	2.5024	43.6	2.4947
27.7	2.5253	31.7	2.5176	35.7	2.5099	39.7	2.5022	43.7	2.4945
27.8	2.5251	31.8	2.5174	35.8	2.5098	39.8	2.5020	43.8	2.4943
27.9	2.5249	31.9	2.5173	35.9	2.5096	39.9	2.5018	43.9	2.4941
28.0	2.5247	32.0	2.5171	36.0	2.5094	40.0	2.5017	44.0	2.4939
28.1	2.5245	32.1	2.5169	36.1	2.5092	40.1	2.5015	44.1	2.4937
28.2	2.5244	32.2	2.5167	36.2	2.5090	40.2	2.5013	44.2	2.4935
28.3	2.5242	32.3	2.5165	36.3	2.5088	40.3	2.5011	44.3	2.4934
28.4	2.5240	32.4	2.5163	36.4	2.5086	40.4	2.5009	44.4	2.4932
28.5	2.5238	32.5	2.5161	36.5	2.5084	40.5	2.5007	44.5	2.4930
28.6	2.5236	32.6	2.5159	36.6	2.5082	40.6	2.5005	44.6	2.4928
28.7	2.5234	32.7	2.5157	36.7	2.5080	40.7	2.5003	44.7	2.4926
28.8	2.5232	32.8	2.5155	36.8	2.5078	40.8	2.5001	44.8	2.4924
28.9	2.5230	32.9	2.5153	36.9	2.5076	40.9	2.4999	44.9	2.4922
								45.0	2.4920

Uncertainty

The repeatability in the method at one standard deviation is 0.0001 g cm^{-3}. If the density of the standard is known to 0.000 01 g cm^{-3} then the standard uncertainty of the method is 0.0001 g cm^{-3}.

9.4 Density of glass and similar materials by hydrostatic weighing

The method is based on ASTM [3] and [4].

9.4.1 Apparatus

A balance of suitable capacity reading up to 0.1 mg or better; beaker of convenient capacity 250 to 750 cm^3 depending upon the size of the glass object; calibrated thermometers range 20 °C to 30 °C with 0.1 °C graduations; nickel chromium or platinum wire with diameter less than 0.2 mm for suspending the specimen either in a basket or in a loop of the same wire; standards of mass of F1 or E2 class of OIML; hydrostatic liquid is freshly boiled distilled water.

9.4.2 Cleaning the suspension wire

The suspension wire should be cleaned before use by degreasing and heating it in vacuum. Another method is to heat the platinum wire in an oxidizing atmosphere until the gases, passing around the wire, emit no colour.

9.4.3 Sample preparation

A sample weighing about 20 g with a minimum of seeds or other defects is taken. If cut from a larger sample it may be in the form of a cylinder or rectangular block. The surface should be smooth and free from cracks or any surface defects. The edges may be slightly rounded off. A gaseous void of 2 mm in diameter may cause an error of 0.05% in the density for a sample of mass 25 g and density 2.5 g cm^{-3}.

The specimen is cleaned properly with hot chromic acid and rinsed in distilled water or in a bath of hot nitric acid or chromic acid fitted with an ultrasonic vibrator. The sample is to be degreased with the appropriate organic solvent and rinsed thoroughly with alcohol and distilled water. After cleaning, it should not be touched by hand—forceps must be used for subsequent operations.

9.4.4 Procedure

(1) Hold the specimen and covered beaker of boiled water near the balance table and ensure that the temperature of the water does not differ from that of the air by more than 0.5 °C. To calculate the density of the ambient air

its temperature, pressure and relative humidity should be measured to the nearest 0.1 °C, 1 mmHg and 10% respectively.

(2) Weigh the glass specimen in air to the nearest 0.1 mg; let the apparent mass be M_a. Note the temperature, pressure and relative humidity of the air and calculate the air density.

(3) Put a suitable mark on the wire and make sure that the wire is submerged exactly to this mark in all subsequent weighings in water. Take the suspension device with the basket or loop to hold the specimen but without the specimen and hang this from the hook of the pan of the balance. Raise the beaker until the water level reaches the specified mark on the wire. Weigh it; let the apparent mass be M_{01}. Apparent mass means the value of the mass of the standards as given in its certificate of calibration.

(4) Place the specimen in the basket or loop of the same wire; adjust the level of the beaker until the water level reaches the mark. Weigh the specimen. Let the apparent mass be M_1. Note the temperature of the water. Let this be t °C.

(5) Remove the specimen, adjust the level of the water up to the mark and weigh the suspension device again. Let this be M_{02}. It should not differ from the previous value by more than 0.1 mg. Take their mean M_0:

$$M_0 = (M_{01} + M_{02})/2. \tag{9.8}$$

(6) Note the temperature, pressure and relative humidity of the air again, calculate the density of the air, and let the mean of the air densities calculated before and after weighing be σ.

(7) For solids with densities less than that of water, a sinker is used. The sinker should be of sufficient weight and density so that the specimen together with the sinker submerge in the water. In steps (3) and (5) the sinker remains attached to either the wire or basket. It should be noted that in this case M_0 will be greater than M_1, hence M_w will be negative.

This method is used to determine the density of plastics in industry.

9.4.5 Calculations

The equilibrium equations are:

$$M_a(1 - \sigma/\Delta) = V(d - \sigma) \tag{9.9}$$
$$M_w(1 - \sigma/\Delta) = V(d - \rho_w). \tag{9.10}$$

Dividing and simplifying we get

$$d = (M_a\rho_w - M_w\sigma)/(M_a - M_w) \tag{9.11}$$

where $M_w = M_1 - M_0$ and d and ρ_w are, respectively, the density of the solid and water at t °C.

When the solids are lighter than water, M_w is given by $M_w = M_0 - M_1$ and is greater than M_w. However, here M_0 is the mass of the suspension device plus that of the sinker, which is much denser than the body.

If d_s is the density of the specimen at the reference temperature t_s, then d_s is given by

$$d_s = d/[1 + 3\alpha(t_s - t)] \tag{9.12}$$

where α is the coefficient of linear expansion of the material from which the solid is made.

Repeatability and uncertainty

In normal circumstances, the repeatability at one standard level is 0.1%. However if proper precautions are taken and necessary corrections are applied with respect to

- dissolved air,
- air bubbles or a fine layer of air adhering around the specimen,
- preventing the suspension wire from submerging,
- variation in the surface tension of water and, finally,
- the correct temperature measurement,

this method is capable of giving a high degree of repeatability, say 0.0001%.

9.5   Density of solids (by making the density of the solution equal to that of the solid by the density gradient method)

This method is normally used for comparatively light materials because of the non-availability of highly dense liquids. The method is based on ASTM [5].

9.5.1   Principle

A comparatively long tube is filled with a solution of varying density so that the denser portion is at the bottom and the density continuously decreases upwards. To find the density of the liquid at any point, calibrated glass floats of known density are used. These spheres depending upon their density will remain suspended at different levels. The height of their centre of gravity is observed with the help of a cathetometer from an arbitrary horizontal line. Either a density height graph is drawn or the density at any height is calculated by interpolation.

9.5.2   Useful liquid mixtures

Combinations of liquids which can be profitably used in this method together with their density ranges are given in table 9.3.

Table 9.3. Useful liquids and their density range.

Mixtures of liquids	density range (g cm^{-3})
Methanol–benzyl alcohol	0.80 to 0.92
Isopropanol–water	0.79 to 1.00
Isopropanol–di-ethylene glycol	0.79 to 1.11
Ethanol–carbon tetrachloride	0.79 to 1.59
Toluene–carbon tetrachloride	0.87 to 1.59
Water–sodium bromide	1.00 to 1.41
Water–calcium nitrate	1.00 to 1.60
Carbon tetrachloride–trimethylene dibromide	1.60 to 1.99
Trimethylene dibromide–ethylene bromide	1.99 to 2.18
Ethylene bromide–bromoform	2.18 to 2.89

9.5.3   Glass floats and their calibration

A number of hollow glass floats, covering the density range, are made. These are well annealed and spherical in shape. The diameter of these floats should be less than one-quarter of the diameter of the cylinder. The surface should be smooth and free from seeds, cracks etc. Prepare a solution with the lowest density of the range. Drop the floats one by one almost from the surface of the solution. Retain those floats which sink slowly. The sinkers which move more rapidly are discarded and retained for higher density solutions. The density of the floats can be adjusted by rubbing the top part on a glass plate with a slurry of 400–500 mesh of silicon carbide.

For calibrating the floats, select two liquids, one denser and the other lighter than the float. Place the lighter liquid in a long cylinder and place it in a thermostatic bath, which maintains the reference temperature within 0.1 °C. Place the float under calibration gently on the liquid surface. It will start to sink. Add the denser liquid drop by drop, until the density of the liquid mixture becomes equal to that of the float. In this situation the float will remain suspended. Instead of waiting for the exact density to be attained by the liquid, the denser liquid should be added until the float starts rising, then add a little of the lighter liquid so that the float just remains stationary or moves downwards but much more slowly. Several trials will attain the objective. Stir the liquid well and ensure the correct reference temperature. The float should not be visibly moving for several minutes. The density of the liquid is determined with the help of Lipkin's bi-capillary pyknometer. All floats are calibrated in this way and a density value assigned to each one.

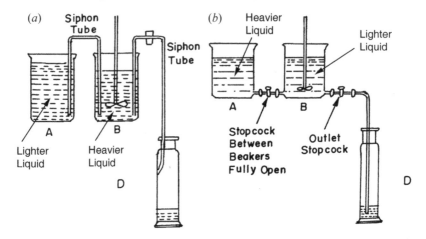

Figure 9.2. (*a*) Arrangement for generating density gradient. (*b*) Arrangement for generating density gradient.

9.5.4   Generating a density gradient

Two beakers A and B of equal diameter are taken and the heavier liquid is placed in beaker B, which is connected through a siphon to the long graduated cylinder D in which the density gradient is to be produced. The cylinder has a ground-glass stopper C. The siphon from beaker B to the cylinder has a stopcock, which can regulate the flow. The end of the siphon reaching cylinder D is just tapered and touches the wall of the cylinder D so that the liquid from beaker B flows along the wall of cylinder D. A stirrer is working in beaker B, so that a well-mixed liquid mixture is transferred to the cylinder. Beaker A, having the lighter liquid, is connected to beaker B again with a siphon tube. The siphon to the cylinder is primed and the stopcock closed. The level in beaker A, containing the lighter liquid, is kept higher than that of the liquid in beaker B, so that the liquid flows from A to B. When the stopcock is opened slowly, the denser liquid starts to flow into the cylinder. Simultaneously the lighter liquid starts arriving at beaker B. Continuous stirring reduces the density of the liquid in beaker B, so the liquid mixture with a continuously decreasing density fills cylinder D. The arrangement is shown in figure 9.2(*a*). Once the cylinder is filled to the desired level it is transferred to the thermostatic bath maintained within 0.1 °C of the reference temperature t_s. Both liquids should be degassed, if necessary, either by slight heating or by creating a slight vacuum. The stopper is fixed to the cylinder and it is kept in a still position for at least 24 h. A linear gradient is thus established, which may remain for quite some time.

In figure 9.2(*b*), beaker B contains the lighter liquid and the delivery tube extends right up to the bottom of the cylinder D, so the lighter liquid starts to flow from B to D. Simultaneously the heavier liquid from beaker A arrives at beaker

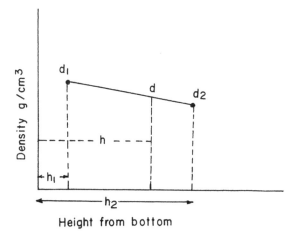

Figure 9.3. Interpolation method.

B and makes it denser due to continuous stirring. This way an increasingly dense liquid flows into the cylinder; the denser liquid pushes up the lighter liquid, hence creating a column of liquid with a continuously varying density.

9.5.5   Measurement of the density of a solid

Floats of the appropriate density are gently introduced. After some time the floats will settle at different heights. When several floats stick together, these should be separated by a wire moved horizontally. Measure the height of the centre of each float from an arbitrary horizontal reference line. Draw a graph of the density (y-axis) against the heights and draw a smooth curve, which normally should be a straight line. If a smooth curve fails to pass through enough points, the solution should be rejected and the whole exercise repeated.

Place gently the specimen on the surface of the liquid and wait until it reaches a stable position and remains there for several minutes. Determine the height from the same reference line and calculate the density from the graph or by interpolation.

Interpolation method

Let d_1 and d_2 be the densities of the floats at heights h_1 and h_2 respectively. Let h be the height of the specimen in between the two floats. Then the density gradient for that region is $(d_1 - d_2)/(h_2 - h_1)$. Referring to figure 9.3, if $h_2 - h$ is the height difference from the float of density d_2, then d, the density of the specimen, is given by

$$d = d_2 + (h_2 - h)(d_1 - d_2)/(h_2 - h_1). \qquad (9.13)$$

This method is quite often used to determine the density of plastic materials.

9.6   Bulk density of manufactured carbon and graphite articles by dimensional measurement [6,7]

9.6.1   Preparation of sample

Machine a test specimen, from the manufactured article, either in the form of a rectangular block or a right circular cylinder. The volume of the rectangular block or cylinder should be no less than 500 mm^3. Furthermore, no dimension of the block should be smaller than five times the length of the largest visible particle and 2000 times the resolution of the measuring device. Similarly the sensitivity of the balance must be better than 0.0005 of the mass of the worked specimen.

No lubricant with a boiling point above 100 °C should be used while machining. All corners, edges and faces of the specimen should be free of chips or gouges. Ensure that the specimen is free from any residue from the machining.

9.6.2   Drying and weighing the specimen

Dry the specimen for at least 2 h in an oven maintained at 110 °C. Allow it to cool to 25 ± 5 °C in a desiccator. The specimen should only be removed from the desiccator just before weighing. Weigh the specimen with an accuracy of 0.05%. Use soft-tipped forceps to lift the specimen.

9.6.3   Measuring the dimensions of the specimen

If the specimen is a rectangular block, make four measurements of the length, from the centre of one face to its opposing face. Measure the width and thickness at each end and at two intermediate points along the length. Determine the mean length, breadth and thickness in the case of a rectangular block. Measure the diameter of the cylinder in mutually perpendicular directions and take the mean. Each dimension is measured with an accuracy of 0.05%. Use an appropriate formulae to calculate the volume of the specimen.

9.6.4   Density of specimen and report

If M is the mass and V the volume of the specimen then its bulk density d is given by

$$d = M/V. \tag{9.14}$$

The report should include the source, grade and shape of the used specimen. The values of the density of individual pieces along with the mean value should be reported.

9.7   Apparent density of free-flowing powders

There are certain solids in powdered form, which behave almost like a liquid when poured from one vessel to another. Examples include soda ash and magnesium oxide. The method described here is suitable for such materials.

9.7.1   Soda ash [8]

Take about 30 g of soda ash and weigh it. Let its apparent mass be M g. Put it in a funnel closing its outlet orifice with a finger. Place a glass volumetric graduated cylinder at a certain distance below it. The distance is specified from the plane of the orifice to the bottom of the measuring cylinder. Open the orifice by removing the finger and start rotating the cylinder at a constant angular velocity. A motor-driven platform may also be used to rotate the measuring cylinder. As soon as the free flow of soda ash ceases, level the surface without any jarring effect. Read the volume V to the nearest 0.1 cm^3. The apparent density d_a is given by

$$d_a = M/V \text{ g cm}^{-3}.$$
\hfill (9.15)

The graduated cylinder is calibrated using the gravimetric method. Pure distilled water is used as the standard of density. The apparent mass of water contained in the cylinder to a specific level is determined. The capacity of the cylinder, up to that level is found by adding a correction C which can be found in the appropriate tables 8.1(a)–(r). In selecting the table, care should be taken to select the correct values for the coefficient of thermal expansion of the glass from which it is made, densities of used weights and the reference temperature. Normally measuring cylinders are made from borosilicate or soda glass. For borosilicate glass the values of correction C are given in tables 8.1(a) and (g). Soda glass has a coefficient of volume expansion of $30 \times 10^{-6}\,^\circ\text{C}^{-1}$ for which the values of C have been given in tables 8.1(o)–(r) for a nominal capacity of 1 cm^3. For other capacities a proportional value of C is used. It should be mentioned that the value of C in mg cm^{-3} is the same as values of C in g per 1000 cm^3 (dm^3) or kg m^{-3}.

9.7.2   Electrical-grade magnesium oxide [9]

The density of electrical-grade magnesium oxide is determined by finding the volume of a given measured mass of it when it fills a measuring cylinder, which is being tapped constantly at a certain specified rate.

Weigh a certain amount of magnesium oxide, say 100.00 ± 0.02 g, start the tapping machine and quickly transfer the magnesium oxide into the hopper of the tapping machine. Stop tapping as soon as the magnesium oxide ceases to flow. Tap the hopper to dislodge any powder adhering to its side. Insert a calibrated plug into the measuring cylinder and tap it five times. Read the volume of the specimen as indicated by the measuring cylinder. Then the tap density is the apparent mass of the magnesium oxide powder divided by the measured volume. The apparatus

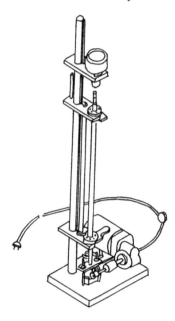

Figure 9.4. Tapping machine.

is shown in figure 9.4. The significance of this method is due to the fact that this is the way in which magnesium oxide is filled in insulation tubular heating and insulation elements.

9.7.3   Activated carbon [10]

The apparent density of granular activated carbon with more than 90% of its particles larger than an 80 mesh is determined in a similar way to that just described. In this method instead of tapping the graduated jar, activated carbon particles are vibrated and made to fall through a feed funnel of specified size.

Dry the activated granular carbon in an oven maintained at $150 \pm 5\,°C$. Keep on weighing at regular time intervals, until the apparent mass of the specimen becomes constant within 1 mg. Add the specimen to a reservoir funnel with a blocked orifice.

The specimen then falls from the reservoir funnel R onto a metal vibrator with a specific frequency. The inclination of the vibrating plate M can be adjusted to suit the rate of flow with which the sample is to be fed through F the feeder funnel of specific dimensions into a calibrated measuring glass cylinder C. Read the volume as indicated by the measure. It is preferable for the volume of specimen to be 100 cm^3 and that it is delivered at the rate of 0.75–1 cm^3 s^{-1} into the measuring cylinder. To get the desired flow-rate, if necessary, the inclination of the vibrator M can be suitably changed. The specimen powder

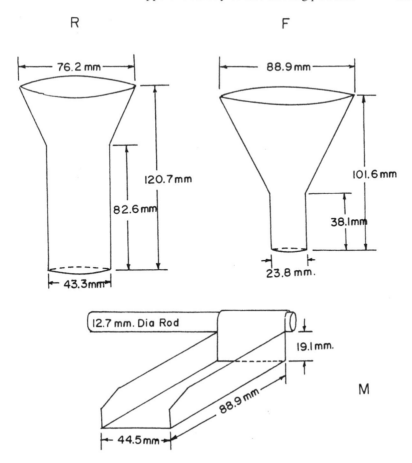

Figure 9.5. Reservoir and feed funnels with vibrator. R is the reservoir funnel, while F is the feed funnel.

is then transferred to the pan of the balance and weighed to the nearest 0.1 g. The apparent density d_a is given by

$$d_a = (\text{apparent mass/volume}) \text{ g cm}^{-3}.$$

The repeatability of the method is about 0.007 g cm^{-3}. The dimensions of the reservoir, feed funnels and vibrator as specified in ASTM D 2854-70 (re-approved 1976) [10] are given in figure 9.5.

The full arrangement is shown in figure 9.6. The powder from the funnel R falls onto the vibrator M and slips down to the feed funnel F and finally to the measuring cylinder C.

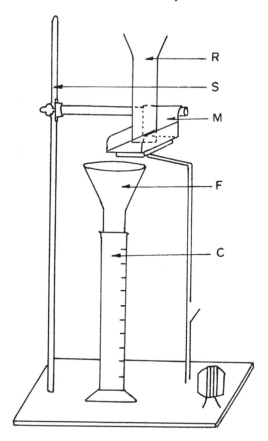

Figure 9.6. Arrangement for measuring the apparent density of granular activated carbon.

9.7.4 Metal powders

The method for determining the apparent density of metal powders is similar to that described in section 9.7.1. Essentially it consists of allowing the powder to flow though a funnel with specified orifice into a cup of given volume. The capacity of the cup is calibrated against water as the standard of density and the gravimetric method is used.

Free-flowing metal powder [11]

The metal powder in question, if necessary, is dried in an oven maintained at 102–107 °C for at least 1 h. A cup of volume 25 ± 0.05 cm^3 is placed on a fairly vibration-less table and the dried powder is made to flow through a 60° funnel with an orifice of 2.5 mm internal diameter. The powder is allowed to fall until it starts overflowing the periphery. The funnel is turned around immediately so that

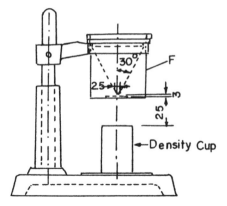

Figure 9.7. Funnel and cylindrical cup for free-flowing powders.

any excess powder does not create unnecessary pressure. It is levelled carefully with a spatula or straight edge. After levelling, the cup is slightly tapped so that there is no slippage while transferring it to the balance. The net apparent mass is determined within 10 mg and the apparent density d_a is given by

$$d_a = \text{apparent mass/volume of the cup.}$$

If the apparent mass is measured within ± 10 mg, then the repeatability of the method in measuring the density is 0.01 g cm^{-3}.

Non-free-flowing metal powders [12]

The method is the same except the diameter of the orifice of the funnel is 5 mm and, if necessary, a wire 3 mm in diameter may be used to push the powder through the orifice. Care should be taken to prevent the wire from going inside the cup.

Pour density of pelleted oxide powder (carbon black) [13]

The powder is poured into the centre of a pre-weighed stainless steel cylindrical cup of capacity 624 cm^3 at 20 °C. The cup has no lip and its walls are straight and smooth from inside. The internal diameter is 100 ± 5 mm. The cup is calibrated by the gravimetric method using water as the density standard.

The pouring is continued until the powder forms a large enough cone above the rim of the cup. It is levelled in one stroke by a straight edge or spatula. The edge is firmly held normal to the surface of the rim and the powder is levelled in a single sweep. The cup is weighed and the apparent mass W is calculated from the difference in the two weighings. Pour density d_p is given by

$$d_p = W/\text{volume of the cup.}$$

9.7.5   Density by specific gravity bottle (pyknometer) method

Density of beryllium oxide powder [14]

The density of beryllium powder is determined by the method described in section 6.1.3. As beryllium oxide reacts with water, toluene is used. The density of toluene is determined with the help of the same specific gravity bottle by using water as the density standard. The density d of beryllium oxide is given by

$$d = (\rho_1 - \sigma)(M_1 - M_0)/[(M_3 - M_0) - (M_2 - M_1)] + \sigma \text{ g cm}^{-3} \text{ at } 20\,^{\circ}\text{C} \quad (9.16)$$

where M_0 is the mass of the empty RD bottle, M_1 is the mass of the RD bottle plus powder, M_2 is the mass of the RD bottle + powder + toluene required to completely fill the RD bottle at $20\,^{\circ}\text{C}$, M_3 is the mass of the RD bottle + toluene required to completely fill the RD bottle at $20\,^{\circ}\text{C}$, ρ_1 is the density of toluene at $20\,^{\circ}\text{C}$ and σ is the density of air at $20\ ^{\circ}\text{C}$.

The density ρ_1 of toluene is given by

$$\rho_1 = \rho_{\text{w}}(M_{\text{L}} - M_0)/(M_{\text{w}} - M_0) \quad (9.17)$$

where M_0 is the mass of the empty RD bottle, M_{L} is the mass of the RD bottle + toluene required to completely fill it at $20\,^{\circ}\text{C}$ and M_{w} is the mass of the RD + water required to completely fill it at $20\,^{\circ}\text{C}$. It should be mentioned here that the unit of density d in this case will be the same as the unit of density in which σ and ρ_{w} are expressed. Of course the same unit is to be taken for both. The method can also be used for flexible cellular material—vinyl chloride polymers and co-polymers (open cell foam) [15].

9.7.6   Density of solid pitch and asphalt [16]

The specific gravity bottle (pyknometer) method already described in detail in section 9.7.5 can also be used to determine the density of pitch and asphalt. The temperature of the bath is maintained at $25 \pm 0.1\,^{\circ}\text{C}$ and water is used as the displaced liquid.

9.8   Refractory materials

Some refractory materials are attacked by water and some are not, so the methods for determining their densities and other associated properties will differ. Furthermore, some tests are quite often to be carried out in the field where there may be little technical help so field methods have to be simpler.

9.8.1   Field method for measuring the bulk density of refractory brick and insulating firebrick [17]

This method essentially consists of measuring the dimensions within 0.5 mm. A 30 cm steel scale with graduations of 0.5 mm is used for measuring the length and breadth of the brick. The scale has a rigid hardened steel hook consisting of a piece at right angles to the length of the scale and the zero of the scale starts from its inner face.

To measure the thickness, several bricks are stacked together and their total height at the middle point of the length of the stack is measured with a similar scale but 1 m in length and graduated in mm or at best 0.5 mm. The thickness of each brick is determined by dividing by the number of bricks taken. For insulating bricks, the thickness of each individual brick may be measured almost at the middle point of its length and the average is taken for several bricks. The actual number of bricks to be taken would depend upon the particular specification. Normally ten bricks are sufficient but they should be a representative sample of the whole lot.

Several bricks are dried at about 110 °C for several hours, cooled and weighed. W, the average weight for one brick, is calculated. Then the apparent bulk density is given by

$$\text{Bulk density (apparent)} = W/LBH$$

where L, B and H are, respectively, the average length, breadth and height of one brick.

9.8.2 Bulk density, water absorption and associated properties of refractory bricks [18]

This method is good for materials which are not attacked by water. To obtain representative results, at least ten bricks should be selected using a good sampling method. Bricks of nominal size $228 \times 114 \times 76 \text{ mm}^3$ are divided into four pieces by cutting along the plane parallel to the $228 \times 114 \text{ mm}^2$ face and then cutting it along the $114 \times 76 \text{ mm}^2$ plane. This way each piece will have four faces from the original mould.

The samples are dried in an oven maintained at 105–110 °C until each sample attains a constant weight. All weighings are to be carried out when the samples have attained room temperature and they should be within 0.1 g. If necessary a dry air enclosure may be used for the cooling. Let D represent the dry weight of the sample. Place the samples in water and boil for 2 h. During boiling all samples should remain covered with water but they should not be in contact with the bottom of the vessel.

Determine the apparent weight S of the sample in distilled water, with the help of a pre-weighed suspension device. Care should be taken for the water levels in the two weighings to reach the same mark on the suspension wire.

For determination of the saturated weight W of the sample, take it out and place it on a cloth saturated with water so that only water adhering to the sample is removed and no absorbed water escapes due to the capillary action of the cloth fibre. Then weigh it; let W be the value; this is known as the saturated weight of the sample. If ρ_w is the density of water, then the exterior volume V of the sample is given by

$$V = (W - S)/\rho_w. \qquad (9.18)$$

If ρ_w is taken to be unity, it will only contribute an error of about 0.3%.

From these data, the bulk density B, the apparent relative density T, porosity P, water absorption A and the volume of open pores V_O can be determined as follows.

$$B = D/V \qquad (9.19)$$
$$T = D/(D - S) \qquad (9.20)$$
$$P \text{ (the apparent porosity in \%)} = 100(W - D)/V \qquad (9.21)$$
$$A \text{ in \%} = 100(W - D)/D \qquad (9.22)$$
$$V_O = W - D \qquad (9.23)$$

and

$$V_{imp} = D - S \qquad (9.24)$$

where V_{imp} is the volume of the impervious portion of the solid.

Almost the same method is used for fired white-ware products [19]. Samples of at least 50 g in mass (apparent) are taken and dried in an oven maintained at 150 °C to a constant weight within 0.01 g. The samples are boiled in water for 5 h and kept in the same water for another 24 h. Then the same procedure and formulae are used to determine S, the weight in water, W, the saturated weight, and D, the weight of the dried sample.

9.8.3   Bulk density, liquid absorption and associated properties of refractory bricks using vacuum and pressure techniques [20]

The bulk density, liquid absorption and associated properties of refractory bricks can be determined by applying vacuum and then pressure to saturate the specimen either with water, if the material is unaffected by it, or otherwise with another liquid, e.g. mineral spirit.

As described in the previous section, make four pieces from $228 \times 114 \times 76$ mm^3 bricks. For refractory materials in other shapes, cut, drill or break the specimen to have samples of volume 410–490 cm^3.

Take at least five samples, which represent the whole batch. As mentioned in section 9.8.2, dry the specimen samples in an oven maintained at 105–110 °C until the weight becomes constant within 0.1 g. Determine the dry apparent mass D. Normally the dry weight is determined before the mass is saturated or suspended. If it seems that part of the specimen has been broken or lost during any of the procedures after drying, the dry weight may be re-determined.

Saturating the sample with water or mineral spirit

Choose a suitable vessel, which can be closed, secured and able to withstand a vacuum of 3.4 kPa on the lower side and 483 kPa on the higher side without any deformation. Place the sample in it, close it and connect it to a vacuum pump reducing the pressure to 6.4 kPa. Maintain this pressure for 30 min. Allow water to enter the vessel if the material does not react with it, otherwise use mineral spirit. Maintain the low pressure for another 5 min. Close the vacuum line and pressurize the vessel up to 207 kPa and keep this pressure up for another hour or so. Reduce the pressure to atmospheric. The saturated sample is then ready to be weighed. Weigh the sample within 0.1 g. Let the saturation weight be W.

Suspended weight

Weigh the samples in the same liquid with which it has been saturated with a pre-weighed loop or any other suspension device. Care should be taken to submerge the suspension wire to the same mark on the wire. Let the suspended weight be S.

Remove the sample from the water or mineral spirit, as the case may be, and remove any adhering liquid with a moist lint-free towel. Alternatively roll the specimen lightly over the wet cloth, which has been pre-saturated with the liquid. Weigh the sample to the nearest 0.1 g. An accuracy of 0.1 g is only valid if the apparent mass of the sample is 500 g or more.

Sample parameters

The exterior volume, V, of the specimen is given by

$$V = (W - S)/\rho_w \text{ or } \rho_{liquid} \tag{9.25}$$

$$\text{Volume of open pores} = (W - D)/\rho_w \text{ or } \rho_{liquid} \tag{9.26}$$

$$\text{Volume of impervious portion} = (D - S)/\rho_w \text{ or } \rho_{liquid} \tag{9.27}$$

$$\text{Apparent porosity } P \text{ in } \% = 100(W - D)/(W - S) \tag{9.28}$$

$$\text{Liquid absorption } A \text{ in } \% = 100(W - D)/D \tag{9.29}$$

$$\text{Apparent relative density (specific gravity) } T = D/(D - S). \tag{9.30}$$

If the sample is weighed in mineral spirit, then T is given by

$$T = [D/(D - S)]\rho_{liquid}. \tag{9.31}$$

The bulk density B is given by

$$B = D/V. \tag{9.32}$$

Whenever mineral spirit is used instead of water, replace ρ_w by ρ_{liquid}.

9.8.4 Bulk density and volume of solid refractory materials by wax immersion [21]

This is a quick method of finding the bulk density, B, and exterior volume of a refractory material, so it is quite often used during the production process. It comprises the following steps:

- choose a representative sample;
- choose some paraffin wax of good quality of known density;
- apply molten wax to a pre-weighed sample of weight W;
- find P, the apparent mass of the sample with coated wax; and
- the suspended weight of the sample in water.

The paraffin wax is to be heated only a few degrees above its melting point. It should be heated through a water bath so that the temperature may be controlled. Vapours from molten wax ignite spontaneously above 200 °C, so its vapours should not be allowed to come into contact with the heating element or flame. Therefore it should be safely heated through a thermocontrolled water bath at 60 °C. A thin layer of wax is applied evenly to the sample. All precautions mentioned in earlier sections should be observed while finding the suspended apparent mass, S, of the sample in water. If P is the weight of the sample with a wax coating and ρ_{wax} is the density of wax, then V_1, the volume of the sample with wax coating, is given by

$$V_1 = (P - S)/\rho_w. \tag{9.33}$$

The volume, v, of the wax is $(P - W)/\rho_{wax}$, giving the exterior volume of the sample V as

$$V = V_1 - v \tag{9.34}$$

and hence the bulk density B as

$$B = W/V. \tag{9.35}$$

As in the refractory items discussed in sections 9.8.1–9.8.4, ρ_w, the density of water, is taken to be unity. The error due to this will amount to only 0.3%.

9.8.5 Density of refractory materials either in the form of granules or their cut pieces

The size of the granules or pieces is chosen depending upon the specific method to be used to find the volume of the sample, which consists of these granules or pieces. The method for finding the volume depends on whether the material reacts with water or not. If the material does not react with water, there are two methods—by pyknometer/specific gravity bottle or by a burette where the volume of water displaced by the sample is measured. If the refractory material reacts with water then the volume of mercury displaced by the sample is measured.

True specific gravity of refractory materials [22]

Sample preparation

Take two walnut-sized samples such that the original surface is excluded. Crush the pieces into smaller pieces of no more than 3 mm. Thoroughly mix the crushed material and reduce it by quartering or riffling to a test sample of 50 g. Grind the entire 50 g sample so that it passes through a 150 μm sieve. Special care should be taken while grinding not to exclude any portions that are difficult to grind.

Transfer the sample to a glass bottle with a stopper. Dry it in an oven maintained between 105 and 110 °C, until the weight becomes constant.

Procedure

Take a pyknometer or a specific gravity bottle P and dry it at 105–110 °C. Cool P in a desiccator and weigh it, transfer the powder into it (P) and weigh it (P), then fill it (P) with distilled water completely and weigh it. Remove the material and water, clean and dry it (P) again and find the mass of P and fill it with distilled water and again weigh it (P). From these data the mass of the sample m, and the mass of water with the same volume as that of the sample can be determined. The ratio of the two will give the relative density (specific gravity) of the sample. Details have already been discussed in previous sections. The water used in this process should be between 15 and 24 °C.

Bulk density of granular refractory materials [23]

In this case, the representative sample consists of pieces sized between 6.7 mm and 2.36 mm. Blow any sample that has been calcined and cooled not more than 2 hours beforehand with clean dry air to remove all dust from the sample. Wash any other material with a stream of tap water for at least 5 min until all dust has been removed. Dry it in an oven at about 105 °C. Separate the sample into four parts and find the bulk density of each part separately to obtain a good mean. Each part has a volume of about 25 cm^3. Weigh the sample to the nearest 0.01 g. Let the weight be W.

Place the sample in a beaker of water and boil the contents for at least 1 h. There should be sufficient water to keep the sample covered during the boiling process. Cool the contents to room temperature by running cold water. Decant the excess water and spread the sample onto a damp towel. Blot the surface of the grains lightly so that only the adhering surface water is removed and no water from the pores is removed by capillary action. None of the grains in this case will adhere to each other.

Take a cleaned 50 cm^3 burette and fill it to about the 25 cm^3 level. Wait for any water to drain out and read the level of water in the burette to within 0.1 cm^3. Pour the grains into the burette, taking care that no grain is left on the towel. Shake well so that no air bubbles are sticking to the grains. Read the level of water in the burette carefully. The difference in the two readings will give the volume of the grains. Let this be V cm^3.

Then the bulk density B is given by

$$B = W/V \text{ g cm}^{-3}. \tag{9.36}$$

9.8.6   Bulk density and porosity of granular refractory materials [24]

In this method the representative sample consists of material sized between 2.00 and 4.75 mm, i.e. those pieces which can pass through a no 4 sieve (mesh size 4.75 mm) but not through sieve no 10 (mesh size 2.00 mm).

Mercury volume meter and porosity apparatus

This has the following parts.

Sample container

The sample container with a hollow stopper with a capillary and hook at its top is shown in figure 9.8. The container is made from borosilicate glass and is shaped of frustums of two cones joined base to base. At the lower end of the lower frustum there is a standard stopcock. A polythene tube can be slipped on the outlet of the stopcock. The uppermost part ends in a ground-glass standard female joint no 16. A small hook is also provided to enable it to be coupled through a rubber band with a similar hook on the stopper. The hollow stopper of total length 140 mm has a bore of 2.5 mm. Its lower end is a ground-glass standard tapered no 16 stopper, which can fit into the top of the sample container. It also has a hook on top. A rubber band holds the two hooks on the stopper and the sample container together when the vibrator is attached. Towards the middle of the capillary a permanent mark is etched and coloured for better visibility. The other dimensions are shown in figure 9.8.

Calibrated burette

There is a standard 50 cm^3 burette with 0.1 cm^3 graduations connected at the top to a reservoir of capacity of at least 100 cm^3. The reservoir can be connected to a vacuum pump. The stopper end of the burette is connected to the sample container with a polythene tube. The burette is calibrated by the gravimetric method against water as the density standard.

Stand and vibrator

The burette and sample container are held in position by a common stand. The sample container can be smoothly moved up or down easily with a boss and holder. A vibrator to vibrate the sample container is also fixed to the rod holding the sample container. The arrangement is shown in figure 9.9.

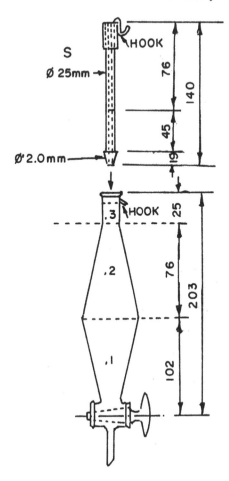

Figure 9.8. Sample container (all dimensions in mm).

Procedure

About 100 g of the sample is dried and kept in a desiccator containing phosphorus penta oxide (P_2O_5) or quick lime. The sample is weighed within 0.05 g.

Add enough mercury to the system and adjust the level of the sample container so that the mercury level is at the specified mark of the hollow stopper and also on the readable portion of the burette near the 50 cm³ mark. Note the reading of the burette to the nearest 0.1 cm³ and let it be V_1. Raise the sample container C. By creating a low pressure with the vacuum pump, transfer most of the mercury to the reservoir R and the burette. Close the stopcock B of the container and release the vacuum. Now both sides of the apparatus are at atmospheric pressure. Transfer the sample to the sample container and open its

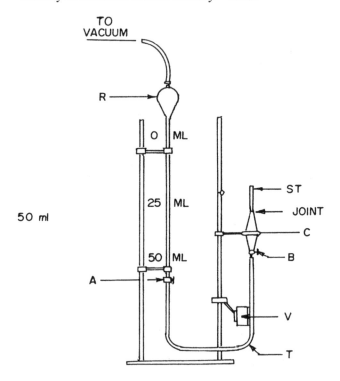

Figure 9.9. Porosity apparatus or mercury volume meter.

stopcock until the mercury flows back to the stopper ST up to the reference mark on it. Read the new mercury level in the burette. Close the stopcock A.

Start the vibrator. Adjust the elevation of the sample container occasionally to bring the mercury back to the reference mark as required. Continue until no further change in the mercury level takes place.

Stop the vibration. With the stopcock open and the mercury level at the reference mark on the hollow stopper, note the volume reading V_2 on the burette to the nearest 0.1 cm^3.

Calculation

Calculate the bulk density as follows.

$$B = W/(V_2 - V_1). \qquad (9.37)$$

Find the true specific gravity separately by some other method. If A is the true relative density then

$$A = W/V_{\text{imp}}. \qquad (9.38)$$

However

$$V_2 - V_1 = \text{total volume} = V_{\text{imp}} + V_O \qquad (9.39)$$

where V_{imp} is the volume of the impervious portion and V_O the volume of the open pores of the solid. Therefore

$$B = W/(V_{\text{imp}} + V_O). \qquad (9.40)$$

From these equations

$$(A - B)/A = [W/V_{\text{imp}} - W/(V_{\text{imp}} + V_O)]/(W/V_{\text{imp}})$$

giving us

$$(A - B)/A = (V_{\text{imp}} + V_O - V_{\text{imp}})/(V_{\text{imp}} + V_O) = V_O/(V_{\text{imp}} + V_O) \quad (9.41)$$

which equals the volume of the open pores/total volume, which is the porosity P. Hence

$$\text{Porosity in } \% = 100(A - B)/A. \qquad (9.42)$$

9.9   Density or relative density of insulating material

9.9.1   Density of insulating paper [25]

Knowledge of the density is useful in the design of electrical insulating systems and in determining the economic aspects of the use of paper as an insulating material. Some physical and electrical properties of paper are related to its density.

Conditioning [26]

The density of insulating paper can be determined under two conditions—wet–wet conditions or dry–dry conditions.

The paper is cut into the form of a rectangle. Each edge is square to the other two. The size of the paper depends upon its availability and the readability of the balance. It should be such that its mass may be determined with a relative accuracy of 0.1% from a balance with a readability of 0.1 mg.

Mass

The mass of paper with a reasonable size is determined on a balance giving a relative accuracy of 0.1%. The length and breadth of the paper is normally measured with a steel ruler graduated in mm. Half of the graduation is estimated when the length/breadth is determined. The length or breadth is measured parallel to the edge at least at three positions—two near the edges and the third in the middle. However, any measurement is taken 1 cm in from the edge. The mean of the three measurements is taken.

Thickness

For films or plastic sheets thicker than 0.25 mm, a micrometre with a ratchet or frictional thimble is used.

To avoid any backlash error, it is customary to close the micrometer on the specimen outside the area to be measured. Open the micrometer by ten divisions from the initial reading and bring it to the first measurement area. Start closing it slowly so that the number of graduations passing the reference mark can be counted. Continue the motion at the same rate until the ratchet has clicked three times or the frictional thimble has slipped, and only then take the reading. To measure another area, again move back ten divisions and repeat the same procedure for the thickness measurement at the second measurement area.

If the micrometer also has a vernier scale, open the micrometer by ten divisions and close it slowly so that it just touches the paper and note the final reading taking into account the reading of the vernier scale.

For papers from 0.25 down to 0.025 mm in thickness, a dial gauge with a broader foot is used. The size of the foot or height from which it is dropped depends upon the specification being used.

While using a dial gauge, place it on a clean and level table with no excessive vibrations. Close the contact surfaces of the pressure foot and anvil. Adjust the zero of the instrument. Lift the pressure foot and place the specimen under it. Lower the foot onto the specimen at a location outside the area under measurement. Then raise the foot slightly, position the specimen to bring the area being measured underneath it, raise the pressure foot seven to ten divisions higher and then drop it. Note the reading and then follow the same procedure for other areas of interest.

The thickness is measured along a line at right angles to the machined direction of the paper.

For paper less than 0.05 mm thick, it is better to use a motorized dial gauge. For still thinner papers the paper pieces are stacked together to make a pad and their combined thickness is measured with a dial gauge. The thickness of an individual piece is determined by dividing the combined thickness by the number of stacked pieces.

Density

Quite often, the mass per unit area in g per m^2 is reported. However, dividing it by the thickness gives the density in g per cm^3. The length, breadth and thickness are expressed in cm.

9.9.2   Insulating boards [27]

The density of insulating boards is generally determined by finding the mass of a cleanly cut rectangular board. Every edge should be square to the other two. The length is measured at three points on the broad side, with 0.5 cm being left from the edges. One reading is necessarily taken at the middle of its breadth. The lengths should not differ from each other by more than twice the least count of the measuring instrument. A larger difference shows a deviation from the rectangular shape. The breadth is measured in the same way. The thickness is measured with a micrometer.

As a rule of thumb, for linear measurements, the following instruments may be used:

- a steel scale within 0.5 mm for dimensions greater than 100 mm;
- a vernier calliper within 0.1 mm for dimensions in between 30 and 100 mm;
- a micrometre within 0.01 mm for dimensions between 3 and 30 mm; and
- a dial gauge within 0.002 mm for dimensions less than 3 mm.

9.10 PVC and similar materials

9.10.1 Apparent density of plastic materials [28]

The two methods given here cover the determination of the apparent density of plastic materials in such different forms ranging from fine powders and granules to large flakes and cut fibres.

Fine powders

This method is similar to the ones discussed in section 9.7.4 for powders. The equipment consists of a 100 cm^3 measuring cup, calibrated against water as the standard using a gravimetric method and a funnel with a specified orifice diameter. It should be noted that the diameter of the orifice is 9.5 ± 0.8 mm. The measuring cup is placed below the funnel at a distance equal to the diameter of the cup. The assembly is shown in figure 9.10. About 115 ± 5 cm^3 of the powder is placed in the funnel keeping the orifice closed. The powder is allowed to fall freely from the funnel. If caking occurs at the funnel, the material is loosened with a glass rod. After all the material has passed through the funnel, the excess powder is scraped with a straight edge without shaking the cup. The cup is weighed to the nearest 0.1 g. Find the apparent mass of the powder. The density is obtained by dividing the mass of powder by the volume of the cup. The capacity of 100 cm^3 should be known to within ± 0.5 cm^3.

Coarse, granular material, including dice and pellets

The method is used for those materials which cannot be poured easily through a normal funnel. The equipment is the same apart from its size and dimensions. The cup has a capacity of 400 cm^3 and the funnel has an opening of 25.4 mm. The equipment is arranged as shown in figure 9.11. Fill the funnel with about 500 cm^3 of material while the outlet is closed with a flat strip. It is then allowed to fall suddenly into the pre-weighed measuring cup. As soon as the material

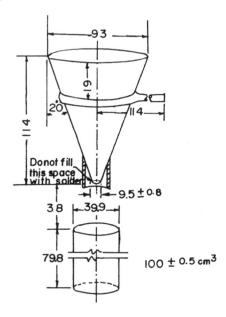

Figure 9.10. Funnel and cup assembly.

stops falling, the excess material is scraped with a straight edge and the cup weighed. Find the apparent mass of the material in the cup; density is calculated from knowledge of the capacity of the cup.

Material in the form of coarse flakes, chips, cut fibres or strand

Materials in the form of coarse flakes, chips, cut fibres or strands cannot be poured through a funnel. In addition as these materials are bulky in their normal form and can be compressed to great extent even by hand, it is logical to measure their density under a certain pressure. The equipment consists of a cylinder of capacity $1000 \ cm^3$ and a plunger whose outer surface is well machined and smooth. The plunger is in the form of an open cylinder with an outer diameter which is slightly smaller than the inner diameter of the measuring cylinder. The plunger has a scale graduated in mm on its outside. The plunger and measuring cylinder are shown in figure 9.12.

Place the measuring cylinder on a paper and take the material of apparent mass 60 ± 0.2 g and loosely drop it into the measuring cylinder from a height almost equal to that of the cylinder. Care should be taken that no material is lost in dropping it. If any fibre or flake falls outside put it back in the measuring cylinder. Before applying the pressure through the plunger, level the material and measure the height of the material in the cylinder in cm. Let this be h_1. Put some lead shots in the plunger cylinder such that plunger cylinder weighs 2300 ± 20 g.

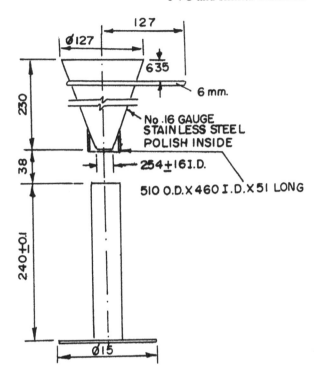

Figure 9.11. Funnel and measuring cylinder.

Lower the plunger slowly until it rests entirely on the material. Wait for a minute and measure the height of the material from the scale on the plunger to the nearest 0.1 cm. Let this height be h_2.

Calculate the apparent density from both the values h_1 and h_2. The volume of the material, V, is given by

$$V = Ah$$

where A is the area of the measuring cylinder and h is the mean of h_1 and h_2. Three observations for each sample are taken and the mean is reported.

9.10.2 Density of rubber chemicals [29]

The specific gravity bottle (pyknometer) method already described in detail in section 9.7.5 is used to determine the density of rubber chemicals. The temperature of the bath is maintained at $25 \pm 0.1\,°C$ and liquids which do not react with rubber chemicals are used as the displacement liquids.

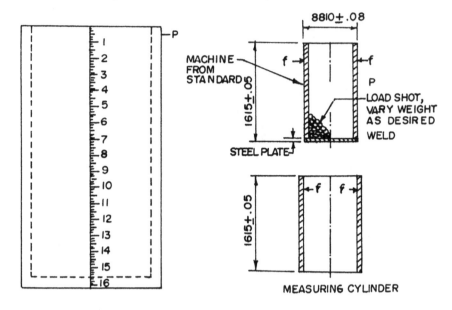

Figure 9.12. Measuring cylinder and its plunger.

9.10.3 Sodium carboxy methyl cellulose [30]

A weighed amount of the substance is transferred to a measuring cylinder. The measuring cylinder has capacity of about 250 cm^3 and is vibrated for 3 min to settle the powder down. The volume of the powder is measured to within 0.1 cm^3. The measure is calibrated against water by using the gravimetric method at the specified reference temperature.

9.10.4 Hydroxy ethyl cellulose [31]

The same method as that described in section 9.10.3 is also employed for this material.

9.10.5 Hydroxy propyl methyl cellulose [32]

The same method as described in section 9.10.3 is also employed for this material.

9.10.6 Flexible cellular materials—vinyl chloride polymers and co-polymers (closed-cell sponge) [33]

The mass is determined with a suitable balance of appropriate capacity and readability 1 mg, while the volume is measured by determining the dimensions as discussed in section 9.9.2. Repeatability is only 1%.

9.10.7   Flexible cellular materials—vinyl chloride polymers and co-polymers (open-cell foam) [15]

The mass is determined with a suitable balance of appropriate capacity and readability 1 mg, while the volume is measured by using the specific gravity bottle method as described in section 9.7.5. Repeatability is only 1%.

9.10.8   Apparent density of leather [34]

Cut a rectangular piece of leather cleanly. Care should taken that each edge is square to the other two adjacent edges. The length and breadth are measured with a steel scale within 1 mm. The area is calculated to the second decimal place of cm^2. The thickness is measured with a disc-type micrometer at four quadrants to within 0.25 mm. Repeatability is around 4%.

9.10.9   Density of high modulus fibres [35]

The density of high modulus fibres is determined either by hydrostatic weighing as described in section 9.4 or by using a sink–float method as described here.

Sink–float method

The sample is placed in a borosilicate glass bottle, which contains liquid that will thoroughly wet the sample and is lighter than it. A liquid of higher density is added slowly and mixed simultaneously. A stage will be reached when the sample will start to float in the liquid and starts rising. By careful addition of the lighter liquid drop by drop, the sample is just suspended in the liquid mixture. In this situation the density of the solid fibre becomes equal to that of the liquid. The density of the liquid mixture is determined separately with the help of a pyknometer, whose volume is calibrated against water at the specified reference temperature.

9.10.10   Polyethylene [36]

Cut a strand, boil it in water for 30 min. Let it cool down to room temperature. Cut 5 mm from the thick end and discard it. At least three pieces are cut from the next 10 mm of the strand. Use a sharp blade so that it can cut the ends cleanly and square to the axis.

Use appropriate liquids, which do not react with polyethylene. Establish a density gradient in a long cylinder, insert the spherical floats and place the specimen strands in the cylinder one by one; measure the height from the middle point of the strand to the reference line from which the heights of the standard floats of known density have been measured. Interpolate the density from the heights and calculate the density. This method has been described in detail in section 9.5.

References

[1] ASTM F 77-69 (re-approved 1979); Apparent density of ceramics for electron device and semi-conductor application

[2] ASTM C 729-80; Density by sink–float comparison

[3] ASTM C 693-74 (re-approved 1980); Density of glass by buoyancy

[4] ASTM D 792-66 (re-approved 1979); Specific gravity and density of plastics by displacement method

[5] ASTM D 1505-68 (re-approved 1979); Density of plastics by the density gradient technique

[6] ASTM C 559-79; Bulk density of manufactured carbon and graphite articles by dimensional measurement

[7] ASTM C 838-80; Bulk density of as manufactured carbon and graphite shapes

[8] ASTM D 501-67 (re-approved 1977); Sampling and chemical analysis of alkaline detergent (density of soda ash)

[9] ASTM D 3347-74 (re-approved 1979); Flow rate tap density of electrical grade magnesium oxide for use in sheathed-type electric heating elements

[10] ASTM D 2854-70 (re-approved 1976); Apparent density of activated carbon

[11] ASTM B 212-76; Apparent density of free-flowing metal powders

[12] ASTM B 417-76; Apparent density of non-free flowing metal powders

[13] ASTM D 1513-80; Pour density of pelleted carbon black

[14] ASTM C 699-79; Chemical, mass spectrometric and spectro-chemical analysis of and physical tests on beryllium oxide powder

[15] ASTM D 1565-76; Flexible cellular material—vinyl chloride polymers and copolymers (open cell foam)

[16] ASTM D 71-78; Density of solid pitch and asphalt

[17] ASTM C 134-70 (re-approved 1977); Size and bulk density of refractory brick and insulating firebrick

[18] ASTM C 20-80; Apparent porosity, water absorption, apparent specific gravity and bulk density of burned refractory brick and shapes by boiling water

[19] ASTM C 373-72 (re-approved 1977); Water absorption, bulk density, apparent porosity and apparent specific gravity of fired whit-ware products

[20] ASTM C 830-79; Apparent porosity, liquid absorption, apparent specific gravity and bulk density of burned refractory brick, and shapes by vacuum pressure

[21] ASTM C 914-79; Bulk density and volume of solid refractories by wax immersion

[22] ASTM C 135-66 (re-approved 1976); True specific gravity of refractory materials by water immersion

[23] ASTM C 357-70 (re-approved 1976); Bulk density of granular refractory materials

[24] ASTM C 493-70 (re-approved 1976); Bulk density and porosity of granular refractory materials by mercury displacement

[25] ASTM D 202-81; Density of untreated paper used for electrical insulation

[26] ASTM D 644-55 (re-approved 1976); Moisture content in paper and paperboard by oven drying

[27] ASTM D 3394-81; Density of electrical insulating board

[28] ASTM D 1895-69 (re-approved 1979); Apparent density, bulk factor and pourability of plastic materials

[29] ASTM D 1817-81; Rubber chemicals

[30] ASTM D 1439-72 (re-approved 1978); Sodium carboxy methyl cellulose

[31] ASTM D 2364-75 (re-approved 1979); Hydroxy ethyl cellulose
[32] ASTM D 2363-79; Hydroxy propyl methyl cellulose
[33] ASTM D 1667-76; Flexible cellular materials—vinyl chloride polymers and co-
      polymers (closed cell sponge)
[34] ASTM D 2346-68 (re-approved 1978); Apparent density of leather
[35] ASTM D 3800-79; Density of high-modulus fibres
[36] ASTM D 2839-69 (re-approved 1980); Melt box index strand for determination of
      density of polyethylene

Some relevant Indian Standards

- IS 8543 (part 1/section 2): 1979 Methods of testing plastics: determination of apparent density of solid plastics
- IS 10441: 1991 Metallic powders—determination of apparent density of oscillating funnel method
- IS 10837: 1984 (re-affirmed 1995) Mould and accessories for determination of cohesion-less soil
- IS 1167: 1986 Method for determination of apparent density of metallic powders by Scott volumeter
- IS 8426 (part 3): 1977 (re-affirmed 1989) equivalent to IEC 556-1963 Methods of measurement for properties of gyro-magnetic materials for use in micro-wave frequencies—permittivity, apparent density and Curie temperature

Appendix A

Glossary of terms

Abundance ratio: The ratio of the number of isotopic atoms of specific mass number, present in a given volume, to the number of atoms of the normal atomic mass number. For example oxygen has isotopes of mass number 18 and 17, while its normal atomic mass number is 16. Then the abundance ratio R_{18} is the ratio of the number of atoms of mass number 18 to those of mass number 16, present in the given volume. Similarly the abundance ratio of isotopes of water with oxygen of mass number 18 or hydrogen of mass number 2, i.e. deuteron (D), will respectively be $R_{18} = n(^{18}O)/n(^{16}O)$ and $R_D = n(D)/n(H)$.

Adhesion—force of: A force that acts to hold two molecules of dissimilar substances together.

Alcohol degree by mass: The percentage of alcohol by mass present in a pure alcohol water mixture.

Alcohol degree by volume: The percentage of alcohol by volume present in a pure alcohol water mixture.

Alcoholometer: A hydrometer used for measuring the percentage of alcohol by either mass or volume in an alcohol water mixture.

API degree: This equals $141.5/S - 131.5$, where S is the specific gravity of liquid at 60/60 °F.

Artefact: An artefact of a unit is the embodiment of its value, which is either realized from the very definition of the unit or otherwise as it is defined. For example the International Prototype of the kilogram at BIPM is the embodiment of the unit of mass as it is defined.

Avogadro's constant: The number of particles or entities contained in the 1 g molecule of the substance. Avogadro's constant has the unit mole^{-1} and its approximate numerical value is 6.02×10^{23}.

Baume degree: For liquids denser than water: the Baume degree is $145 - 145/S$; for liquids lighter than water: the Baume degree is $140/S - 130$, where S is the specific gravity of liquid at 60/60 °F. Here water has degree zero from the first formula and degree 10 from the second.

Baume scale: There are two arbitrary scales due to Baume, one for liquids lighter than water and another for liquids denser than water.

Beattie's formula: $\rho = \rho_0[1 + a_0(t \, °C^{-1}) + a_1(t \, °C^{-1})^2 + a_2(t \, °C^{-1})^3 + a_3(t \, °C^{-1})^4]^{-1}$, where $a_0 = 1.815\,868 \times 10^{-4}$, $a_1 = 5.458\,43 \times 10^{-9}$, $a_2 = 3.4980 \times 10^{-11}$, $a_3 = 1.5558 \times 10^{-14}$ and $\rho_0 = 13\,595.076$ kg m^{-3}. This value has been derived by using Beattie's formula and the value of the density of mercury at 20 °C as 13 545.848 kg m^{-3}.

Brix degree: The percentage of sucrose by mass in a pure sucrose solution at 20 °C. Zero degree Brix corresponds to pure water at 20 °C.

Centre of buoyancy: The centre of buoyancy of a body fully or partly submerged in a fluid is the centroid of the volume of the displaced fluid; it is also the point at which the resultant buoyant force acts on the body.

Certificate correction (CC): Correction to be applied at the point of observation of a calibrated instrument as given by the Calibration Laboratory in its certificate of calibration.

Coefficient of compressibility: The fractional change in volume of a substance for unit change in pressure. The unit is Pa^{-1} or multiples of it.

Coefficient of cubic/volume expansion: The increase in volume per unit volume for 1 K or 1 °C rise in temperature. The unit is K^{-1} or °C^{-1}.

Coefficient of linear expansion: The increase in length per unit length for 1 K or 1 °C rise in temperature. The unit is K^{-1} or °C^{-1}.

Coherent radiation: The radiation that emanates from a monochromatic source which has a definite phase relationship is coherent radiation.

Coherent sources: Two sources of monochromatic light of the same wavelength are said to be coherent, if the phase difference between the two light waves remains unaltered with respect to time.

Cohesion: The force of attraction between the molecules of a liquid that enables drops and thin films to be formed.

Collimated beam: A beam of particles with a narrow cross section that has little or no spreading angle.

Compressibility: The extent to which a material reduces its volume when it is subjected to compressive stresses.

Compressible diver: An object which floats in a liquid but is constructed in such a way that its volume decreases with hydrostatic pressure within the liquid.

Conservation of energy: Energy cannot be created or destroyed, but it can be changed from one form to another; no violation of this principle has been found.

Conservation of mass: Mass cannot be created or destroyed; this principle generally holds true in larger contexts but can be violated at the microscopic level, as in a nuclear reaction.

Conservation of mass–energy: A fundamental principle stating that energy cannot be created or destroyed, and that mass carries an inherent energy equal to the mass multiplied by the square of the speed of light in a vacuum. This is the combination of the previous two principles (general theory of relativity).

Constant-mass hydrometer: A hydrometer which floats to different submerged volume depending upon the density of the liquid. Its mass remains constant.

Constant-volume hydrometer: A hydrometer which is made to float in liquids of different density with a constant submerged volume by varying its mass.

Content method: For determining density; measures the mass of the substance contained in a measure of known volume.

Density (of gases): The mass of unit volume of a gas at stated pressure and temperature. Mostly the density of a gas is stated at normal pressure (101 325 Pa) and temperature (0 °C). It should be expressed as n kg m^{-3} at x Pa and y °C. Here also n, x and y are pure numbers. Instead of the pascal other units of pressure like the bar or mm Hg may also be used.

Density (of solids and liquids): The mass of unit volume of material at a specific temperature. The SI unit of density is kg m^{-3}, but g cm^{-3} or g ml^{-1} may also be used. For solids, the volume shall be that of an impermeable portion. It should be expressed as n kg m^{-3} at x °C. Here n and x are pure numbers. As liquids and solids are very much less compressible, pressure is not mentioned in industrial measurements. But when precision is demanded, all density values are given at atmospheric pressure (101 325 Pa).

Density (of solids and liquids), apparent: The apparent mass (weight) of a unit volume of a material at a specified temperature. For solids, the volume shall be that of an impermeable portion only. It should be expressed as n kg m^{-3} at x °C. Here n and x are pure numbers.

Density (of solids), bulk: The apparent mass (weight) of a unit volume of a permeable material (it includes both impermeable and permeable voids normal to the material) at a stated temperature. It should be expressed as n kg m^{-3} at x °C. Here n and x are pure numbers.

Density of a body: The ratio of the mass of a body to its volume. The SI unit of density is kg m^{-3}. However its sub-multiples like kg dm^{-3}, g cm^{-3} or mg cm^{-3} may also be used. The use of g cm^{-3} is more frequent. It should be noted that the numerical value of density in kg m^{-3} or mg cm^{-3} is the same.

Density of matter: The basic property of matter, it is the concentration of matter in a given space, measured as mass per unit volume, and is expressed in such units as kg m^{-3}, g cm^{-3}.

Deviation of isotopic composition of water: The difference from unity of the ratio of the abundance ratios of the water sample to that of SMOW. The multiplying factor of 1000 is normally used to avoid smaller fractions. For example

$$\delta D = (R_D(\text{sample})/R_D(\text{SMOW})-) \times 10^3$$
$$\delta\,^{18}O = (R_{18}(\text{sample})/R_{18}(\text{SMOW})-) \times 10^3.$$

Dilatation of water: Change in volume per unit volume for unit change in temperature. The unit is K^{-1} or °C^{-1}.

Dimensions: There are three fundamental dimensions namely mass, length and time. These are respectively represented by M, L and T. For convenience, at most two more—say temperature and charge—may be taken and may be represented by K and C, respectively. A unit representing any physical quantity can now be expressed in terms of these dimensions.

Displacement method: For determining the density of a liquid; measures the mass of the liquid displaced by a solid of known volume.

Dry–dry condition: The sample is dried in an oven maintained at 105 ± 3 °C for such a time that its weight becomes constant. Weigh the sample after regular intervals until its apparent mass becomes constant within the prescribed limits.

Equation with dimensional parity: In any equation there are various quantities; each quantity will have its dimensions in terms of length, mass, time, charge and temperature. A dimensionally balanced equation is one in which every term has the same dimension. If a quantity is expressed as a function of many other quantities then each term should have the same dimensions as that of the quantity.

Equatorial plane: The horizontal symmetry plane, if one exists, through a body revolving about a vertical axis of symmetry.

Expansion of mercury: $\rho = \rho_0[1 + a_0(t\,^\circ\mathrm{C}^{-1}) + a_1(t\,^\circ\mathrm{C}^{-1})^2 + a_2(t\,^\circ\mathrm{C}^{-1})^3 + a_3(t\,^\circ\mathrm{C}^{-1})^4]^{-1}$, where $a_0 = 1.815\,868 \times 10^{-4}$, $a_1 = 5.458\,43 \times 10^{-9}$, $a_2 = 3.4980 \times 10^{-11}$ and $a_3 = 1.5558 \times 10^{-14}$ and $\rho_0 = 13\,595.076\,\mathrm{kg\,m^{-3}}$. This value has been derived by using Beattie's formula and the value of the density of mercury at $20\,^\circ\mathrm{C}$ as $13\,545.848\,\mathrm{kg\,m^{-3}}$.

Fabry–Perot fringes: A series of light and dark concentric rings observed when monochromatic light is passed through a Fabry–Perot interferometer.

Fabry–Perot interferometer: A type of interferometer in which a light beam is reflected between two glass or quartz plates a number of times before it is transmitted.

Fizeau fringes: The pattern that results from the interference of the two beams of light coming from the top and bottom surfaces of a thin layer or wedge.

Fizeau interferometer: An interferometer in which a beam of light is split, reflected a number of times, and then recombined in order to produce thin, multiple-beam interference fringes.

Flatness: A surface is said to be flat if a plane defined by any three points on the surface contains every other point on it. To measure the flatness, the coordinates of a large number of points with a certain frame of reference are determined and a regression plane is passed through the points. Then the square root of the mean of the squares of the distances of all other points not lying on the plane gives the measure of the flatness of the surface.

Fractional fringe order: An optical path difference of half a wavelength is equivalent to a change of one fringe. So the distance between the two points can be expressed as the sum of a whole number of fringes and a part of it. The fraction of the fringe is the fractional fringe order and is estimated from the interferogram. The integer part of the fringe order is calculated either by actual fringe counting from a reference point or by the method of exact fractions by using radiations of at least two wavelengths.

Fringe order: The fringe for which the path difference is zero is termed the central fringe and this fringe is assigned zero order. On either side of it the fringe order is increased by one for each fringe passed. If N_0 is the fringe order it means n fringes have been passed from the central fringe.

Fringes: A set of alternating bright and dark bands that are produced by interference or diffraction of light waves. The fringe is the locus of equal optical path difference.

Fundamental constants: The constants of physical data whose values are determined by experiment, such as the speed of light in a vacuum, the mass of an electron and the charge of a proton.

Gravitational system of units: A system of physical units in which length, time and force are regarded as fundamental units. Note, in this system force is the base unit instead of mass. The unit of force is established as the gravitational force on a standard body at a standard location on the earth.

Interference phenomenon: The interaction of two or more wave motions affecting the same part of a medium so that the instantaneous disturbances in the resultant wave are the vector sum of the instantaneous disturbances in the interfering waves. Therefore surfaces at which every point has an amplitude equal to the sum or difference of the amplitudes of two wave motions are formed. Thus the phenomenon of interference is the redistribution, in space, of light energy due to the interaction of two coherent beams of light waves. The phenomenon gives a pattern of alternate dark and light bands.

Isothermal compressibility of mercury: $\beta = \beta_0(1 = b_1 p + b_2 p^2 + b_3 p^3)$, where $\beta_0 = 40.25 \times 10^{-12}$ Pa, $b_1 = -3.730\,11 \times 10^{-10}$ Pa^{-1} , $b_2 = 1.938\,77 \times 10^{-19}$ Pa^{-2} and $b_3 = -7.299\,26 \times 10^{-29}$ Pa^{-3}.

Lactometer: Special name given to a hydrometer used for measuring the density or purity of liquid milk.

Mass: A fundamental property of an object, which makes it resist acceleration and which determines its gravitational attraction. This property can be generally regarded as equivalent to the amount of matter in the object. Mass and weight are not synonymous; mass is not affected by the forces acting on an object, but weight is a relative property that can change according to the gravitational force exerted on the object.

Meniscus correction: The meniscus correction in respect of a hydrometer is equal to $1000(TR/D\rho Lg)[(1 + 2gD^2\rho/T)^{1/2} - 1]$, where T is the surface tension in N m^{-1}; R is range and ρ is density in kg m^{-3}; D is the diameter of the stem, L is the scale length in metres and g the acceleration due to gravity in m s^{-2}.

Monochromatic radiation: An electromagnetic radiation with one wavelength or with a narrow range of wavelengths.

Newton's rings: These are interference fringes formed by the two interfering beams in which one beam has a plane wavefront (parallel beam) and the other has a spherical wavefront (converging beam). To be specific they are the bright and dark rings that are formed around the point of contact of a slightly convex lens and a flat sheet of glass, caused by interference in the air between the surfaces, which are observable in monochromatic light.

Open-scale hydrometer: A hydrometer which has very small scale interval with comparatively smaller range. It measures much finer differences in density.

Optical axis: The imaginary line that passes through the centres of curvatures of both optical surfaces of a single lens or the centres of all surfaces of a correctly centred lens system.

Phase: The phase of a wave is the value of the argument of the sine function; if the amplitude is defined as being equal to $A_0 \sin(\theta)$, then θ is the phase of the wave. The phase consists of two components, namely temporal and spatial. If ω is the frequency in radians per second of the wave and λ is its wavelength then the temporal phase is the product of the frequency ω and t, the time taken from some arbitrary point. The spatial component is due to the distance travelled from an arbitrary point and is expressed as $2\pi/\lambda$ times x, the distance travelled from the arbitrary point. Both are expressed in radians. In general: (1) a particular stage or point in a sequence through which time has advanced, measured from some arbitrary starting point; (2) the relative angular displacement of one sinusoidal quantity with respect to a reference angle or to another sinusoidal varying quantity of the same frequency.

Phase angle: The relative angular displacement between a periodic quantity and a reference angle or between two sinusoid varying quantities of identical frequencies

Point source: A source (of radiation, mass, or charge) whose dimensions are small compared to the distance to the observation point.

Porosity: The ratio of the total amount of void space in a material (due to pores, small channels, and so on) to the bulk volume occupied by the material. Generally porosity P is expressed in %.

Primary standard: The top of the hierarchy of standards—it is only second to the prototype. A primary standard is also an embodiment of the unit realized through the phenomenon by which the unit has been defined.

Pyknometer: A vessel of fixed and precisely known capacity.

Range of a hydrometer: The difference between the maximum and minimum density or any measuring quantity, which it is capable of measuring. Normally it is the difference between the values of extreme graduations of its scale.

Reference temperature: The temperature at which the instrument is supposed to give correct readings or is intended to be used.

Relative density or specific gravity (of solids and liquids): The ratio of the mass of unit volume of a material at a stated temperature to the mass of the same volume of gas-free distilled water at a stated temperature. The two stated temperatures might be the same or different. If the material is a solid, the volume shall be that of an impermeable portion. It should be expressed as n at $x\,°C/y\,°C$. Here n, x and y are pure numbers.

Relative density: The ratio of the density of a substance to that of the reference (standard) substance (normally water). It is independent of units of measurement of density but it is qualified by the temperature of the density of the substance and the temperature at which the density of the reference substance has been taken. $S\ t_1/t_2\,°C$. t_1 and t_2 may either be different or the same.

Relative density (apparent) or specific gravity (of solids and liquids): The ratio of apparent mass (weight) of a certain volume of a material at a stated temperature to the apparent mass (weight) of same volume of gas-free distilled water at the stated temperature. If the material is a solid, the volume shall be that of an impermeable portion. It should be expressed as n at $x\,°C/y\,°C$. Here n, x and y are pure numbers.

Relative density (bulk) or specific gravity (of solids): The ratio of the apparent mass (weight) of a certain volume of a permeable material (including both the impermeable and permeable voids normal to the material) at a stated temperature to the apparent mass (weight) of the same volume of gas-free distilled water at the stated temperature. If the material is a solid, the volume shall be that of impermeable and permeable portions. It should be expressed as n at $x\,°C/y\,°C$. Here n, x and y are pure numbers.

Relative density bottle: A measure of fixed volume in the shape of a bottle having a ground-glass stopper with an axial capillary bore.

Relative density of water: The ratio of the density of water at a given temperature to its maximum density, which occurs at a temperature very close to $4\,°C$.

Relative density or specific gravity (of gases): The ratio of the mass of a certain volume of a gas under the observed temperature and pressure to the mass of the same volume of dry air containing 0.04% of CO_2 at the same or normal temperature and pressure. It should be expressed as n at $x\,°C/y\,°C$. Here n, x and y are pure numbers.

Sample holder: A device to hold a number of samples under test and capable of bringing samples onto the pan of a balance one at a time but in all possible combinations. There are two types of sample holder discussed in this book. In a vertical sample holder all the samples are in one vertical plane, i.e. one above the other. A separate lifting device is provided for each sample to bring it onto the pan and movement of the sample is only in the vertical direction and that only by 1 mm, so disturbance to the hydrostatic liquid is at a minimum. However, as the samples are in different horizontal planes, a correction due to the vertical gradient of gravity has to be applied. In a horizontal sample holder all the samples are placed in one horizontal plane and are brought onto the pan by rotating the sample holder about its vertical axis. Hence all samples are moved, which causes more disturbance to the hydrostatic liquid, but no gravity gradient correction is required. In this case, a small temperature gradient in the vertical plane is immaterial.

Scale interval: The difference in the value of the quantity the hydrometer is intended to measure, between two consecutive graduated marks.

Scale length: The length between the graduation marks representing the range of the hydrometer.

Scale of hydrometer: A set of graduations which indicate the quantity the hydrometer is intended to measure. These are marked on a paper which is then inserted in the stem so that the axis of the scale is parallel to the axis of the hydrometer.

Sikes scale: An arbitrary scale to measure the strength of spirits. Zero degree corresponds to 70% of over proof of an alcohol–water mixture and 100% to pure water.

Sinker: A body of known volume denser than the liquid in which it is intended to be used.

SLAP: Standard Light Antarctic Precipitation is the second reference standard of water; it is obtained from melted ice from the Antarctic from Plateau station, which is at an altitude of 3700 m and at $79°15'$S and $40°30'$E. SLAP has the lowest value of R_D. The values are $R_D = (89.02 \pm 0.05) \times 10^{-6}$ and $R_{18} = (1893.9 \pm 0.45) \times 10^{-6}$.

SMOW: Standard Mean Ocean Water is the first reference standard of pure water with the following isotopic composition:

$$R_{18} = (2005.2 \pm 0.45) \times 10^{-6}$$
$$R_D = (155.769 \pm 0.05) \times 10^{-6}$$

and

$$R_{17} = (371.0 \pm 0.45) \times 10^{-6}.$$

Solid angle: The ratio of the area of a surface of a sphere bounded by arcs each equal to that of the radius of the sphere to the square of its radius.

Solid-based density standard: A solid of known volume and mass, e.g. a silicon sphere of known geometry whose diameter and variation in its diameter is known within a few nm and mass within a few μg.

Specific gravity: The ratio of the density of a material at a given temperature and pressure to the density of some standard material.

Standard deviation: The square root of the mean of the squares of deviations from the mean value. If x_i are the n variables following normal distribution and x_m is the mean value of the variables, then $x_m = \sum x_i/n$ and standard deviation $\sigma = \sqrt{\sum(x_i - x_m)^2/n}$. The estimate of σ for samples of finite size is $s = \sqrt{[\sum(x_i - x_m)^2/(n-1)]}$.

Standard uncertainty: The square root of the sum of the variances due to all component uncertainties due to all the influencing quantities (all sources of error). But the relative standard uncertainty is the ratio of standard uncertainty and the quantity itself.

Summation method: Finding the value of the quantity by fragmenting it into small fractions, measuring or finding the value of each fragment and taking the algebraic sum of all such fragments.

Surface tension: Due to cohesive forces in between the molecules of a liquid, its surface is stretched, like a stretched rubber membrane; the surface tension is the result of this tension and is expressed as force per unit length. The SI unit of surface tension is $N\,m^{-1}$. However it is usually expressed in terms of $mN\,m^{-1}$.

Surface tension category of hydrometers: There are three categories of hydrometers from the point of view of surface tension, namely low, medium and high. For example, low surface tension hydrometers are meant to measure the density of liquids whose surface tension varies from 15 to 35 $mN\,m^{-1}$.

Surface tension correction: $10^3\pi\rho D(T - T_s)/mg$, where ρ is the density in $kg\,m^{-3}$, D is the diameter of the stem in metres, m is the mass of the hydrometer in kg, g is the acceleration due to gravity in $m\,s^{-2}$ and T is the surface tension in $N\,m^{-1}$. The unit of surface tension correction is $N\,m^{-4}\,s^2$, i.e. $kg\,m^{-3}$.

Surface tension factor: $10^6\pi\rho D/mg$, where ρ is the density in $kg\,m^{-3}$, D is the diameter of the stem in metres, m is the mass of hydrometer in kg and g is the acceleration due to gravity in $m\,s^{-2}$. The unit of surface tension factor is $m^{-3}\,s^2$.

Temperature of maximum density of water: Water contracts as its temperature is increased from $0\,°C$, acquires a minimum and then starts expanding from that temperature onward. The temperature of water at which the given mass of water acquires its minimum volume is the temperature of its maximum density. The latest value of this temperature is $3.983\,035\,°C$ and the value of maximum density of water is $999.974\,950 \pm 0.000\,84\,kg\,m^{-3}$.

Thermal dilatation: The ratio of the volume of a given mass of water at a given temperature to its volume at 4 °C.

Tuning fork densitometer: A densitometer with a tuning fork as its vibrating element.

Twaddle degree: Zero degree Twaddle corresponds to the specific gravity of 1 and one Twaddle degree interval is equivalent to 0.005 of specific gravity at 60/60 °F

$$T \text{ degree Twaddle} = 200(S60/60\,°F - 1).$$

Uncertainty: A measure of the inherent variability of repeated measurements of a quantity. A prediction of the probable variability of a result, based on the inherent uncertainties in the data, found from a mathematical calculation of how the data uncertainties would, in combination, lead to uncertainty in the result. In particular, it is the range about the measured value in which the true value is most likely to lie. Also it is the range in which any future measured value is likely to lie. If U is the semi-range of uncertainty and the reported value of the quantity is X with a level of confidence say 95%, then the chance that the true value of the measured quantity lies between $X - U$ and $X + U$ is 95%. Alternatively if anybody else measures the same quantity, then the measured value should lie in between the limits $X - U$ and $X + U$, 95 times out of 100.

Vertical gradient of gravity: Rate of variation in acceleration due to gravity in the vertical direction. The gradient is negative as the value of acceleration due to gravity decreases as we go higher up. Its value in space is 3×10^{-7} m s^{-2}. (The line joining the point under consideration to the centre of the earth is the vertical at that point.)

Vibrating element densitometer: An instrument to measure the density of a fluid by measuring the frequency of a vibrating element. The square of the resonant frequency is linearly related to the density of the medium.

Volume of stem: The volume of the stem of a hydrometer between the two graduation marks representing the maximum and minimum quantity the hydrometer is intended to measure.

Volume of the bulb: The volume of the bulb of a hydrometer up to the lowest graduation mark.

Wavefront: A surface at which every point acts as a secondary source and has the same phase.

Wavelength: The spatial distance between adjacent points with the same phase.

Weight of a body: The gravitational force experienced by a body on the earth's surface or in some other gravitational field. Weight and mass are not synonyms; weight is related to force and thus can change according to position. For example, an object in outer space will have less weight than it does on earth, because of its distance from the earth's gravitational field, but its mass will be unaffected.

Wet–wet condition: The sample is kept in air maintained at $23 \pm 2\,°C$ with relative humidity of $50 \pm 2\%$ for a minimum period of 4 h. The sample is supported in such a way that air circulates freely around it.

Young's two-slit interference: A method used by Thomas Young to disprove Newton's corpuscular theory of light; a beam of light from a single source was split into two beams by closely spaced parallel slits and the resulting interference was observed on a screen.

Appendix B

List of symbols

α	Coefficient of linear thermal expansion
Δ	Density of standard weights
σ	Density of air
γ	Coefficient of cubic thermal expansion of material of measures
α	Angle of contact
ρ	Density of liquids
β	Coefficient of isothermal compressibility
χ	Level of air saturation
λ	Wavelength of light
ω	Angular velocity of light
ε	Fraction of a fringe
μ	Refractive index
$\sum$	Summation
π	Ratio of the length of the circumference of a circle to its diameter
$\Delta\rho$	Change in density of water
$\delta\varphi$	Difference in phase angle
$\Delta\mu$	Change in refractive index
$\rho_{(p,t)}$	Density of water at pressure p (Pa) and temperature t (°C)
$\rho_{(SMOW)}$	Density of SMOW
$\rho_{(V\text{-}SMOW)}$	Density of V-SMOW. It may be noted that $\rho_{(V\text{-}SMOW)} \equiv \rho_{(SMOW)}$
δ, θ	Phase angles
$\delta^{17}O$	$\delta^{17}O = 0.5\delta^{18}O$
$\delta^{18}O$	Deviation of the ratio $R_{18(sample)}/R_{18(SMOW)}$ from unity, equal to $[R_{18(sample)}/R_{18(SMOW)} - 1] \times 10^3$; a multiplication factor of 1000 is used to avoid smaller fractions
δD	Deviation of the ratio of $R_{D(sample)}/R_{D(SMOW)}$ from unity, equal to $[R_{D(sample)}/R_{D(SMOW)} - 1] \times 10^3$; a multiplication factor of 1000 is used to avoid smaller fractions
γ_f	Compressibility of solids

$\delta f / \delta x$	Partial differential coefficient of function f with respect to an independent variable x
ω_{ij}	Solid angle subtended by an elementary area formed at the centre of the sphere bounded by the points (i, j), $(i, j - 1)$, $(i + 1, j - 1)$ and $(i + 1, j)$
$\Delta L / L$	Fractional change in length
ρ_m	Density of mercury
σ_p	Density of air at pressure p
δV	Change in volume
$\delta V / V$	Fractional change in volume
γ_w	Compressibility of water
ρ_w	Density of water
$\rho_{w(t,P)}$	Density of water at pressure P and temperature t
A	Atmospheric pressure
A_{ij}, a_{ij}	Constants or coefficients of various powers of a variable
B	Bulk density of solids
BC	Buoyancy correction
b_{ij}	Constants or coefficients of various powers of a variable
C_ρ	Change of density with respect to temperature
cm^3	A unit of volume
C_p	Specific heat at constant pressure
D	Mass of sample dried until mass becomes constant
D	Diameter of stem of hydrometer
D	Diameter of wire
d	Relative density of water
df / dh	Differential coefficient of f with respect to h (height)
d_g	Density of glass
d_0	Diameter of sphere at $0\,°C$
d_s	Density of solids
d_t	Diameter of sphere at $t\,°C$
d_w	Density of water
f	Focal length of the lens
h	Height of a point inside a liquid from arbitrary reference line
h	Depth or height of water column
I	Density interval of a hydrometer
I_1, I_2	Indications by a damped balance
L	Scale length of hydrometer
M, m or W	Mass
MC	Meniscus correction
n	Frequency
$n(D)$	Number of atoms, present in given volume, of deuteron of mass number 2
$n(H)$	Number of atoms, present in given volume, of hydrogen of mass number 1

$n(^{16}O)$	Number of atoms, present in given volume, of oxygen of mass number 16
$n(^{17}O)$	Number of atoms, present in given volume, of oxygen of mass number 17
$n(^{18}O)$	Number of atoms, present in given volume, of oxygen of mass number 18
N_0	Integer number of fringes or fringe order
°F	Degrees Fahrenheit
P	Porosity
P	Pressure in Pa
ppm	Parts per million = one part in 10^6
R	Range of a hydrometer
R_1, R_2	Rest points as calculated from the turning points of a freely oscillating balance
R_{17}	$n(^{17}O)/n(^{16}O)$ = abundance ratio of water molecules having oxygen of mass number 17
R_{18}	$n(^{18}O)/n(^{16}O)$ = abundance ratio of water molecules having oxygen of mass number 18
R_D	$n(D)/n(H)$ = abundance ratio of water molecules having heavy hydrogen (deuteron) of mass number 2
S_{ij}	The difference between the jth diameter of the ith meridian and its equatorial diameter d_{i1}
$St_1/t_2\,°C$	Specific gravity of a liquid equal to the density of liquid at t_1 and density of water at $t_2\,°C$
STC	Surface tension correction
STF	Surface tension factor of a hydrometer
t_0	Reference temperature of the capacity measure
T	Surface tension of the liquid
TC	Temperature correction
t_s	Reference temperature for the standard hydrometer
t_u	Reference temperature for hydrometer under test
V	Volume
V	Volume of bulb
v	Volume of stem
Var	Variance = square of standard deviation
V_{imp}	Volume of impervious portion of the solid
V_t	Volume at $t\,°C$
w	Apparent mass.

Index

T - #0086 - 101024 - C0 - 234/156/20 [22] - CB - 9780750308472 - Gloss Lamination